A Structural Perspective on Respiratory Complex I

Leonid Sazanov
Editor

A Structural Perspective on Respiratory Complex I

Structure and Function of NADH:ubiquinone oxidoreductase

Springer

Editor
Leonid Sazanov
Medical Research Council Mitochondrial Biology Unit
Wellcome Trust/MRC Building, Hills Road
Cambridge CB2 0XY, UK

ISBN 978-94-007-4137-9 ISBN 978-94-007-4138-6 (eBook)
DOI 10.1007/978-94-007-4138-6
Springer Dordrecht Heidelberg New York London

Library of Congress Control Number: 2012938257

Printed on acid-free paper

Springer is part of Springer Science+Business Media (www.springer.com)

Foreword

Complex I (NADH:ubiquinone oxidoreductase) is the first enzyme of the respiratory chain in mitochondria and bacteria. It is one of the largest and most elaborate membrane protein assemblies known. It plays a central role in cellular energy production, providing about 40% of the proton flux required for ATP synthesis. Complex I dysfunction has been implicated in many human neurodegenerative diseases and mutations in its subunits are the most common human genetic disorders known. Complex I is also a major source of reactive oxygen species in mitochondria, which may lead to Parkinson's disease and could be involved in aging. The enzyme transfers two electrons from NADH to quinone, coupling this process to the translocation of four protons across the membrane out of the mitochondrial matrix, by a mechanism as yet not fully established. Mitochondrial complex I consists of 45 different subunits, whilst the prokaryotic enzyme is simpler, consisting of 14 "core" subunits with a total mass of about 550 kDa. The mitochondrial and bacterial enzymes contain equivalent redox components (flavin and 8–9 Fe-S clusters) and have a similar, rather unusual, L-shaped structure. The hydrophobic arm is embedded in the membrane and the hydrophilic peripheral arm protrudes into the mitochondrial matrix or the bacterial cytoplasm. The "core" subunits exhibit a high degree of sequence conservation, which suggests that the complex I mechanism is likely to be the same throughout all species. Hence, the bacterial enzyme is used as a 'minimal' model of human complex I in order to understand its structure and mechanism. Recent years have been marked by spectacular progress in the structural characterization of complex I, which now finally allows us to begin to understand the mechanics of this large molecular machine, making this book very timely.

Until about 5–6 years ago structural information on complex I was absent, and so understanding of it was very limited, especially compared to other enzymes of the respiratory chain. Complex I used to be known as a notorious "monster" enzyme, the "black box" of bioenergetics. In 40 or so years since its discovery it was established that complex I most likely pumps four protons per two electrons transferred from NADH to quinone. Electron transfer was known to occur via flavin mononucleotide (FMN) and series of at least 6 iron-sulfur (Fe-S) clusters, which were detected by electron paramagnetic resonance (EPR). Not all the clusters were observed experimentally, since the presence of 8–9 Fe-S clusters was predicted on the basis of sequence analysis. The sequence of events during electron transfer was unknown and the mechanism of proton translocation was even more enigmatic. Two possible mechanisms of coupling between electron transfer and proton translocation have been vigorously discussed: direct (redox-driven, akin to the Q-cycle) and indirect (conformation-driven). However, in the absence of structural information, they were mostly speculative.

All started to change in 2005–2006, when we solved the first crystal structure of the hydrophilic domain of complex I, using the enzyme from *Thermus thermophilus*. It established the electron transfer pathway from NADH, through flavin mononucleotide (FMN) and seven conserved Fe-S clusters, to the quinone-binding site at the interface with the membrane domain. In 2010–2011 we have solved the structure of the membrane domain of *E. coli* complex I and determined the architecture of the entire *T. thermophilus* enzyme at lower resolution. Thus, the atomic structure of only one "core" subunit, Nqo8/NuoH (*Thermus/E. coli* nomenclature), found at the interface of the two main domains, remains unknown. Additionally, low-resolution X-ray analysis of the mitochondrial enzyme from *Yarrowia lypolityca* was published in 2010, indicating a similar arrangement of the "core" subunits, surrounded by many supernumerary subunits. The membrane-spanning part of the enzyme lacks covalently bound prosthetic groups, but our structures show how proton translocation through the three largest hydrophobic subunits of complex I, homologous to each other and to the antiporter family, may be driven by a long α-helix, akin to the coupling rod in a steam engine. This and other features of the structure strongly suggest that electron transfer in the peripheral arm is coupled to proton translocation in the membrane arm purely by long-range conformational changes. Mutations causing human diseases are found near key residues involved in proton transfer, explaining their effects on activity.

Not all the details of the mechanism are clear yet, but we are now operating on a completely different level of knowledge than just a few years ago. This led to the idea of summarizing in book form current knowledge of complex I, taking into account structural information. No books on complex I have been published previously, and the last special issue of a journal devoted to complex I was published in 2001, when it was still known as a "black box". Therefore, it is hoped that this book will provide the reader with a timely and comprehensive review of current state-of-the-art research on complex I.

In Chap. 1, current knowledge of the structure of complex I is reviewed, starting from the peripheral domain, followed by a detailed description of the new structure of the membrane domain, and ending with implications for the mechanism. In Chap. 2, the binding of substrates, the role of individual Fe-S clusters (in particular those away from the main pathway) and the mechanism of proton translocation are discussed on the basis of data from site-directed mutagenesis, EPR and FTIR spectroscopy, as well as other studies. In Chap. 3, current knowledge of the characteristics and roles of each Fe-S cluster in complex I is overviewed. Chapter 4 provides a review of many specific inhibitors of complex I, the use of which has been very informative in characterisation of the quinone-binding site and the terminal electron transfer step.

In Chap. 5, some of the earliest studies on complex I, in particular EPR spectroscopy leading to the first identification of Fe-S clusters, are summarised.

Complex I has an intricate evolutionary history, originating from the unification of hydrogenase and transporter modules. In Chap. 6, the evolutionary relationship with [Ni-Fe]-hydrogenases is analysed and mechanistic implications are derived from comparisons of known crystal structures. In Chap. 7, the emphasis is on the relationship with the Mrp antiporter family and it is proposed that antiporter-like subunits in modern complex I may have different functions.

Mutations in complex I subunits, both mitochondrially- and nuclear-encoded, lead to a range of human diseases. Many of these mutations have been reproduced in bacterial systems for mechanistic studies. Chapter 8 provides a review of site-directed mutagenesis studies that helped in identifying residues essential for structural integrity, cofactor ligation, substrate binding, electron transfer and proton translocation. In Chap. 9, a comprehensive overview of the cellular consequences of pathological mtDNA-encoded mutations in complex I subunits is provided.

Mitochondrial complex I contains, in addition to the "core" subunits, up to 31 "supernumerary" subunits, with poorly understood roles. Chapter 10 describes an intricate process of assembly of the complex in several stages, involving distinct functionally and evolutionarily conserved modules, and requiring a number of chaperones. In Chap. 11, the similarities and peculiarities of the subunit composition of mitochondrial complex I in plants and the complex I analogue in chloroplasts are described.

In the respiratory chain of mitochondria complex I appears not to exist on its own, but as part of even larger assemblies, or "supercomplexes". These involve complexes I, III and IV, as described in Chap. 11, and may promote substrate channelling.

Thus, combined, the chapters cover a wide range of topics which should provide the reader with an up-to-date review of research on complex I in these exiting times, when the molecular basis for its mechanism is finally starting to become clear.

Leonid Sazanov
Medical Research Council
Mitochondrial Biology Unit
Wellcome Trust/MRC Building, Hills Road
Cambridge, UK

Contents

Part I
Structure and Mechanism of Complex I

Chapter 1
Structure of Complex I

Rouslan G. Efremov and Leonid Sazanov

Abstract Complex I is the first enzyme of the respiratory chain and plays a central role in cellular energy production. It has been implicated in many human neurodegenerative diseases, as well as in ageing. One of the biggest membrane protein complexes, varying in size from 0.5 to 1 MDa, it is an L-shaped assembly consisting of hydrophilic and membrane domains. Previously, we determined the structure of the hydrophilic domain in several redox states. It established the pathway for electron transfer from NADH to quinone via seven Fe-S clusters. Recently, we solved the structure of 6 out of 7 membrane domain subunits and described the architecture the entire bacterial complex I. This progress in structural characterization of the enzyme finally allows us to begin to understand the mechanism of this large molecular machine. The proposed mechanism of coupling between electron transfer and proton translocation involves long-range conformational changes, coordinated in part by a long α-helix, akin to the coupling rod of a steam engine.

Keywords NADH: ubiquinone oxidoreductase (complex I) • Respiratory chain • Antiporters • X-ray crystallography • Fe-S cluster • Electron transfer • Proton translocation • *E. coli* • *T. thermophilus*

R.G. Efremov
Max-Planck-Institute for Molecular Physiology, Otto-Hahn Str. 11,
Dortmund 44227, Germany

L. Sazanov (✉)
Medical Research Council Mitochondrial Biology Unit, Wellcome Trust/MRC
Building, Hills Road, Cambridge CB2 0XY, UK
e-mail: sazanov@mrc-mbu.cam.ac.uk

L. Sazanov (ed.), *A Structural Perspective on Respiratory Complex I: Structure and Function of NADH:ubiquinone oxidoreductase*, DOI 10.1007/978-94-007-4138-6_1,

1.1 Introduction

Complex I is a main entry-point for electrons to the electron transport chain. It catalyses reversible oxidation of NADH by ubiquinone, coupled to translocation of four protons across the inner mitochondrial membrane (in eukaryotes) or cytoplasmic membrane (in bacteria), with a maximum rate of about 200 cycles per second (Walker 1992; Yagi and Matsuno-Yagi 2003; Sazanov 2007; Brandt 2006). It is also considered as the main source of reactive oxygen species (ROS) in mitochondria, which can damage mtDNA and cause Parkinson's disease (Dawson and Dawson 2003) and possibly aging (Balaban et al. 2005). Mutations in nucleus and mitochondria encoded subunits have been associated with several neurodegenerative diseases (Sazanov 2007; Schapira 1998). Complex I has an intricate evolutionary history, representing a chimera of hydrogenases and cation-proton antiporters (reviewed in (Friedrich 2001; Moparthi and Hagerhall 2011)). The complex is present in many bacteria and in the mitochondria of most eukaryotes, including animals, plants and fungi. Modified versions of the enzyme, utilizing different electron inputs and reducing various quinone analogues, have an even broader spread, encompassing chloroplasts and archaea (Moparthi and Hagerhall 2011).

Complex I is one of the biggest membrane protein assemblies known. The total molecular weight is close to 1 MDa for the mitochondrial enzyme and about 550 kDa for the bacterial version. Complex I composition differs between organisms, numbering from a minimal 14 subunits in many bacteria up to 45 subunits in the bovine enzyme (Carroll et al. 2006). The core 14 subunits are conserved between all organisms and none of them can be removed without compromising enzyme function, suggesting that all complexes I share a similar mechanism. Electron microscopic reconstructions of the enzyme structure in negative stain and in vitreous ice established its overall L-shaped appearance in all organisms studied (Clason et al. 2010), with a peripheral arm protruding into the bacterial cytoplasm/mitochondrial matrix and a membrane embedded arm. The mass of the enzyme is approximately equally distributed between peripheral and membrane arms, each of which is around 180 Å long (including the junction). The 14-subunit bacterial enzymes represent a 'minimal model' that began acquiring supernumerary subunits before the endosymbiotic event that lead to the origin of mitochondria and creation of eukaryotic cell (Yip et al. 2011).

Direct (redox-driven), indirect (conformation driven) and mixed mechanisms of coupling have been suggested (Brandt 2006; Friedrich 2001; Sazanov 2007; Yagi and Matsuno-Yagi 2003). However, in the absence of high resolution structural information, they were largely speculative.

As is common for large and fragile protein complexes, determination of the structure of bacterial complex I was tackled by purifying and crystallizing its more stable fragments. The junction of the peripheral and membrane arms is especially fragile in the bacterial complex (Hinchliffe et al. 2006; Hinchliffe and Sazanov 2005; Leif et al. 1995). Crystallization of the mitochondrial complex I, generally more stable than bacterial enzyme, is complicated by a number of post translational modifications and compositional heterogeneity (Hunte et al. 2010). First, in 2006, the molecular structure of the peripheral arm from *Thermus thermophilus* was determined at 3.3 Å resolution (Sazanov and Hinchliffe 2006); later improved up to 3.1 Å (Berrisford and Sazanov 2009). The last 2 years were marked with a great progress: 3.9–4.5 Å resolution structures of the membrane domain and the intact bacterial complex were solved (Efremov et al. 2010), as well as the crystallographic electron density map of eukaryotic complex I from yeast *Yarrowia lipolytica* being reported at 6.3 Å resolution (Hunte et al. 2010). Very recently, the 3.0 Å resolution molecular structure of six membrane subunits has been determined (Efremov and Sazanov 2011b), nearly completing the puzzle; only the structure of membrane subunit NuoH/Nqo8, at the junction between two main domains, remains unknown.

In this chapter we give an account of our current structural understanding of complex I, focusing on functional aspects and a plausible mechanism.

1.2 Overall Structure

Subunit nomenclature is different for complex I from different organisms. Complex I is encoded by the Nuo operon (NADH-ubiquinone oxidoreductase) in *E. coli* (subunits NuoA-L) and the Nqo operon (NADH-quinone oxidoreductase) in *T. thermophilus* (subunits Nqo1-14, with Nqo15 outside this operon). Mitochondrial complex I is composed of nuclear and mitochondrially encoded subunits named differently in human, bovine and yeast enzymes (Brandt 2006). Both *E. coli* and *T. thermophilus* naming will be used throughout the text.

Seven core subunits constitute the peripheral arm of bacterial complex I and another seven its membrane arm. The peripheral arm provides a rigid scaffold harbouring eight to ten iron-sulfur clusters, seven of which constitute a conserved electron transfer pathway between the NADH binding site at the tip of the domain (distant from the membrane) and the ubiquinone binding site located 20–25 Å above membrane surface (Efremov et al. 2010; Hunte et al. 2010) (Fig. 1.1a). The peripheral arm sits on top of membrane subunit NuoH (Nqo8). This likely provides the major interaction surface between peripheral and membrane arms, with additional contributions from small trans-membrane subunits NuoA/J/K (Nqo7, 10, 11). An 11-helix bundle of subunits NuoA/J/K separates the peripheral arm from the three membrane antiporter-like subunits NuoN/M/L. These are arranged linearly, like the carriages of a train, attached to the NuoA/J/K bundle (Fig. 1.1a, b). A notable structural element, the 110 Å long amphipathic helix from the carboxy-terminal part of NuoL, spans nearly the entire length of the membrane domain, stabilizing it and likely playing important mechanistic role (Efremov et al. 2010).

Analysis of the electron density map from eukaryotic complex I showed that the fold of the core 14 subunits is indeed highly conserved between bacterial and eukaryotic enzymes, as is their ternary organization. Only slight re-arrangement of core subunits (as conserved rigid bodies) has occurred during billions of years of evolution. The analysis also allowed visualization of positions of supernumerary subunits, distributed around the conserved catalytic core and likely playing stabilizing and regulatory roles (Efremov and Sazanov 2011a).

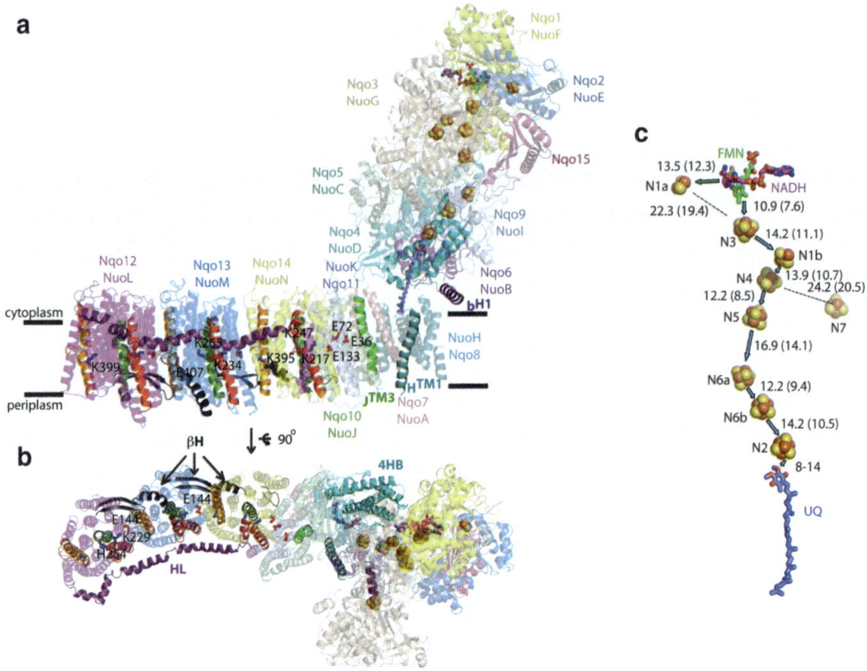

Fig. 1.1 Structure of bacterial complex I. (**a, b**). Structure of peripheral arm with α-helical model of subunit NuoH from *Thermus thermophilus* (PDB code 3M9S), aligned (via membrane domain) to high resolution structure of membrane domain from *E. coli* (PDB code 3RKO). Subunits are shown in different colors. FMN, bound NADH and iron-sulfur clusters are shown as *spheres*, as is the modeled position of ubiquinone. Functionally important structural elements are highlighted and labeled in *bold* (see text for details). Helices TM7, TM8 and TM12 from antiporter-like subunits are shown in *red*, *green* and *orange*. Charged amino acid residues, crucial for proton translocation and coupling, are shown as *sticks*. (**c**), Positions of the redox cofactors in complex I. Position of ubiquinone is modeled based on its expected distance from cluster N2 and its location close to the surface of the membrane, facing the cavity formed between subunits NuoB/D. *Blue arrows* show the main electron transfer pathway between FMN and UQ. *Green arrow* shows electron transfer pathway to cluster N1a, likely serving for temporary storage of electrons and thus reducing ROS production

1.3 Structure of the Peripheral Arm

The peripheral arm of complex I from *T. thermophilus* in itself is a remarkably stable assembly, although it dissociates easily from the membrane arm during purification (Hinchliffe et al. 2006; Hinchliffe and Sazanov 2005). Its crystallographic structure revealed the molecular architecture of this 280 kDa subcomplex of eight subunits, seven of which are core subunits and one, Nqo15, is organism specific (Sazanov and Hinchliffe 2006).

The peripheral arm contains the NADH binding site, formed within subunit Nqo1 (NuoF) (Berrisford and Sazanov 2009), at least a part of quinone binding site (not completely resolved yet), and all redox centres, including flavin mononucleotide (FMN), eight redox active iron-sulfur clusters (conserved between all enzymes), and an additional cluster (found in some bacterial complexes I) (Fig. 1.1c). The latter cluster, N7, is separated by more than 20 Å from the chain of redox active clusters, and hence is not involved in electron transport. It likely presents an evolutionary remnant (Sazanov and Hinchliffe 2006). The NADH and quinone binding sites are separated by the distance of nearly 100 Å. A non-covalently bound FMN, coordinated by subunit Nqo1, lies at the deep end of the solvent-exposed cavity containing the NADH-binding site. During the catalytic cycle two electrons are transferred from NADH to FMN as a hydride ion. Upon binding, the nicotinamide ring of NADH forms a stacking interaction with the isoalloxazine ring of FMN, thus providing a favourable geometry for fast hydride transfer between C^{4N} of NADH and N^5 of FMN (Berrisford and Sazanov 2009). Further, electrons are transferred one by one to quinone along the chain of clusters N3→N1b→N4→N5→N6a →N6b→N2 (Fig. 1.1c). From cluster N2 electrons tunnel to the quinone, most likely bound in the crevice formed between subunits Nqo4 and 6 (NuoD and B). The distances between neighbouring redox centres in the chain are within 14 Å, the maximal distance of physiological electron transfer (Page et al. 1999). Most EPR-visible clusters of the chain are equipotential (E_{m7} −250 mV), with the exception of high-potential cluster N2 (E_{m7} −100 mV), while the one-electron redox potentials of FMN are −300 mV ($FMNH_2$/flavosemiquinone) and −390 mV (flavosemiquinone/oxidized flavin) (Sled et al. 1994). Clusters N5 and N6b (Fig. 1.1c) are EPR-silent due to their low potentials, resulting in an alternating energy landscape along the chain (Roessler et al. 2010). The geometrical arrangement of the cofactors, combined with the favourable values of the redox potentials of neighbouring centres, allows electrons to tunnel between FMN and cluster N2 in the microsecond time range demonstrated experimentally (Verkhovskaya et al. 2008) and is consistent with theoretical estimates (Hayashi and Stuchebrukhov 2010).

Binuclear cluster N1a, coordinated by subunit Nqo2 (NuoE), does not belong to the main redox chain (Fig. 1.1c). It is, however, conserved in complex I from all species, which suggests it has a functional role. Found in an hydrophobic environment, cluster N1a is 12.3 Å away from FMN, has a one-electron potential of −370 mV (in bovine complex I) and can thus reduce flavosemiquinone (FSQ) efficiently (Sazanov and Hinchliffe 2006). It was suggested that N1a plays an important role in reducing ROS production by complex I (Sazanov 2007; Sazanov and Hinchliffe 2006). Under physiological steady state conditions all EPR-visible iron-sulfur clusters are reduced (Kotlyar et al. 1990). Both NADH and quinone are two electron donors, thus at any time complex I carries an even number of electrons. Because there are seven clusters in the main chain (5 reducible, i.e. EPR-visible), in the absence of N1a, one electron would nearly always reside on FSQ, which is exposed to the solvent after NAD^+ dissociates (Berrisford and Sazanov 2009). Solvent-exposed FSQ is an efficient electron donor to cytoplasm-dissolved oxygen and hence the source of ROS. Cluster N1a can temporary store electrons and, unlike

FMN, it is shielded from the solvent by protein, preventing electron leak. Flavosemiquinone redox potential would depend on the redox state of nearby cluster N3, and so as soon as the main redox chain is oxidised by quinone, N1a can 'release' its electron via flavosemiquinone to the higher potential clusters (Sazanov 2007; Sazanov and Hinchliffe 2006).

Details of the subunits' folds are described in (Sazanov and Hinchliffe 2006). The origin of the peripheral arm can be traced back to hydrogenases (Friedrich 2001), which are often built like combinations of lego blocks (Vignais et al. 2001). The evolutionary ancestors include different types of ferredoxins (subunits Nqo2 and Nqo9), FeFe-hydrogenases (N-terminus of subunit Nqo3), molybdopterin-containing enzymes (C-terminus of subunit Nqo3) and NiFe-hydrogenases (subunits Nqo4 and Nqo6). Apparently, individual subunits or subcomplexes have been added to the original core of Ni-Fe hydrogenase scaffold in the course of evolution, 'adjusting' to suit available electron donors, and resulting in an unusually long electron transfer chain (see also Chaps. 6 and 7 in this book).

A notable structural feature is the unusual coordination of cluster N2 by tandem cysteines, Cys45 and Cys46 of subunit Nqo6 (NuoB) in *T. thermophulus*, which is very rare among iron-sulphur containing proteins. Apart from complex I, the Protein Data Bank (PDB) contains a single entry, APS reductase (Chartron et al. 2006), with a cubane iron-sulfur cluster coordinated by tandem cysteines. Interestingly, similar consecutive cysteines are also specific to oxygen-resistant Ni-Fe hydrogenases. These are evolutionarily related to complex I (as well as to non oxygen resistant hydrogenases), where a total of six cysteines, present around a proximal cluster (equivalent of N2), play an important role in oxygen resistance (Goris et al. 2011). In complex I, the unusual coordination leads to a strained conformation of the cysteines and is suggested to play a functional role (Berrisford and Sazanov 2009). Indeed, crystallographic structures of the peripheral arm reduced by NADH and/or dithionite suggest either Cys45 or Cys46 (depending on the redox state) disconnects from the cluster upon reduction, thus possibly playing a role in generating structural changes and/or protonation of bound quinone (Berrisford and Sazanov 2009). Reduction of the peripheral arm leads to structural changes at the interface with the membrane domain: the four-helix bundle of subunit Nqo4/NuoD shifts by about 1 Å towards the membrane and helices H1 and H2 from Nqo6/NuoB move "sideways", likely playing an important role in the coupling mechanism.

1.4 Structure of the Membrane Arm: Fold and Proton Translocation Pathways

Comprising seven core subunits in *E. coli* complex I, the membrane arm spans the membrane with a total of 63 helices (Fig. 1.1a, b) and presents one of the largest membrane-residing protein assemblies (Efremov et al. 2010). It includes subunits NuoH/A/J/K/N/M/L (Nqo8/7/10/11/14/13/12). A major part of the domain, subunits

NuoK/L/M/N, is related to multi-subunit Mrp cation/H^+ antiporters (Mathiesen and Hagerhall 2003).

NuoH likely forms the major interaction surface between membrane and peripheral arms (Efremov et al. 2010) (Fig. 1.1a). NuoH loops facing subunits NuoB/D have well conserved sequences. It is at the interface of these subunits that the likely quinone binding site is formed and it is at this site where the major part of the redox energy is transformed to conformational changes. NuoH (Nqo8) is the only subunit for which a molecular model is missing, although the arrangement of its 8 TM helices was revealed at resolution of 4.5 Å (Efremov et al. 2010). Six of these helices are highly tilted (by up to about 40°) relative to the lipid bilayer normal, consistent with a plausible role for NuoH in conformational coupling.

1.4.1 Subunits NuoA, J and K

Small subunits NuoA, J and K, spanning the membrane with 3, 5 and 3 helices, respectively, are arranged in a compact intricate bundle (Figs. 1.1a, b and 1.2c, d) separating NuoH (Nqo8) from antiporter-like subunits. A plausible proton translocation pathway is formed inside the bundle and at its interface with subunit NuoN by acidic residues and cavities likely filled with water (Fig. 1.2c, d). It includes conserved residues, among which are functionally important $_K$Glu36, $_K$Glu72 (prefix denotes subunit name) (Kao et al. 2005; Kervinen et al. 2004), and a fragment of $_J$TM3 with a proline-less π-bulge in the middle of the membrane, rendering this helix flexible and likely functionally important (Efremov and Sazanov 2011b).

1.4.2 Architecture of Antiporter-Like Subunits

Three antiporter-like subunits, NuoL, M and N, share a similar fold of 14 TM helices, unique among membrane proteins of known structure. The remote subunit NuoL contains an additional carboxy-terminal extension beginning with TM15 (the most distal helix of the membrane domain), followed by the amphipathic helix HL, residing on the cytoplasmic surface of the membrane, and ending with TM16, harboured at the interface with subunits NuoJ, K and N (Fig. 1.1a, b). The helix HL is likely similarly arranged across species, since residues contacting other subunits are relatively well conserved, unlike the rest of the helix (Efremov and Sazanov 2011b), as might be expected for a purely mechanical structural element.

The assembly of antiporter-like subunits is additionally stabilized, on the opposite side of the domain, by extended and well-ordered β-hairpins contacting neighbouring subunits *via* carboxy-terminal amphipathic helices. This unexpected β-hairpin-helix element (termed the βH element, Fig. 1.1b) extends over the entire length of the antiporter-like subunits, thus contributing to the stability of the complex.

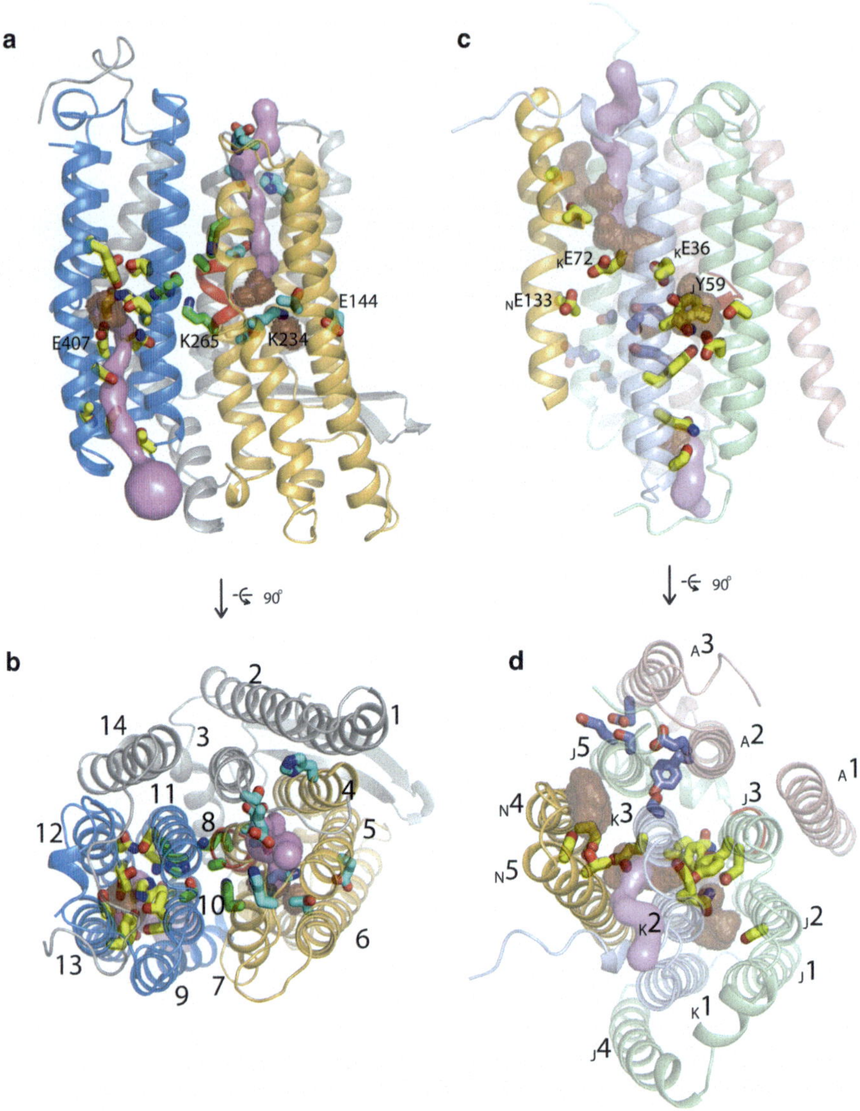

Fig. 1.2 Proton translocation pathways in: (**a**) and (**b**), antiporter-like subunits, shown with subunit NuoM as an example; (**c**) and (**d**), the forth channel formed by small hydrophobic subunits NuoK/J/A and the surface of subunit NuoN. Polar residues forming proton-translocation pathway are shown as *sticks*, hydrophilic cavities (calculated in program VOIDOO (Kleywegt and Jones 1994)) surrounded by polar and charged residues constituting the channels, are shown in *brown*. The tunnels connecting clusters of polar residues and cavities to the cytoplasmic and periplasmic surfaces of the protein, calculated in CAVER (Petrek et al. 2006), are shown in *pink*. Regions where TM helices are disrupted by π-bulge are colored in *red*. In (**a**) and (**b**), symmetry related core helices of NuoM, TM 4–8 and TM9-13, are shown in *wheat* and *marine*, respectively. Other helices are in *grey*. Residues constituting the cytoplasmic half-channel, connecting and periplasmic half-channel are shown in *cyan*, *green* and *yellow* respectively. Key residues are labeled. In (**a**), TM9 is omitted for clarity. In (**c**) and (**d**), subunits are color coded as in Fig. 1.1a, b and helices are labeled. Residues in the main channel are in *yellow* and in the alternative pathway in *purple*. In (**c**), some helices of subunits NuoN/K/J are omitted for clarity

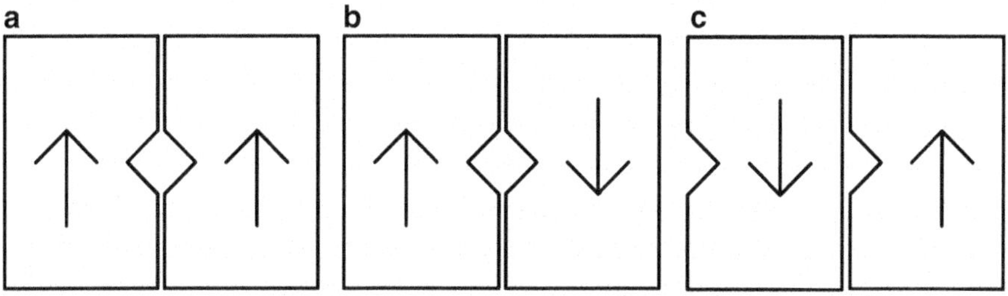

Fig. 1.3 Internal symmetry in membrane proteins. Schematic representation of arrangement of symmetry-related domains in membrane proteins with two structural repeats. (**a**) and (**b**), previously described mutual arrangements of the domains and (**c**), novel arrangement found in antiporter-like subunits of complex I

The fold of 14 conserved helices can be subdivided into a highly-conserved ten trans-membrane (TM) helical core (TM helices 4–13) and the less conserved TM1-3 and TM14 (Figs. 1.1 and 1.2). In higher metazoans $_{N}$TM1-3 are absent (Birrell and Hirst 2010; Mathiesen and Hagerhall 2002) while some insects and worms lack $_{L}$TM1 (Efremov and Sazanov 2011b). In the conserved core two sets of five helices are related to each other by a unique symmetry transformation along a pseudo-two-fold screw axis. Namely, TMs 4–8 can be superimposed on TMs 9–13 by a rotation of about 180° along the axis lying in the membrane plane and a shift directed along the long axis of the domain. Symmetry-related sets of helices are common in transporters. They have been observed in a parallel or anti-parallel fashion (Vinothkumar and Henderson 2010) (Fig. 1.3a, b), but have always been found in a face-to-face orientation. However, in complex I, they are oriented face-to-back, representing a novel arrangement (Fig. 1.3c). The symmetry of the helical sets suggests also that the core was formed by a gene duplication (Vinothkumar and Henderson 2010).

The symmetry related helices TM7 and TM12 are interrupted in the middle of the bilayer by an extended loop of 5–7 residues. The tips of these loops contain a proline that is conserved between all three antiporter-like subunits. Such helices are considered as functionally important for proton or ion transport because they introduce flexibility and charge to the middle of the membrane (Screpanti and Hunte 2007; Vinothkumar and Henderson 2010). The broken helices are strategically located: TM7s contact helix HL, while TM12s are placed at the interfaces of antiporter-like subunits. In addition to these broken helices, TM8, found at the interface of symmetry related domains, is partly unwound in the middle by a proline-less kink disrupting local secondary structure, similarly to $_{J}$TM3. Such π-bulges (Cooley et al. 2010) are usually found at protein functional sites, pointing towards a functional importance of TM8.

1.4.3 Proton Translocation Channels in Antiporter-Like Subunits

Each symmetry-related set of five helices contains an apparent half-channel for proton translocation: TM4-8 – cytoplasmic half; TM9-13 – periplasmatic half

(Efremov and Sazanov 2011b) (Fig. 1.2a, b). The half-channels are formed by combinations of conserved polar residues and polar cavities likely filled with water molecules, some of which were identified by crystallography (Efremov and Sazanov 2011b).

At the bottom of each half-channel, roughly in the middle of the membrane, there are functionally indispensable lysine residues (with a single exception of Glu407 in the NuoM periplasmic half-channel) that are conserved between all complexes I and Mrp antiporters. In the cytoplasmic half-channel these lysines are the last residue of the periplasmic half of discontinuous TM7: $_{L}$Lys 229, $_{M}$Lys234 and $_{N}$Lys217, termed LysTM7. In the periplasmic half-channel the key residues are $_{L}$Lys399, $_{N}$Lys395 and $_{M}$Glu407 on TM12 (termed Lys/GluTM12). Positions of the side chains of LysTM7 and Lys/GluTM12 are approximately related by interdomain symmetry.

Unexpectedly, a strictly conserved and functionally essential glutamate in the middle of TM5 ($_{L}$Glu144, $_{M}$Glu144 and $_{N}$Glu133, termed GluTM5 here), suggested earlier to play a central role in proton translocation (Efremov et al. 2010; Torres-Bacete et al. 2007; Euro et al. 2008; Nakamaru-Ogiso et al. 2010), is wedged at the interface of TMs 5 and 6, and is exposed to both the cytoplasmic half-channel and the interface with the adjacent subunit. GluTM5 is just 5–6 Å away from LysTM7 of the same subunit and around 3 Å further from Lys/GluTM12 of the neighbouring subunit (in NuoN it is close to $_{K}$Gly72). Thus, it has the capacity to approach these two functionally important lysines alternatively in the course of the catalytic cycle.

The half-channels are connected by conserved charged and polar residues in the middle of the membrane (Fig. 1.2a, b). The link is most obvious in NuoN: LysTM7 – W (observed water molecule) – Lys247 – W – His305 – W – LysTM12. Although the distances between ionisable residues and crystallographically resolved water molecules are 4–6 Å, there are no obstacles between them. This should allow for efficient proton transfer due to conformational flexibility and the likely presence of additional water molecules. Some waters may be coordinated (also in NuoL and M) by the exposed backbone carbonyls from the π-bulge of TM8. Additionally, Tyr231 and Tyr333 nearby may participate in proton transfer, as suggested for a conserved tyrosine in cytochrome *c* oxidase (Belevich et al. 2010). Importantly, central $_{N}$Lys247 (Lys265 in NuoM) found on the TM8 π-bulge is invariant and essential for activity (Amarneh and Vik 2003; Euro et al. 2008; Torres-Bacete et al. 2007). In NuoM, the pathway between the channels involves His248, Lys265, His348 and invariant His322, as well as resolved and putative water molecules. In NuoL, an analog of $_{N}$Lys247 is absent, but in this area there is His254 and also Lys342, both invariant. Therefore, the likely pathway between the channels involves His254, Lys342, His338, His334 and water molecules.

A complete proton translocation pathway through each antiporter-like subunit is formed by two half-channels linked in the middle of the membrane. Additionally, cavities at the interfaces NuoL/M, NuoM/N and NuoN/K/J might be used for 'side-entry' of protons *via* GluTM5. However, such a pathway is less likely compared to the cytoplasmic half-channel, since the NuoL/M and M/N interfaces are not extensive and are less suited for proton transport. The pathway through the cytoplasmic half-channel is also more likely since residues lining it are more conserved than those at the interfaces, and it is consistent with the internal symmetry of the protein. The overall design, with two interacting anti-symmetrical half-channels, involves complete subunits in proton translocation, rather than a single isolated channel of 3–4 helices. This would allow high coupling efficiency between protein conformation and proton motive force.

The suggested proton translocation pathway is unusual and novel, which prompts us to discuss potential alternatives. First, is a complete single proton channel formed in one of the two symmetry-related domains? It is less probable, because both half-channels contain crucial Lys or Glu residues as well as conserved polar residues. One can hypothesize that the ion pair GluTM5/LysTM7 acts as a conformational switch inducing proton-translocating structural changes in the second channel. However, the presence of conserved polar residues and cavities linked to the cytoplasm in the first half-channel then remains unexplained. Second, may both channels function as proton pumps? In this case complex I would be potentially able to translocate at least six protons per cycle, but this stoichiometry has never been observed and is not thermodynamically feasible. Homology modelling provides an additional argument supporting the model of the single proton channel consisting of two half-channels. Models of Mrp antiporter subunits MrpA and MrpD (NuoL and NuoM homologues, respectively) also contain a similar proton translocation pathway comprising two half-closed channels connected by charged residues (Efremov and Sazanov 2011b), whilst cation antiport probably involves other subunits or subunit interfaces.

Sequence similarity suggests that antiporter-like subunits in chloroplast Ndh complexes and membrane-bound hydrogenases are also likely to have a similar design.

1.4.4 Antiporter-Like Subunits Do Not Contain Quinone Binding Sites

The molecular structure of the membrane domain provides no evidence for the presence of quinone binding site(s) in any of the antiporter-like subunits, as has been discussed widely in literature. The arguments for existence of such sites were as follows. First, photoaffinity labelling experiments with analogues of specific hydrophobic inhibitors showed labelling of ND2 (Nakamaru-Ogiso et al. 2010a), ND4 (Gong et al. 2003) and ND5 (Nakamaru-Ogiso et al. 2003) subunits of bovine complex I, homologous to *E. coli* NuoN, M and L, respectively. Second, the presence of the quinone-binding signature motif (L/A-X_3-H-$X_{2/3}$-L/T/S) (Fisher and Rich 2000) was suggested in sequences of the antiporter-like subunits. The signature motif itself is weak and more indicative of a true quinone-binding site only when part of a highly conserved region. Sequence motifs centred on $_{L}His^{334}$, $_{L}His^{338}$, $_{M}His^{241}$, $_{M}His^{322}$, $_{M}His^{348}$ and $_{N}His^{224}$ have been discussed as potential quinone binding sites (Fisher and Rich 2000; Amarneh and Vik 2003; Nakamaru-Ogiso et al. 2010b).

Third, Amarneh and Vik (2003) observed inhibition of NADH oxidase activity by decylubiquinone in several mutants, including $_{N}His^{224}$.

The structure shows that the majority of the above mentioned histidines are, in fact, buried deep inside the protein and are parts of putative proton translocation channels. Only $_{M}His^{241}$ and $_{N}His^{224}$ (structurally and sequentially conserved) are located on TM7 pointing outside the subunit. However, they interact directly with helix HL, which is likely the primary reason for their conservation. Importantly, inhibition or lack of activation with decylubiquinone were observed for mutations of other surface residues interacting with HL, namely $_{N}Lys^{158}$ and $_{N}Tyr^{300}$ (Amarneh and Vik 2003). Recently, analogous residues in NuoL and NuoM ($_{L}Lys^{169}$, $_{M}Lys^{173}$, $_{L}Gln^{236}$ and $_{M}His^{241}$), all interacting with HL, were mutated, and the mutants display similar behaviour for all three antiporter-like subunits (Michel et al. 2011). Consequently, the effect of these mutations cannot be attributed to disruption of quinone binding sites. Rather, it is due to interference with conformational coupling, which is likely dependent on interaction between helix HL and the antiporter-like subunits. Both proton-pumping and oxidoreductase activities were significantly affected in these mutants (Michel et al. 2011), confirming the essential coupling role of helix HL. Labelling with photoaffinity inhibitor analogues may have been unspecific, due to the presence of hydrophobic crevices at the interfaces between subunits. Global conformational changes upon enzyme reduction or inhibitor binding would explain the effects observed in labelling experiments (Gong et al. 2003; Nakamaru-Ogiso et al. 2003).

1.4.5 Does NuoN Translocate Protons?

The structure indicates that all antiporter-like subunits perform active proton transport. The suggestions made by several groups that NuoN does not pump protons and/or contains bound quinone cofactor (Q_{Ns}) (Ohnishi et al. 2010a, b; Birrell and Hirst 2010) are not supported by the structure. The arguments in favour of different role of NuoN are as follows. One tightly bound quinone molecule (Shinzawa-Itoh et al. 2010), as well as semiquinone radicals have been observed in bovine complex I (Ohnishi 1998). EPR signals from two semiquinone species were detected: fast-relaxing semiquinone (Q_{Nf}), sensitive to the membrane potential and interacting with cluster N2, and slow-relaxing semiquinone (Q_{Ns}), insensitive to trans-membrane potential and not interacting with N2 (Ohnishi 1998; Ohnishi et al. 2010b). Additionally, mutations of GluTM5 in NuoN do not affect activity as drastically as those in NuoL and M (Amarneh and Vik 2003) and, as noted above (Mathiesen and Hagerhall 2002; Birrell and Hirst 2010), helices TM1-3 are absent in NuoN from higher metazoans.

However, TMs1-3 are found at the periphery of antiporter-like subunits and do not belong to the conserved functional core (helices 4–13). Furthermore, the environment of NuoN is fully preserved in the structure (all subunits contacting it are present), but no bound cofactors are observed, while some ordered lipid chains and bound detergent molecules could be clearly seen in the density. Also, complex I from *Y. lipolytica* contains only 0.2–0.4 mol/mol of tightly bound ubiquinone (Drose et al. 2002) and *T. thermophilus* enzyme does not contain any (Minhas and Sazanov, unpublished), but these enzymes are fully active. Hence, one semiquinone species observed by EPR (Q_{Ns}) can represent the population of substrate quinone molecules bound to complex I but fully embedded in the membrane (thus far away from cluster N2; these molecules may have escaped from active site before reduction reaction was completed), while the other species (Q_{Nf}) can represent quinone bound in the Q-site and interacting with cluster N2.

The difference in effects of GluTM5 mutations can be explained if GluTM5 plays important role in communicating conformational changes between antiporter-like subunits at the interfaces of NuoN-NuoM and NuoM-NuoL (as discussed below). However, in NuoN, the equivalent $_N$Glu133 does not face another antiporter-like subunit. Moreover, conserved $_K$Glu72 is located close to $_N$Glu133 and can probably partially compensate for the absence of $_N$Glu133 in the mutants. Consequently, mutations of $_N$Glu133 may not impede overall conformational change and catalytic activity, even though proton pumping by NuoN might be compromised in the mutant, resulting in a drop of stoichiometry from 4 to 3, which is difficult to measure experimentally. $_N$Glu^{133} is not conserved in worms (Birrell and Hirst 2010; Michel et al. 2011) which, however, show other sequence deviations and also lack $_K$Glu^{72} and $_J$Tyr^{59}. It is possible that channel 4, involving all three residues, is not functional in worms. The three crucial lysines (217, 247 and 395) are conserved in NuoN from these species, suggesting that this subunit is still involved in proton translocation. Mutations of any of these lysines in NuoN completely abolish activity in *E. coli* (Amarneh and Vik 2003), advocating the role of NuoN in active proton translocation.

In summary, the structure does not provide support for the presence of any additional quinone-binding sites in antiporter-like subunits, nor for proposals that subunit NuoN is functionally different from NuoL/M. The presence of a single Q-site at the interface of the two main domains, involving subunit NuoH (Fig. 1.1a), is consistent with all available functional and mutagenesis data.

1.5 Implications for the Coupling and Proton-Pumping Mechanisms

Advances in resolving high-resolution structure allow us now to comprehend many controversial aspects of the mechanism of complex I, although raising simultaneously new questions. In combination, all the structural features indicate unambiguously that complex I operates purely by a conformation-driven mechanism.

Based on the available structural data the following sequence of conformational changes can be envisaged. Reduction of the hydrophilic domain by NADH induces shifts of helix $_B$H1 and the four-helix bundle from NuoD (Berrisford and Sazanov 2009), located at the interface with the membrane domain (Figs. 1.1, and 1.4).

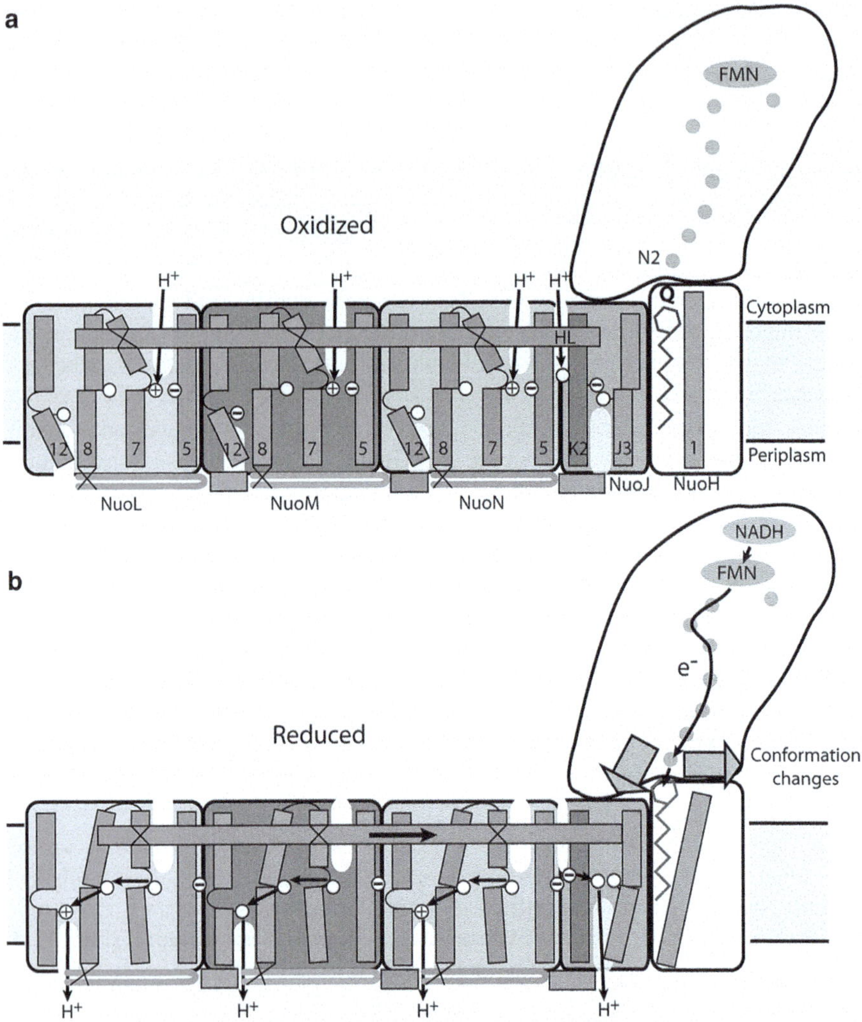

Fig. 1.4 Suggested mechanism of coupling and proton translocation in complex I. (**a**), Oxidised state. (**b**), Reduced state. Crucial charged residues (GluTM5, LysTM7, Lys/GluTM12, Lys/HisTM8 from NuoL/M/N, as well as $_{K}Glu^{72}$ and $_{K}Glu^{36}$) are indicated by *circles* showing charge of the residues. In NuoL/M/N, LysTM7 from the first half-channel is assumed to be protonated in the oxidised state. Conformational changes upon ubiquinone reduction are transmitted from the Q-site to antiporter-like subunits by helix HL (cytoplasmic side) and the βH element (periplasmic side). They move GluTM5 away from LysTM7, forcing lysine to donate its proton into the link between the two half-channels and eventually to Lys/GluTM12. Upon return to the oxidised state, GluTM5 moves back, LysTM7 is protonated from the cytoplasm and the pump is loaded again, whilst Lys/GluTM12 ejects its proton into the periplasm. The fourth proton per cycle is translocated at the interface of NuoN, K and J

These structural changes can induce conformational changes in subunits NuoA/J/K, through either direct contact (NuoD to NuoA) or *via* NuoH. Consistently, cross-links between NuoA and J disappear upon reduction of complex I (Berrisford et al. 2008). Additionally, structural changes must be generated in the Q-site, most probably by subunit NuoH, where most of the energy of electrons fed from NADH is released.

Communication of the conformational changes may proceed in part *via* the long rigid helix TM1 of NuoH (Efremov et al. 2010) (Fig. 1.1), which approaches close to TMs1-3 of NuoJ with one end and, with the other end, close to helix $_{B}$H1 linked to Fe-S cluster N2 (Berrisford and Sazanov 2009). The importance of helix HL for the complex assembly and stability was demonstrated by recent biochemical studies using constructs with HL and TM15 or 16 truncated in several positions as well as with a complete NuoL deletion mutant (Steimle et al. 2011; Torres-Bacete et al. 2011). Steime et al. (Steimle et al. 2011) constructed a plasmid containing all complex I subunits, including a truncated NuoL, and over expressed it in *E. coli*. The group of Yagi employed site directed mutagenesis of *E. coli* chromosomal DNA to produce the mutants (Torres-Bacete et al. 2011). In the first case, truncation of HL or deletion of NuoL resulted in fully assembled enzyme with high oxidoreductase activity but with lower $H^+/2e^-$ stoichiometry, while in the second case elimination of part of TM16, as well as various truncations of HL and TM15, lead to incomplete complex assembly and total loss of NADH-ubiquinone oxidoreductase activity. In spite of strong controversy over the results, likely arising from the different strategies used to construct the mutants, the studies agree on the importance of the functional role of the helix HL. Further experiments need to be performed to resolve the ambiguity.

Because helix HL interacts with subunits NuoJ and K *via* $_{L}$TM16, conformational changes in NuoJ and K can move HL, piston-like, along the membrane surface of the domain. A shift of helix HL will drag flexible TM7 of NuoL/M/N, with which it interacts strongly, altering in turn the environment of LysTM7 and changing its distance from GluTM5. Lysine is an unusual proton translocator due to its high pKa (~10) which, however, can be lowered by 3–4 units in a membrane buried residue, as in the ApcT transporter (Shaffer et al. 2009). Movement of GluTM5 relative to LysTM7 will alter the pKa of the latter.

The βH element, together with helix HL, may coordinate conformational changes. It interacts with TM12 through the C-terminus of TM14 (linked to helix CH and hydrogen bonded to semi-conserved $_{L}\mathrm{Trp}^{67}/_{M}\mathrm{Trp}^{71}$ from the hairpin) (Fig. 1.1b). Additionally, two half-channels from neighbouring subunits can interact via an invariant proline in the TM12 intra-helical loop ($_{M}\mathbf{Pro}^{399}$ and $_{N}\mathbf{Pro}^{387}$), which contacts GluTM5 from subunit NuoL and M, respectively. Conformational changes may press the proline against TM5, leading to a change in distance between GluTM5 and LysTM7. The flexibility of TM5 can be provided by another proline (conserved in NuoM, Pro^{149}), which introduces a slight kink in the helix. Although this proline is absent in NuoL, $_{L}$TM5 contains three conserved glycines.

The interaction between two half-channels within an antiporter-like subunit can be mediated by TM8, found between half-channels and connected to the coupling elements. Invariant ${}_{N}Lys^{247}$, ${}_{M}Lys^{265}$ and ${}_{L}His^{254}$, connecting two channels, are located on TM8 near its flexible kink. The βH element includes conserved (also in Mrp antiporters) salt bridges between the hairpin (${}_{M}\mathbf{Asp^{84}}$) and the C-terminus of TM8 (${}_{M}\mathbf{Arg^{273}}$), while the N-terminus of TM8 is connected to TM7 by a very short rigid loop containing a conserved proline (${}_{M}\mathbf{Pro^{252}}$). In this way, TM8 is linked to both HL and βH coupling elements.

The conformational changes coordinated by both the βH element and TM8, can modulate the pK_a of Lys/GluTM12 *via* interactions with the exposed C-terminus of TM12a or with invariant charged residues nearby – ${}_{N}His^{305}$, ${}_{M}His^{322}$ and ${}_{L}His^{334}$. Additionally, interaction of GluTM5 with Lys/GluTM12, both exposed to the inter-subunit cavities, can facilitate switching between the two conformations (Fig. 1.4). In the first, observed in the current structure and probably representing the oxidised state, GluTM5 is closer to LysTM7 of the same subunit. Upon enzyme reduction, GluTM5 can approach Lys/GluTM12 on the nearby subunit, increasing its pKa and leading to its protonation via the first channel of that subunit. This way, protonation of the crucial residue in the opposite half-channels would be achieved in different parts of the catalytic cycle, as required for the directionality of the pump. It is not exclusively necessary, as LysTM12 in NuoL clearly operates without such a partner. Mutagenesis studies support a "switch" possibility. When GluTM5 in NuoM is shifted one helix turn up or down (Torres-Bacete et al. 2009), the activity is retained. Only in these positions is the residue mutated to glutamate close to LysTM7 and is also at the interface with NuoN. In other cases activity was lost even if the residue is close to LysTM7 (E144A/M146E). The suggested translocation cycle is shown in Fig. 1.4, assuming that LysTM7 is protonated and that Lys/Glu TM12 is deprotonated in the oxidized state.

A mechanism recently proposed by Brandt (Brandt 2011) postulates only two functioning proton pump modules in complex I, one in each of the two halves of the membrane domain (divided as NuoHAJKN and NuoLM sub-complexes), while the catalytic cycle includes two "strokes", each translocating two protons. This model implies that either subunit NuoM or NuoL is not functional, which is highly unlikely in view of the many conserved and essential charged residues found in the proton translocation channels in both subunits. In our opinion, four protons are translocated through four channels simultaneously in a single "stroke". Even though it involves a single large drop in energy, it is important to note that this "stroke" is effectively divided into four parallel steps in four channels (as in a parallel electrical circuit), and so there is no contradiction with the general principles of bioenergetics, in which large energy drops are usually broken into smaller intermediate steps.

The structure of the complex I is now nearly complete. Step by step it is revealing a surprising and previously unthinkable intricate architecture of the enzyme. We are slowly approaching a true understanding of how redox energy from NADH, binding at the tip of the hydrophilic domain, is used in the membrane domain, at distances of up to ~300 Å away, to generate trans-membrane potential.

References

Amarneh B, Vik SB (2003) Mutagenesis of subunit N of the *Escherichia coli* complex I. Identification of the initiation codon and the sensitivity of mutants to decylubiquinone. Biochemistry 42:4800–4808

Balaban RS, Nemoto S, Finkel T (2005) Mitochondria, oxidants, and aging. Cell 120:483–495

Belevich I, Gorbikova E, Belevich NP, Rauhamaki V, Wikstrom M, Verkhovsky MI (2010) Initiation of the proton pump of cytochrome c oxidase. Proc Natl Acad Sci USA 107:18469–18474

Berrisford JM, Sazanov LA (2009) Structural basis for the mechanism of respiratory complex I. J Biol Chem 284:29773–29783

Berrisford JM, Thompson CJ, Sazanov LA (2008) Chemical and NADH-induced, ROS-dependent, cross-linking between subunits of complex I from *Escherichia coli* and *Thermus thermophilus*. Biochemistry 47:10262–10270

Birrell JA, Hirst J (2010) Truncation of subunit ND2 disrupts the threefold symmetry of the antiporter-like subunits in complex I from higher metazoans. FEBS Lett 584:4247–4252

Brandt U (2006) Energy converting NADH:quinone oxidoreductase (complex I). Annu Rev Biochem 75:69–92

Brandt U (2011) A two-state stabilization-change mechanism for proton-pumping complex I. Biochim Biophys Acta 1807:1364–1369

Carroll J, Fearnley IM, Skehel JM, Shannon RJ, Hirst J, Walker JE (2006) Bovine complex I is a complex of 45 different subunits. J Biol Chem 281:32724–32727

Chartron J, Carroll KS, Shiau C, Gao H, Leary JA, Bertozzi CR, Stout CD (2006) Substrate recognition, protein dynamics, and iron-sulfur cluster in *Pseudomonas aeruginosa* adenosine 5′-phosphosulfate reductase. J Mol Biol 364:152–169

Clason T, Ruiz T, Schagger H, Peng G, Zickermann V, Brandt U, Michel H, Radermacher M (2010) The structure of eukaryotic and prokaryotic complex I. J Struct Biol 169:81–88

Cooley RB, Arp DJ, Karplus PA (2010) Evolutionary origin of a secondary structure: pi-helices as cryptic but widespread insertional variations of alpha-helices that enhance protein functionality. J Mol Biol 404:232–246

Dawson TM, Dawson VL (2003) Molecular pathways of neurodegeneration in Parkinson's disease. Science 302:819–822

Drose S, Zwicker K, Brandt U (2002) Full recovery of the NADH:ubiquinone activity of complex I (NADH:ubiquinone oxidoreductase) from Yarrowia lipolytica by the addition of phospholipids. Biochim Biophys Acta 1556:65–72

Efremov RG, Sazanov LA (2011a) Respiratory complex I: 'steam engine' of the cell? Curr Opin Struct Biol 21:532–540

Efremov RG, Sazanov LA (2011b) Structure of the membrane domain of respiratory complex I. Nature 476:414–420

Efremov RG, Baradaran R, Sazanov LA (2010) The architecture of respiratory complex I. Nature 465:441–445

Euro L, Belevich G, Verkhovsky MI, Wikstrom M, Verkhovskaya M (2008) Conserved lysine residues of the membrane subunit NuoM are involved in energy conversion by the proton-pumping NADH:ubiquinone oxidoreductase (complex I). Biochim Biophys Acta 1777:1166–1172

Fisher N, Rich PR (2000) A motif for quinone binding sites in respiratory and photosynthetic systems. J Mol Biol 296:1153–1162

Friedrich T (2001) Complex I: a chimaera of a redox and conformation-driven proton pump? J Bioenerg Biomembr 33:169–177

Gong X, Xie T, Yu L, Hesterberg M, Scheide D, Friedrich T, Yu CA (2003) The ubiquinone-binding site in NADH:ubiquinone oxidoreductase from *Escherichia coli*. J Biol Chem 278:25731–25737

Goris T, Wait AF, Saggu M, Fritsch J, Heidary N, Stein M, Zebger I, Lendzian F, Armstrong FA, Friedrich B, Lenz O (2011) A unique iron-sulfur cluster is crucial for oxygen tolerance of a [NiFe]-hydrogenase. Nat Chem Biol 7:310–318

Hayashi T, Stuchebrukhov AA (2010) Electron tunneling in respiratory complex I. Proc Natl Acad Sci USA 107:19157–19162

Hinchliffe P, Sazanov LA (2005) Organization of iron-sulfur clusters in respiratory complex I. Science 309:771–774

Hinchliffe P, Carroll J, Sazanov LA (2006) Identification of a novel subunit of respiratory complex I from *Thermus thermophilus*. Biochemistry 45:4413–4420

Hunte C, Zickermann V, Brandt U (2010) Functional modules and structural basis of conformational coupling in mitochondrial complex I. Science 329:448–451

Kao MC, Nakamaru-Ogiso E, Matsuno-Yagi A, Yagi T (2005) Characterization of the membrane domain subunit NuoK (ND4L) of the NADH-quinone oxidoreductase from *Escherichia coli*. Biochemistry 44:9545–9554

Kervinen M, Patsi J, Finel M, Hassinen IE (2004) A pair of membrane-embedded acidic residues in the NuoK subunit of *Escherichia coli* NDH-1, a counterpart of the ND4L subunit of the mitochondrial complex I, are required for high ubiquinone reductase activity. Biochemistry 43:773–781

Kleywegt GJ, Jones TA (1994) Detection, delineation, measurement and display of cavities in macromolecular structures. Acta Crystallogr D Biol Crystallogr 50:178–185

Kotlyar AB, Sled VD, Burbaev DS, Moroz IA, Vinogradov AD (1990) Coupling site I and the rotenone-sensitive ubisemiquinone in tightly coupled submitochondrial particles. FEBS Lett 264:17–20

Leif H, Sled VD, Ohnishi T, Weiss H, Friedrich T (1995) Isolation and characterization of the proton-translocating NADH: ubiquinone oxidoreductase from *Escherichia coli*. Eur J Biochem 230:538–548

Mathiesen C, Hagerhall C (2002) Transmembrane topology of the NuoL, M and N subunits of NADH:quinone oxidoreductase and their homologues among membrane-bound hydrogenases and bona fide antiporters. Biochim Biophys Acta 1556:121–132

Mathiesen C, Hagerhall C (2003) The 'antiporter module' of respiratory chain complex I includes the MrpC/NuoK subunit – a revision of the modular evolution scheme. FEBS Lett 549:7–13

Michel J, Deleon-Rangel J, Zhu S, Van Ree K, Vik SB (2011) Mutagenesis of the L, M, and N subunits of complex I from *Escherichia coli* indicates a common role in function. PLoS One 6:e17420

Moparthi VK, Hagerhall C (2011) The evolution of respiratory chain complex I from a smaller last common ancestor consisting of 11 protein subunits. J Mol Evol 72:484–497

Nakamaru-Ogiso E, Sakamoto K, Matsuno-Yagi A, Miyoshi H, Yagi T (2003) The ND5 subunit was labeled by a photoaffinity analogue of fenpyroximate in bovine mitochondrial complex I. Biochemistry 42:746–754

Nakamaru-Ogiso E, Han H, Matsuno-Yagi A, Keinan E, Sinha SC, Yagi T, Ohnishi T (2010a) The ND2 subunit is labeled by a photoaffinity analogue of asimicin, a potent complex I inhibitor. FEBS Lett 584:883–888

Nakamaru-Ogiso E, Kao MC, Chen H, Sinha SC, Yagi T, Ohnishi T (2010b) The membrane subunit NuoL (ND5) is involved in the indirect proton pumping mechanism of *Escherichia coli* complex I. J Biol Chem 285:39070–39078

Ohnishi T (1998) Iron-sulfur clusters/semiquinones in complex I. Biochim Biophys Acta 1364:186–206

Ohnishi ST, Salerno JC, Ohnishi T (2010a) Possible roles of two quinone molecules in direct and indirect proton pumps of bovine heart NADH-quinone oxidoreductase (complex I). Biochim Biophys Acta 1797:1891–1893

Ohnishi T, Nakamaru-Ogiso E, Ohnishi ST (2010b) A new hypothesis on the simultaneous direct and indirect proton pump mechanisms in NADH-quinone oxidoreductase (complex I). FEBS Lett 584:4131–4137

Page CC, Moser CC, Chen X, Dutton PL (1999) Natural engineering principles of electron tunnelling in biological oxidation-reduction. Nature 402:47–52

Petrek M, Otyepka M, Banas P, Kosinova P, Koca J, Damborsky J (2006) CAVER: a new tool to explore routes from protein clefts, pockets and cavities. BMC Bioinformatics 7:316

Roessler MM, King MS, Robinson AJ, Armstrong FA, Harmer J, Hirst J (2010) Direct assignment of EPR spectra to structurally defined iron-sulfur clusters in complex I by double electron–electron resonance. Proc Natl Acad Sci USA 107:1930–1935

Sazanov LA (2007) Respiratory complex I: mechanistic and structural insights provided by the crystal structure of the hydrophilic domain. Biochemistry 46:2275–2288

Sazanov LA, Hinchliffe P (2006) Structure of the hydrophilic domain of respiratory complex I from *Thermus thermophilus*. Science 311:1430–1436

Schapira AH (1998) Human complex I defects in neurodegenerative diseases. Biochim Biophys Acta 1364:261–270

Screpanti E, Hunte C (2007) Discontinuous membrane helices in transport proteins and their correlation with function. J Struct Biol 159:261–267

Shaffer PL, Goehring A, Shankaranarayanan A, Gouaux E (2009) Structure and mechanism of a Na+-independent amino acid transporter. Science 325:1010–1014

Shinzawa-Itoh K, Seiyama J, Terada H, Nakatsubo R, Naoki K, Nakashima Y, Yoshikawa S (2010) Bovine heart NADH-ubiquinone oxidoreductase contains one molecule of ubiquinone with ten isoprene units as one of the cofactors. Biochemistry 49:487–492

Sled VD, Rudnitzky NI, Hatefi Y, Ohnishi T (1994) Thermodynamic analysis of flavin in mitochondrial NADH:ubiquinone oxidoreductase (complex I). Biochemistry 33:10069–10075

Steimle S, Bajzath C, Dorner K, Schulte M, Bothe V, Friedrich T (2011) Role of subunit NuoL for proton translocation by respiratory complex I. Biochemistry 50:3386–3393

Torres-Bacete J, Nakamaru-Ogiso E, Matsuno-Yagi A, Yagi T (2007) Characterization of the NuoM (ND4) subunit in *Escherichia coli* NDH-1: conserved charged residues essential for energy-coupled activities. J Biol Chem 282:36914–36922

Torres-Bacete J, Sinha PK, Castro-Guerrero N, Matsuno-Yagi A, Yagi T (2009) Features of subunit NuoM (ND4) in *Escherichia coli* NDH-1: topology and implication of conserved Glu144 for coupling site 1. J Biol Chem 284:33062–33069

Torres-Bacete J, Sinha PK, Matsuno-Yagi A, Yagi T (2011) Structural contribution of the C-terminal segments of NuoL (ND5) and NuoM (ND4) subunits of complex I from *E. coli*. J Biol Chem 286:34007–34014

Verkhovskaya ML, Belevich N, Euro L, Wikstrom M, Verkhovsky MI (2008) Real-time electron transfer in respiratory complex I. Proc Natl Acad Sci USA 105:3763–3767

Vignais PM, Billoud B, Meyer J (2001) Classification and phylogeny of hydrogenases. FEMS Microbiol Rev 25:455–501

Vinothkumar KR, Henderson R (2010) Structures of membrane proteins. Q Rev Biophys 43:65–158

Walker JE (1992) The NADH – ubiquinone oxidoreductase (complex I) of respiratory chains. Q Rev Biophys 25:253–324

Yagi T, Matsuno-Yagi A (2003) The proton-translocating NADH-Quinone oxidoreductase in the respiratory chain: the secret unlocked. Biochemistry 42:2266–2274

Yip CY, Harbour ME, Jayawardena K, Fearnley IM, Sazanov LA (2011) Evolution of respiratory complex I: "supernumerary" subunits are present in the alpha-proteobacterial enzyme. J Biol Chem 286:5023–5033

Chapter 2
On the Mechanism of the Respiratory Complex I

Thorsten Friedrich, Petra Hellwig, and Oliver Einsle

Abstract The energy-converting NADH:ubiquinone oxidoreductase, respiratory complex I, couples the electron transfer from NADH to ubiquinone with a proton translocation across the membrane. Electron microscopy revealed the two-part structure of the enzyme complex. A peripheral arm, composed of globular subunits, extends into the aqueous phase. The arm contains the cofactors for the electron transfer reaction, namely one flavin mononucleotide and up to ten iron-sulfur (Fe/S) clusters. The other arm, the membrane arm, is embedded in the lipid bilayer and thus necessarily involved in proton translocation. The (ubi)quinone binding site is most likely located at the interface of the two arms. The oxidation of one NADH is coupled with the translocation of four protons (current consensus value). In this chapter, the binding of the substrates NADH and (ubi)quinone, the role of individual Fe/S clusters and the mechanism of proton translocation are discussed in the light of recent data obtained from our laboratories. We propose a model for the respiratory complex I, in which the electron transfer is coupled with the translocation of two protons by the (ubi)quinone redox chemistry and the residual two protons by conformational changes within the membrane arm.

Keywords *Escherichia coli* • EPR spectroscopy • FTIR spectroscopy • Iron-sulfur clusters • NADH:ubiquinone oxidoreductase • Proton translocation • (ubi)quinone

---

T. Friedrich (✉) • O. Einsle
Institut für Organische Chemie und Biochemie, Albert-Ludwigs-Universität, Albertstr. 21, 79104 Freiburg, Germany
e-mail: thorsten.friedrich@uni-freiburg.de

P. Hellwig
Faculté de chimie, Laboratoire de spectroscopie vibrationelle et électrochimie des biomolécules, Université de Strasbourg, 1 rue Blaise Pascal, 67008 Strasbourg, France

L. Sazanov (ed.), *A Structural Perspective on Respiratory Complex I: Structure and Function of NADH:ubiquinone oxidoreductase*, DOI 10.1007/978-94-007-4138-6_2,

2.1 Introduction

The enzyme complexes of respiratory chains transform the energy released by a redox reaction into the energy of an ion gradient across the membrane, in which they are embedded. This gradient is used for energy-consuming processes such as ATP synthesis, active transport and motion. The first enzyme complex of many electron transfer chains is the energy-converting NADH:ubiquinone oxidoreductase, also called respiratory complex I (Weiss et al. 1991; Walker 1992; Friedrich et al. 1995; Ohnishi 1998; Vinogradov 1998; Yagi and Matsuno-Yagi 2003; Brandt 2006; Friedrich and Pohl 2007; Sazanov 2007). Complex I couples the transfer of two electrons from NADH to (ubi)quinone with the translocation of four protons across the membrane (current consensus value). Homologues of the complex are present in the inner mitochondrial membrane of eukaryotes and the cytoplasmic membrane of bacteria. While ubiquinone is generally the substrate of the mitochondrial complex (but see also: Van Hellemond et al. 2003), the bacterial complex also reacts with other quinone species such as menaquinone (Unden and Bongaerts 1997). Due to this, the term (ubi)quinone is used in this chapter when the quinone species is not further specified. Cyanobacteria and chloroplasts also contain a complex I homologue, but they most likely use an electron donor other than NADH (Friedrich and Scheide 2000). This chapter deals solely with the complex I from bacteria and mitochondria.

Electron microscopy revealed the two-part structure of the complex consisting of a peripheral arm extending into the aqeous phase and a membrane arm embedded in the lipid bilayer (Guenebaut et al. 1998; Grigorieff 1999; Peng et al. 2003; Baranova et al. 2007; Clason et al. 2010). All known redox cofactors, namely one non-covalently bound flavin mononucleotide and – depending on the organism – up to ten iron-sulfur (Fe/S) clusters, are located within the peripheral arm (Ohnishi 1998; Sazanov and Hinchliffe 2006). The spatial arrangement of the cofactors became evident from the structure of the peripheral arm of *Thermus thermophilus* complex I, which was solved at molecular resolution (Sazanov and Hinchliffe 2006). The substrate (ubi)quinone is most likely directly reduced by one of the Fe/S clusters. The exact location of the (ubi)quinone-binding site is not yet known, but based on (ubi)quinone-site inhibitor studies and inhibitor resistance of site-directed mutants it is expected to be located at the interface between the peripheral and the membrane arm (Darrouzet et al. 1998; Schuler et al. 1999; Murai et al. 2009; Tocilescu et al. 2010).

The membrane arm does not contain a conserved motif for binding any further cofactors. However, the presence of two quinone radicals was demonstrated by EPR spectroscopy (Ohnishi and Salerno 2005). These quinone species are most likely involved in proton translocation (see below). The three major subunits of the membrane arm are homologous to monovalent cation/proton antiporters (ATS; antiporter type subunits). This, together with the recently published X-ray structure of the entire complex from *T. thermophilus* (Efremov et al. 2010) and *Yarrowia lipolytica* (Hunte et al. 2010) suggests that this arm is involved in proton translocation most

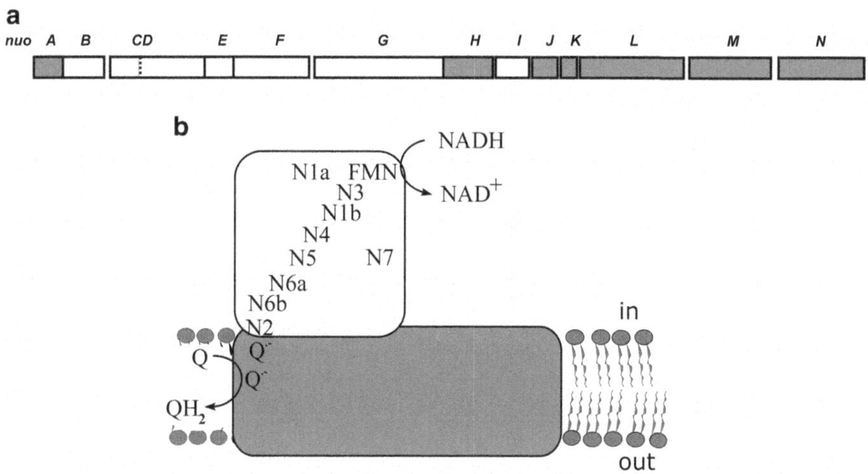

Fig. 2.1 Scheme of (a) the *E. coli nuo*-operon and (b) of complex I. (**a**) shows the name and the sequence of the *nuo* genes on the *E. coli* chromosome. Genes coding for globular proteins constituting the peripheral arm are shown in *white*, genes coding for polytopic proteins constituting the membrane arm are shown in *grey*. (**b**) represents a model of complex I with the peripheral arm shown in *white* and the membrane arm in *grey*. The position of the substrate binding sites and the cofactors is given. Q stands for (ubi)quinone, Nx for the Fe/S clusters (Modified from: Friedrich and Pohl 2007)

likely indirectly driven by the (ubi)quinone redox chemistry. Surprisingly, one of the ATS contains a long horizontal helix aligning the membrane arm. It was suggested that this amphipatic helix acts as a 'piston' transmitting the energy released by the redox reaction to the ATS (Efremov et al. 2010; Ohnishi 2010).

Bacteria contain a structurally minimal form of an energy-converting NADH:ubiquinone oxidoreductase consisting of 13–15 subunits called either NuoA to NuoN or Nqo1 to Nqo15 (Weidner et al. 1993; Friedrich et al. 1995; Yano et al. 1997; Friedrich 1998). In most cases the corresponding genes are organized in one large operon or gene cluster (Fig. 2.1). The gene order is well-conserved amongst bacteria. Homologues were found in all organisms containing an energy-converting NADH:ubiquinone oxidoreductase. Therefore, the subunits for which they are coding are termed 'minimal' subunits essential to catalyse redox-driven proton translocation (Table 2.1). Noteworthy, the subunits NuoE, F and G, which build up the electron input module of the complex, are either not present or modified in the homologous complex found in cyanobacteria, chloroplasts and in *Helicobacter pylori* and *Campylobacter jejuni* (Smith et al. 2000). The latter complex was shown to react with flavodoxin rather than NADH (Weerakoon and Olson 2008). NuoE, F and G together with subunits NuoB, CD and I represent the seven globular proteins comprising the peripheral arm of the complex. The residual seven, most hydrophobic subunits NuoA, H, J, K, L, M and N build up the membrane arm. They include the ATS NuoL, M and N, which are derived from gene triplication (Kikuno and Miyata 1985; Fearnley and Walker 1992).

Table 2.1 Nomenclature, properties, and proposed functions of subunits of complex I

Subunit			
E. coli	*T. thermophilus*	Localization (TM helices)	Proposed function
NuoA	Nqo7	Membraneous (3)	Q-binding
NuoB	Nqo6	Peripheral	[4Fe4S] 'N2' Q-binding
NuoC	Nqo5	Peripheral	–
NuoD	Nqo4	Peripheral	Q-binding
NuoE	Nqo2	Peripheral	[2Fe2S] 'N1a'
NuoF	Nqo1	Peripheral	NADH-binding FMN-binding [4Fe4S] 'N3'
NuoG	Nqo3	Peripheral	[2Fe2S] 'N1b' 3×[4Fe4S] 'N4', 'N5', 'N7'
NuoH	Nqo8	Membraneous (8)	Q-binding
NuoI	Nqo9	Peripheral	2×[4Fe4S] 'N6a', 'N6b'
NuoJ	Nqo10	Membraneous (5)	–
NuoK	Nqo11	Membraneous (3)	Q-binding
NuoL	Nqo12	Membraneous (16)	H^+-translocation
NuoM	Nqo13	Membraneous (14)	H^+-translocation
NuoN	Nqo14	Membraneous (14)	Q-binding (?)

The genes of NuoC and NuoD are fused in a few bacteria. The tetranuclear Fe/S cluster N7 on NuoG is only present in a few bacteria. (Ubi)quinone is abbreviated with 'Q', the type and name of the Fe/S clusters is given

2.1.1 *Electron Transfer from the NADH Binding Site to the Putative Quinone Binding Site*

The NADH binding site is located near the top of the peripheral arm. NADH delivers its electrons to the non-covalently bound FMN on subunit NuoF. The midpoint redox potential of the FMN/$FMNH_2$ couple was determined to −0.32 V (Sled et al. 1994; Bungert et al. 1999; Kohlstädt et al. 2008). Electrons are passed from the reduced FMN by a chain of 95 Å consisting of seven Fe/S clusters to the (ubi) quinone binding site at the interface between the two arms (Table 2.2; Fig. 2.2). Cluster N3 is located at a 7.6 Å distance from the FMN and represents the entrance of the electron transfer chain consisting of clusters N3, N1b, N4, N5, N6a, N6b and N2. The edge-to-edge distances between the clusters of the chain vary from 8.5 to 14 Å. Thus, all distances are compatible with the physiological electron transfer rate (Moser et al. 2006; but see below). The short distances between the individual Fe/S guarantee an electron transfer rate faster than the physiological turnover. It is noteworthy that the distance between N5 and N6a, with 14 Å, is relatively long indicating that this transition might be a bottle-neck for the electron transfer reaction along the Fe/S cluster. From a calculation of the electron transfer rates it was proposed that a loss of N1b and N3 would not have a significant effect on the overall reaction rate (Moser et al. 2006). The connection of the NADH oxidation site

Table 2.2 Localization and midpoint potential of the cofactors of complex I

Cofactor	Localization	Midpoint potential [mV]
FMN	NuoF	–320
N1a	NuoE	–250 to –330 (Bacteria)
		–370 to –410 (Mitochondria)
N1b	NuoG	–230
N2	NuoB	–220
N3	NuoF	–270
N4	NuoG	–270
N5	NuoG	–250
N6a	NuoI	–270
N6b	NuoI	–270
N7	NuoG	–260

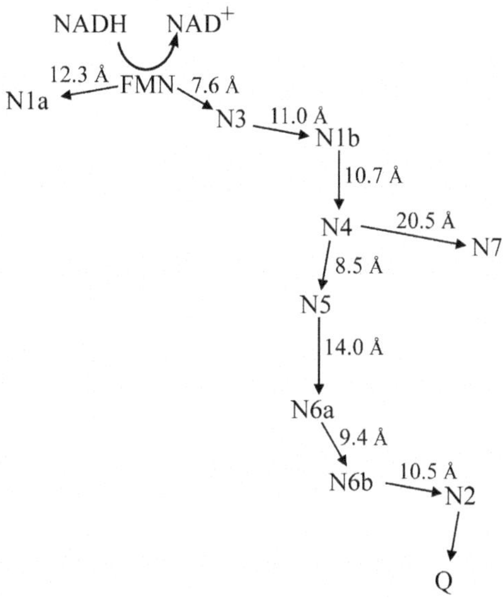

Fig. 2.2 Scheme of the electron pathway in the peripheral arm of complex I. Spatial arrangement of the cofactors within the peripheral arm of complex I. The edge-to-edge distances between the cofactors as deduced from the crystal structure (Sazanov and Hinchliffe 2006) are given

with the (ubi)quinone reduction site by a chain of seven Fe/S clusters spanning a distance of approximately 95 Å seems to be rather unusual. An explanation for this might be that today's respiratory complex I evolved from preexisting modules catalyzing electron transfer and proton translocation (Friedrich et al. 1995; Friedrich and Weiss 1997; Friedrich and Scheide 2000; Friedrich 2001; Mathiesen and Hägerhall 2003).

Most clusters of the chain, namely N3, N1b, N4, N5, N6a and N6b exhibit a more or less identical midpoint redox potential of about –0.25 V (Ohnishi 1998; Yagi and Matsuno-Yagi 2003; Sazanov 2007; Hirst 2010). Only N2 has a more positive and pH-dependent midpoint redox potential of about –0.1 V (Ingledew and Ohnishi 1980).

N2 is expected to be the direct reductant for the substrate (ubi)quinone. Two semiquinone species called Q_{Nf} and Q_{Ns} have been detected as essential intermediates during the turnover in the bovine heart complex I by means of EPR spectroscopy (Yano et al. 2000, 2005). The term 'Q_N' indicates their origin as semiquinones bound to complex I, while the subscript 'f(ast)' and 's(low)' reflect their relaxation properties in EPR spectroscopy. The Q_{Nf} signal is only detectable in the presence of a membrane potential. It was shown that N2 interacts with the semiquinone radicals (Ohnishi and Salerno 2005). The distance between N2 and Q_{Nf} was determined to 12 Å, while that between N2 and Q_{Ns} might be larger than 30 Å (Ohnishi et al. 2010a, b). Complex I contains two additional Fe/S clusters named N1a and N7, which are not part of the electron transfer chain mentioned above. The putative role of these clusters is discussed below.

2.1.2 Coupling Electron Transfer with Proton Translocation

The recently determined structures of the bacterial complex I from *T. thermophilus* (Efremov et al. 2010) and the mitochondrial complex from *Yarrowia lipolytica* (Hunte et al. 2010) mark a milestone in complex I research. The models derived from the data confirmed the L-shaped structure of the complex and the presence of all known redox cofactors in the peripheral arm of the complex. Although the distance of the terminal Fe/S cluster N2 to the membrane and the position of the (ubi) quinone binding site(s) remain under debate, it became clear that most redox reactions do not contribute to proton translocation. It is now evident that only the (ubi) quinone chemistry, possibly assisted by the redox reaction of N2 drives proton translocation by both a direct and an indirect mechanism.

The membrane arm is composed of seven 'minimal' subunits. Among them, there are the three major subunits NuoL, M and N, which derive from a common ancestor and which are homologues of monovalent cation/H^+ antiporters (ATS). It is most reasonable to assume that these ATS are involved in proton translocation. The X-ray structure of the *T. thermophilus* complex revealed that the ATS are located at the most distal position of the membrane arm (Fig. 2.3). They consist of 14 transmembraneous (TM) helices arranged in a core of four central helices surrounded by a ring of ten TM helices and include two discontinuous TM helices connected by a peptide loop in the membrane. This type of helix is also found in transporters and channels and they seem to play an important role in ion translocation (Screpanti and Hunte 2007).

In addition, NuoL contains an additional C-terminal domain, which is not present in the other ATS. This domain is made up of the 15th TM helix holding an unusual, 110 Å long amphipathic helix aligned parallel to the membrane arm. This 'horizontal' helix is anchored to the membrane arm by the 16th TM helix of NuoL. A similar 'horizontal' helix was found in the mitochondrial complex, however, at a length of 60 Å. This unique domain is conserved within the family of energy-converting hydrogenases and antiporters (Efremov et al. 2010) and clearly distinguishes NuoL

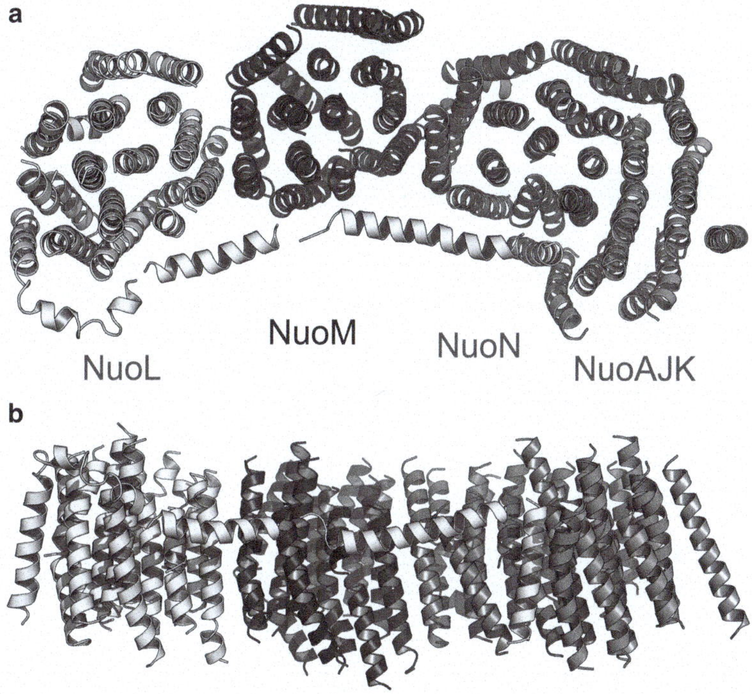

Fig. 2.3 Structure of the membrane arm of the *E. coli* complex I. The integral membrane part of complex I (PDB-ID 3M9C) is dominated by subunits NuoL, NuoM and NuoN that are homologous to Na^+/H^+ antiporters. (**a**) A top view highlights the three transporter subunits, but also the unique C-terminal helix of NuoL that was suggested to be involved in the coupling of electron transfer and proton translocation. (**b**) A side view of (**a**) shows the amphipatic helix of NuoL to be located within the membrane

from its homologues NuoM and N. It was proposed that the quinone redox chemistry leads to conformational changes that are somehow transmitted to the 'horizontal' helix. The helix could act as a piston to transmit the redox energy released in the peripheral arm to proton translocation in the membrane arm. Due to this movement, the three ATS are opened and closed, respectively, leading to a translocation of one proton by each of the subunits. It is proposed that the fourth proton is translocated at the quinone site by another yet unknown mechanism.

2.2 Methodology

To investigate the role of individual amino acid residues for the function of the *E. coli* complex I several mutants were generated. The mutagenesis of the large *nuo*-operon requires special techniques that are described in this paragraph. The variants were characterized -amongst other methods - by Fourier transform infrared (FTIR) spectroscopy. As this technique is not generally available for membrane proteins, its basics will be shortly described in this paragraph.

2.2.1 Mutagenesis of the nuo-Operon

Mutagenesis of the *E. coli nuo*-operon coding the complex I subunits is very laborious due to its enormous size of approximately 16 kb (Weidner et al. 1993). In addition, the insertion of a resistance cartridge in the *nuo*-operon disturbed the assembly of the complex (Schneider et al. 2008). The first attempt to generate site-directed mutants was to introduce unmarked mutations in the *E. coli* chromosome by means of genomic replacement using the *nptI-sacRB* suicide vector (Ried and Collmer 1987; Spehr et al. 1999; Flemming et al. 2003a, b). The suicide vector was inserted in the gene of interest by P1 transduction leading to kanamycin resistance and sucrose sensitivity (Oden et al. 1990). Deletion of the gene and directed integration of the unmarked mutations in the chromosome were achieved by marker exchange eviction mutagenesis (Ried and Collmer 1987; Hamilton et al. 1989; Flemming et al. 2003b). Strains carrying the desired mutation were able to grow in the presence of sucrose. In order to avoid time-consuming chromosomal mutagenesis, strains with an in-frame deletion of the individual *nuo*-genes were constructed using *E. coli* AN387 as parental strain (Wallace and Young 1977). The loss of the corresponding protein was detected by Western blot analysis. The lack of the individual proteins prohibited the assembly of the complex indicated by the absence of complex I activity in the membranes of the mutant strains (Flemming et al. 2003a, b, 2006; Schneider et al. 2008). The *nuo*-deletion strains were transformed with the corresponding *nuo*-gene on the inducible pBAD33 plasmid (Guzman et al. 1995) to complement the chromosomal deletion *in trans*. In that way, complex I was rescued and the desired mutations were introduced into the episomal *nuo*-gene using standard PCR techniques. As an example, 90% of the wild type complex I activity in the membrane were restored in the strain with a chromosomal *nuoB* deletion complemented with pBAD*nuoB* (Flemming et al. 2003a). Sucrose gradient centrifugation of a detergent extract of the cytoplasmic membranes of this strain revealed the presence of a stable complex I. An SDS-PAGE analysis of the preparation of complex I from this strain confirmed all complex I subunits. The enzymatic activity and EPR spectroscopic features of the presence of Fe/S clusters of the preparation were virtually identical to those of the parental strain (Flemming et al. 2003a).

In a third attempt to make the genetic manipulation of the *nuo*-operon easier, we constructed an extrachromosomal expression system containing all *nuo*-genes (Pohl et al. 2007a). The chromosomal DNA of an AN387 derivate strain called ANN003 was isolated, digested and a 27,866 bp fragment containing the entire *nuo*-operon and the T7 promotor region was purified and cloned into a linearized cosmid. The T7 promotor region was replaced by the L-arabinose inducible P_{araBAD} from pBAD33, yielding a 21.3 kb construct named pBAD*nuo* (Pohl et al. 2007a). By applying λ-Red (recombination defective)-mediated recombineering (Datsenko and Wanner 2000) any desired mutation can be introduced in the episomal encoded *nuo*-genes. The 21 kb expression plasmid pBAD*nuo* itself is too large for site-directed mutagenesis. Therefore, the *nuo*-gene of interest is mutated on small plasmids called subclones containing just one *nuo*-gene by standard PCR techniques. Firstly, the *nuo*-gene of interest on the expression plasmid is replaced by the *nptIsacRB* selection cartridge *via* λ-Red-mediated recombination. To prevent unwanted crossover recombination with the chromosomal *nuo*-genes, the chromosomal deletion strain DH5αΔ*nuo* is used for recombineering. In a second recombination step, the selection cartridge is replaced by a linear DNA fragment containing the mutated version of *nuo*-gene, which is amplified from the individual subclone.

Using this technique, a hexahistidine coding sequence was inserted upstream *nuoF*, leading to the production of a stable complex I engineered with a histidine-tag N-terminal on NuoF. This allowed the rapid and efficient purification of *E. coli* complex I and several variants of the complex (Pohl et al. 2007a, b, 2010; Steimle et al. 2011). The complex as prepared by affinity chromatography is pure and has a high lipid content due to the mild and fast purification procedure. To measure the enzymatic activity of the complex and the variants encoded by pBAD*nuo* directly in the membrane, the expression strain BW25113Δ*nuo* (Vranas and Friedrich, unpublished data) can be transformed with the expression plasmid. The chromosomal *nuo* deletion ensures that the mutant strain contains only the episomally encoded variant of the complex.

2.2.2 FTIR Spectroscopy

2.2.2.1 Reaction-Induced FTIR Difference Spectroscopy

Specialized difference FTIR techniques allow the detection of side-chain modifications, of the delocalization of water molecules, or of minute secondary structure changes (Garczarek and Gerwert 2006; Zscherp and Barth 2001). By using reaction-induced FTIR difference spectroscopy, the reorganization of proteins upon induction of the reaction is monitored without contributions from the background. Reaction-induced FTIR difference spectroscopy became an especially important tool for the analysis of enzyme mechanisms.

The success of the approach is based on the possibility of monitoring the vibrational absorption bands of a single -COOH group, or any other residue, in a protein. Data reported is thus typically obtained by cycling the reaction of interest and averaging a large number of scans and cycles. Several approaches have been exploited to obtain reaction-induced difference FTIR spectra. The goal in all cases is to maintain the sample at a constant concentration and path length, while perturbing the state of the sample in a way that is informative. A significant difficulty of the reaction-induced technique is the discrimination of individual contributions in the spectrum and their unequivocal assignment. A number of strategies have been developed over the years, including the variation of external parameters such as pH, temperature, specific and non-specific isotope labeling as well as the use of site-directed mutants (Vogel and Siebert 2000).

For complex I, electrochemically induced FTIR difference spectra have been presented (Hellwig et al. 2000, 2004) revealing signals that derive from the reorganization of the protein upon the induced redox reaction and coupled protonation events. In addition, the spectra contain contributions from the flavin cofactor and bound ubiquinone. The contribution of lipids with the reorganization of the enzyme upon electron transfer was probed (Hielscher et al. 2006). It was shown that in *E. coli* complex I the reduction of N2 is coupled with a deprotonation of glutamate and possibly aspartate residues (Friedrich and Hellwig 2010) and with the protonation of tyrosine residues (Flemming et al. 2003b). These amino acids were identified by site-directed mutagenesis on subunit NuoB (Flemming et al. 2006). In addition, the putative sodium translocation activity of complex I was discussed by means of this technique (Friedrich et al. 2005).

Data for the mitochondrial complex I was also obtained in perfusion-induced ATR-FTIR experiments. In this type of setup, the sample is attached to the surface of a reflexion unit and the interaction with inhibitors, redox partners and substrates can be probed by FTIR spectroscopy (Marshall et al. 2006). This led to the identification of IR properties of ubiquinone bound to the mitochondrial complex I.

2.2.2.2 H/D Exchange

Protein dynamics play an important role in the catalytic efficiency of enzymes. Yet, large domain movements often remain unnoticed. Large conformational changes can be studied by monitoring hydrogen/deuterium ($^1H/^2H$) exchange kinetics at the level of the amide proton in the mid infrared spectral range (Knox and Rosenberg 1980; Gregory and Lumry 1985; Haris and Chapman 1995; de Jongh et al. 1997; Vigano et al. 2004). $^1H/^2H$ exchange was first introduced through the pioneering work of Linderstrøm-Lang and colleagues in the mid-1950s, who, after the discovery of protein α-helices and β-sheets, realized that amide hydrogen exchange rates should reflect the presence of hydrogen-bonded structure (Hvidt and Linderstrom-Lang 1954, 1955). The relationship between the rate of hydrogen exchange and protein dynamics have subsequently been formulated as well as mathematical descriptions of the exchange models (Hvidt and Linderstrom-Lang 1954, 1955; Hvidt and Nieslen 1966). It was suggested that protein structure can be divided in three types of structure characterized by their particular ($^1H/^2H$) exchange dynamics, and that these dynamic structures or domains are functionally and evolutionarily more relevant than those defined on the basis of their secondary structures. The slowly exchanging domain, for example has been related for several proteins to the initial folding core during the sequence of events leading to protein folding (Kim et al. 1993). In addition, medium and fast exchanging domains can be distinguished.

The tool is thus useful for the study of large conformational changes. If the presence of nucleotides in the soluble part of the enzyme is to influence the membrane arm in order to control proton translocation, the overall structure of the enzyme may change together with the accessibility of the different domains. In order to probe this possibility, hydrogen exchange experiments were carried out with complex I from *E. coli* (Hielscher et al. 2011). In general, the initial infrared absorbance spectra of the proteins are characterized by an intense absorption band at 1,650 cm^{-1}, called amide I band that includes ν(C=O) (70–85%) and ν(C=N) (10–20%) vibrational modes. In addition, the amide II band is localized near 1,550 cm^{-1}. It derives from the in-plane N-H bending vibration (40–60%) coupled to the ν(C-N) (20–40%) vibrational mode (Krimm and Bandekar 1986). Upon $^1H/^2H$ exchange, the amide II band intensity at 1,550 cm^{-1} decreases. The in-plane N-H (N-D) bending mode uncouples and appears in the 940–1,040 cm^{-1} region and the ν(C-N) moves near 1,450 cm^{-1}, mixing with other modes to form a new band called amide II', whose intensity increases during the exchange.

2.3 Nucleotide Binding

The binding site for NADH is located on subunit NuoF that contains a domain with a Rossmann-like nucleotide-binding fold and holds the FMN moiety, the primary electron acceptor of complex I (Sazanov and Hinchliffe 2006). This particular Rossmann-type fold consists of 4 parallel ß-strands flanked by a helix and has so far not been detected in any other enzyme. A typical feature of the standard Rossmann-fold is that the turn between the first strand and the following helix contains glycine residues binding one of the phosphate groups of the NAD pyrophosphate and enabling packing of the helix against the ß sheet (Lesk 1995). In NuoF, this loop does not contain glycine residues and interacts with the isoalloxazine ring of the flavin nucleotide.

The structure of the peripheral arm of the *T. thermophilus* complex I with bound NADH revealed the nucleotide binding mode (Fig. 2.4). FMN and NADH are placed in an optimal position for a stereo-specific hydride transfer. The hydride transfer from NADH to FMN was demonstrated by a kinetic isotope effect on the NADH oxidase activity using tritiated NADH (Ernster et al. 1965). Considering the high affinity of NADH to the complex reflected in a K_M of less than 10 μM (Vinogradov 1998; Stolpe and Friedrich 2004), the nucleotide is bound by remarkably few interactions, partially to the protein and partially to the FMN. The adenine moiety is cradled in a hydrophobic pocket of the protein lined by the aromatic side chains of three conserved phenylalanine residues, two of which stack ideally with the purine base. The adenosine ribose is H-bonded to a glutamate side chain from one side and a lysine side chain from the other side. The nicotinamide ring is involved in stacking interactions with the isoalloxazine ring of the FMN as expected. The nicotinamide ribose and the pyrophosphate are involved in H-bonds with the FMN (Fig. 2.4). The glycine-rich loop of the Rossmann-fold in complex I is not contained within

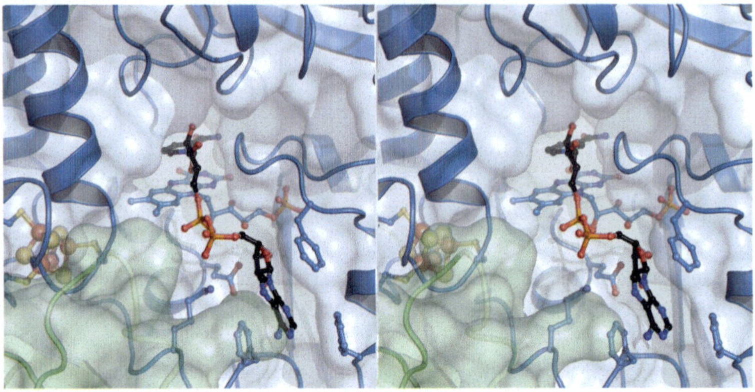

Fig. 2.4 NADH binding in complex I. Stereo representation of the FMN site of the peripheral arm of *T. thermophilus* complex I (PDB-ID 3IAM). NADH is bound in a pocket largely formed by the FMN cofactor and a series of hydrophobic residues that enter into stacking interaction with the adenine moiety of the substrate. The nicotinamide ring of NADH is placed directly above the FMN isoalloxazine, ideally positioned for electron transfer

the standard Rossman-fold and is involved in hydrogen-bonding of the nicotinamide and its ribose moiety (Berrisford and Sazanov 2009).

The structure of NADH bound to subunit NuoF revealed that the carboxylate group of the conserved Glu183^F (*E. coli* nomenclature; the superscript denotes the name of the corresponding Nuo-subunit) is hydrogen-bonded to the hydroxyl groups of the adenosine ribose (Berrisford and Sazanov 2009). The ribose is shielded by the protein in such a way that there is some, but not sufficient space below the O^{2B} atom for the phosphate group of NADPH (Fig. 2.4). This arrangement could explain the low (but detectable) reactivity of complex I with NADPH (Vinogradov 1998; Pohl et al. 2008). Submitochondrial particles exhibit a NADPH oxidase activity with a pH optimum around 6 (Hatefi and Hanstein 1973), while the NADH oxidase activity shows a broad optimum around pH 7. The NADPH oxidase activity was about one-fifth of the NADH oxidase activity at pH 6 (Djavadi-Ohaniance and Hatefi 1975). The NADPH dehydrogenase activity is an intrinsic property of the mitochondrial complex I and not related to the NADPH/NAD transhydrogenase activity. With the given structural information, it was reasonable to assume that NADPH could be a substrate for the complex, but that due to the steric interaction and electrostatic repulsion with Glu183^F the reaction rate with NADPH is at least five-times slower than with NADH as a substrate (Vinogradov 1998; Pohl et al. 2008).

2.3.1 Engineering Complex I to NADPH:Ubiquinone Oxidoreductase

To test the structure-derived hypothesis that position Glu183^F determines the selectivity of complex I to NADH over NADPH this residue was replaced by Asp, Gln,

Table 2.3 Kinetic parameters of the NAD(P)H:decyl-ubiquinone oxidoreductase activity of *E. coli* complex I and several E183X^F variants

	K_M (μM)		k_{cat} (s^{-1})		k_{cat}/K_M ($s^{-1}\cdot\mu M^{-1}$)	
Variant	NADH	NADPH	NADH	NADPH	NADH	NADPH
Parental	13±1	1´870±30	26±2	3±1	2±0.3	0.002±0.001
E183D^F	6±1	390±30	37±2	34±2	6±0.4	0.09±0.006
E183Q^F	12±2	45±4	28±2	14±1	2±0.5	0.32±0.003
E183N^F	14±1	480±50	11±2	11±3	1±0.1	0.02±0.006
E183H^F	6±1	25±2	37±2	10±1	6±0.3	0.4±0.003

Asn, and His using the expression, production and purification system described above (Morina et al. 2011). The mutations were introduced on the episomal *nuo*-operon by site-directed mutagenesis and the variants were purified by means of affinity and anion-exchange chromatography. The kinetic properties of the physiological NAD(P)H:decyl-ubiquinone oxidoreductase activity of the variants were determined after reconstitution in lipids and compared with that of the parental strain (Table 2.3).

The preparation of the parental complex showed a K_M^{NADH} of 13 μM and a v_{max} of 2.9 μmol/(min · mg) when measuring the rate of NADH oxidation. Determination of the rate of decyl-ubiquinone reduction led to the same K_M^{NADH} and a v_{max} of 2.8 μmol/(min · mg). The reaction was inhibited to 95% by an addition of 10 μM piericidin A, a specific complex I inhibitor. Thus, the physiological electron transfer from NADH to quinone was measured in the reconstituted system. The K_M^{NADPH} was determined to 1.9 mM. The v_{max} with NADPH as substrate was eightfold lower than with NADH (Table 2.3). He-temperature EPR spectroscopy revealed that NADPH reduced the Fe/S clusters of the complex to the same amount as NADH (Morina et al. 2011). To answer the question whether NADPH oxidation is capable of driving proton translocation, the preparation was reconstituted in proteoliposomes and the generation of a proton gradient was determined by the fluorescence of the pH sensitive dye ACMA (Morina et al. 2011). An addition of NADH led to an ACMA quench, which was sensitive to the proton uncoupler CCCP and piericidin A. Thus as expected, NADH oxidation drives the complex I-mediated proton translocation. However, no signal was detected after an addition of NADPH. Either the NADPH oxidation is not coupled with proton translocation by complex I or the rate of proton translocation is slower than the proton backflow due to the proton permeability of the proteoliposomes (Morina et al. 2011).

To determine the influence of the conserved Glu183^F on the NAD(P)H:decyl-ubiquinone oxidoreductase activity, the kinetic parameters of the variants were determined (Table 2.3). The NAD(P)H:decyl-ubiquinone oxidoreductase activity of the preparations of all variants was inhibited by piericidin A with an IC_{50} identical to that of complex I from the parental strain. Thus, these variants exhibited true NADPH:ubiquinone oxidoreductase activity (Morina et al. 2011). The Glu183Asn^F and the Glu183Gln^F variants showed a similar reactivity towards NADH as the parental strain. In contrast, the variants Glu183His^F and Glu183Asp^F exhibited a significantly higher affinity to NADH. The Glu183Gln^F variant showed the highest turnover rate with NADH. Noteworthy, all variants showed a better reactivity with NADPH than the complex from the parental strain. The Glu183His^F variant exhibited a 75-fold higher affinity to and a three-fold higher turnover with NADPH. Most strikingly, the Glu183Asp^F variant showed an 11-times higher turnover with NADPH, which is even higher than the turnover of the parental complex with NADH. The affinity of the Glu183Asp^F variant to NADPH was only five-fold greater than that of the parental strain. Although its reactivity with NADH was not significantly changed, the Glu183Gln^F variant showed a 40-fold higher affinity to NADPH and a four-fold enhanced turnover. The catalytic efficiency with NADH was not enhanced in the Glu183Asn^F and Glu183Gln^F variants, but three-fold higher in the Glu183Asp^F and Glu183His^F variants. Remarkably, the catalytic efficiency with NADPH was enhanced 200-fold in the Glu183His^F variant, 160-fold in the Glu183Gln^F variant, 45-fold in the Glu183Asp^F variant, and ten-fold in the Glu183Asn^F variant (Table 2.3).

He-temperature EPR spectroscopy showed that NADH as well as NADPH reduced the Fe/S clusters of all variants (Morina et al. 2011). After reconstitution in proteoliposomes the Glu183His^F variant was capable of proton translocation measured by means of ACMA quenching with either NADH or NADPH as substrate. As described for complex I from the parental strain, the quenching of the ACMA signal was sensitive to an addition of either CCCP or piericidin A demonstrating its origin from the redox driven proton translocation by the variant. Thus, the failure to detect an ACMA quench with the parental complex was due to the low turnover rate of the complex with NADPH as substrate.

By changing a single amino acid residue in an enzyme complex consisting of nearly 5,000 amino acid residues, an energy-converting NADPH:ubiquinone oxidoreductase emerged. This is experimental proof of the assumption that Glu183^F is a key residue in substrate recognition as derived from structural analysis. The strictly conserved Glu183^F seems to assure that complex I only uses NADH but not NADPH. Although the catalytic efficiency of the Glu183His^F variant with NADH is three-fold higher than with the wild type complex (Table 2.3), the mutation has not evolved in any species for which a genome sequence is available, most likely because the higher efficiency is at the cost of the loss of substrate selectivity (Morina et al. 2011). Most recently it was reported that *E. coli* adapts to high concentration of cellular NADPH by a single mutation of either Glu183Ala^F or Glu183Gly^F (Auriol et al. 2011). These mutations lead to a higher catalytic efficiency with NADPH compared to the parental strain. However, the enhancement is less than those obtained with the Glu183His^F, Glu183Gln^F and Glu183Asp^F substitutions (Schulte et al. unpublished results). Thus, NADPH stress in *E. coli* is overcome by adapting the substrate specificity of a single enzyme (Auriol et al. 2011). The usual way to cope with such a problem would be to either increase the capacity of active metabolic pathways or activate the ‘silent’ pathways (Ibarra et al. 2002).

In addition, the oxidation of NADPH by complex I and its variants is always coupled with the generation of reactive oxygen species (ROS), although the

NADPH-induced ROS production is slightly diminished in the variants (Morina et al. 2011). However, the variants also show a slightly enhanced NADH-induced ROS production. The source of ROS evolving during the redox reaction of complex I is still under discussion (Hirst et al. 2008, see below). As the quinone site inhibitor piericidin A had no effect on the ROS-production, which was on the other hand significantly influenced by the mutations on NuoF, the ROS are formed at the flavin site under experimental conditions (Morina et al. 2011). The Glu183Asp^F variant exhibiting a higher turnover with NADPH than complex I with NADH and a very low affinity to NADPH showed an NADPH-induced ROS production. This indicates that the mode of nucleotide binding determines the ROS production at this site. Noteworthy, the Glu183His^F substitution leading to the highest catalytic efficiency with NADPH as substrate is coupled with a higher ROS-production than the substitution Glu183Ala^F (Schulte et al. unpublished results), which evolved by natural selection (Auriol et al. 2011). Thus, it seems that under physiological conditions a lower ROS production is favored over a higher catalytic efficiency.

2.3.2 Conformational Changes Coupled to Nucleotide Binding

It was shown that an addition of nucleotides led to conformational changes in *E. coli* complex I (Mamedova et al. 2004; Pohl et al. 2008). These conformational changes were either detected by a change in the cross-link pattern between individual subunits or by electron microscopy (Mamedova et al. 2004; Pohl et al. 2008; Berrisford and Sazanov 2009). In addition, the nucleotide-induced conformational changes were identified by ATR-FTIR spectroscopy (Pohl et al. 2008).

Infrared spectroscopy enables the determination of secondary structure elements based on the analysis of the amide I band including the ν(C=O) vibration of the protein backbone influenced by its structure-specific hydrogen bonding environment. To analyze the effect of NADH on complex I, the soluble NADH dehydrogenase fragment of the *E. coli* complex I was used for simplicity (Uhlmann and Friedrich 2005). This fragment is made up of the subunits NuoE, F, and G and contains the FMN and six Fe-S clusters as well as the NADH binding site. From the structure of the peripheral arm of the *T. thermophilus* complex the distribution of secondary structure elements of the NADH dehydrogenase fragment in the oxidized state was calculated to 30% alpha helical, 17% beta sheet and 53% not attributed structures (Pohl et al. 2008). Spectral deconvolution of the IR absorbance spectrum of the amide I band of the NADH dehydrogenase fragment revealed a distribution of the secondary structure elements of 46% alpha helical, 26% beta sheet, 10% turns, 12% random structures. Approximately 6% of the signal derived from side chains. After reduction by NADH, the amount of alpha helical and beta sheet structures decreased by 4%, which is within the error of the method. However, signals corresponding to turns were increased by 8% while random type elements decreased by the same amount. Clearly, binding of NADH induced a conformational change in the frag-

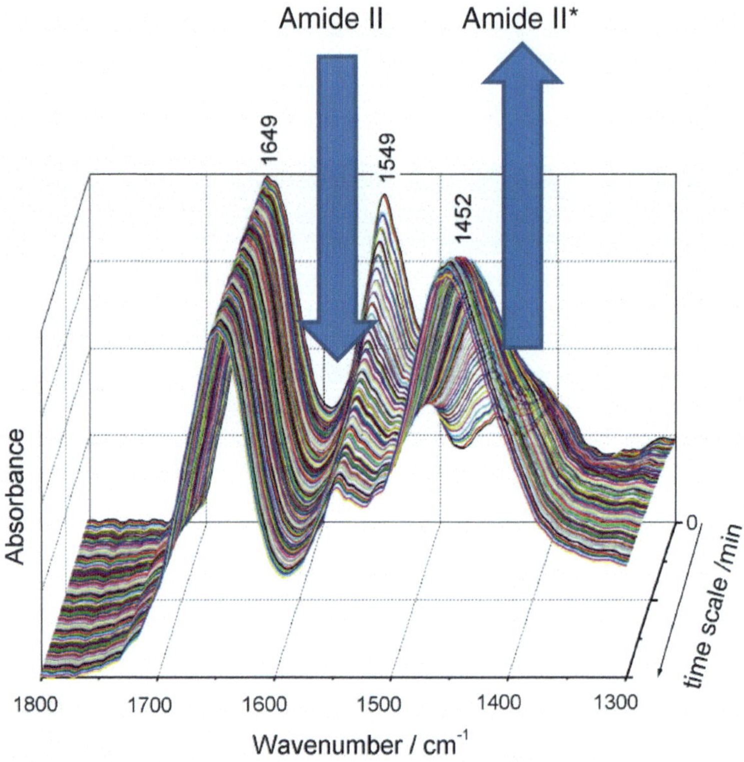

Fig. 2.5 3-D representation of the shifts in the absorbance spectra of the *E. coli* complex I upon ^{1}H/^{2}H exchange

ment, most likely affecting the hydrogen-bond network (Pohl et al. 2008; Hielscher 2009).

As described above, (^{1}H/^{2}H) exchange kinetics provide information on large domain movements. The influence of the nucleotide on the tertiary structure of complex I and the NADH dehydrogenase fragment were examined using this technique (Fig. 2.5). During (^{1}H/^{2}H) exchange a typical complex I spectrum evolves with a decrease of the amide II band intensity at 1,550 cm^{-1} (Hielscher et al. 2011). The in-plane N-H (N-D) bending mode uncouples and appears in the 940–1,040 cm^{-1} region and the ν(C-N) moves near 1,450 cm^{-1}, mixing with other modes to form a new band called amide II', the intensity of which increases during the exchange.

The kinetics of the ^{1}H/^{2}H exchange are interpreted by contributions resulting from a fast exchange (a_1 and τ_1), an intermediate exchange (a_2 and τ_2), and a slow exchange (a_3 and τ_3), with a_x as the percentage of the species of the entire population and τ_x as the lifetime of the corresponding species (Table 2.4). The data lead to the tentative conclusion that NADH binding to the NADH dehydrogenase fragment gives rise to a better accessibility of residues, especially of those initially involved

Table 2.4 $^1H/^2H$ exchange parameters of *E. coli* complex I and its NADH dehydrogenase fragment (NDF)

Sample	a_1/%	τ_1/min	a_2/%	τ_2/min	a_3/%	τ_3/min
NDF	4.0±0.2	1.1±0.1	70.9±3.5	34.6±1.7	26.1±1.3	134±7
NDF+NADH	43.6±2.2	4.0±0.2	18.2±0.9	27.8±1.4	32.0±1.6	743±38
Complex I	45.6±2.3	0.4±0.1	20.9±1.0	22.3±1.1	33.5±1.7	1,226±61
Complex I+NADH	41.4±2.1	2.8±0.1	26.2±1.3	18.2±0.9	31.8±1.6	505±25
Complex I +NADH+decyl-Q	50.0±2.5	0.8±0.1	28.2±1.4	13.2±0.7	21.5±1.1	354±18

The fractions of the various populations are named a_i and are characterized by the time constant τ_i of hydrogen exchange

in hydrogen exchange. NADH binding gives rise to a more 'open' structure. The conformation of the complex in buffer in the presence of NADH and in the presence of NADH and decyl-ubiquinone is not the same, giving evidence for individual domain movements upon binding of either substrate. Again, the structure of the complex with bound substrate seems to be more 'open'.

2.4 Electron Transfer

2.4.1 Importance of Aromatic Amino Acids

The computational predictions of the electron transfer rates between the individual Fe/S clusters of the complex have to consider the overall electron transfer reaction coefficient k_{ET}. This catalytic turnover has been determined experimentally to 100–180 s^{-1} (Ragan and Racker 1973; Vinogradov 1998; Stolpe and Friedrich 2004; Sharpley et al. 2006; Verkhovskaya et al. 2008). The individual electron transfer steps calculated from theoretical models have to be larger or at least equal to that value. Dutton and coworkers used the known distances between the Fe/S clusters to determine the time-dependent distribution of excess electron density along the chain of Fe/S clusters (Moser et al. 2006). Using the described parametrization, rates of 10^6–10^7 s^{-1} are computed for all transfer steps with the exception of the rate limiting k_{ET} of about 10^4 s^{-1} for the for the transfer between N5 and N6a. These clusters are separated by an edge-to-edge distance of 14 Å (Berrisford and Sazanov 2009). However, a sufficiently oxidizing quinone/quinole couple would drag the electrons from N6 *via* N2 to the quinone (Moser et al. 2006). It was also proposed that distances longer than 14 Å can be used to regulate electron transfer (Moser et al. 2006).

In an alternative approach, the k_{ET} between the Fe/S clusters was calculated on a microscopic level using atoms and atomic orbitals as smallest entities of a model for the electronic structure and electron transfer rates (Wittekindt et al. 2009). A semiempirical model was set up using parameters that derive from experimental studies or a careful fit to high-level *ab initio* calculations. Within this model several mechanisms of biological electron transfer such as through-space, through-bond, superexchange or hopping transfer are treated in an unbiased way. Application of this model to determine the electron transfer rates between the Fe/S clusters of complex I led to the unexpected conclusion that neither a through-space, through-bond nor a superexchange theory account for the experimentally determined k_{ET} (Wittekindt et al. 2009). Especially the k_{ET} of 2.2×10^{-4} between N5 and N6a was much too low to correspond to the experimentally determined data.

To overcome the discrepancy between the experimentally determined k_{ET} and those derived from the model, transient localization of charge carriers on aromatic amino acid residues was taken into account (Wittekindt et al. 2009). The use of the hopping hypothesis was motivated by the fact that excess electrons reside in high-energy molecular orbitals dominated by d contributions. Thus, a relatively small energy gap to the lowest unoccupied π orbitals was expected. Positively charged amino acid residues may exist in the vicinity of the postulated aromatic amino acid residues, that would decrease the energy of the excess electrons narrowing the energy gap. In addition, it was shown that backbone dipoles generally generate positive potentials in proteins (Gunner et al. 2000). With this approach seven aromatic amino acids were identified that could potentially be used as stepping stones within the electron pathway through complex I (Wittekindt et al. 2009). One of these is localized on the *T. thermopilus* subunit Nqo15, which is not present in complex I from most other species (Hinchliffe et al. 2006; Pohl et al. 2007c). A subsequent bioinformatic analysis revealed that the residual six aromatic amino acids in question are either conserved or substituted by other aromatic amino acid residues throughout the various species. Three of them (His514^{CD}, Phe515^{CD} and Trp221^{G}) are located between N5 and N6a, the clusters with the longest edge-to-edge distance of the complex I Fe/S clusters. Mutation of these amino acid residues to non-aromatic amino acid residues reduced the overall catalytic turnover by approximately one-half (Kashani-Poor et al. 2001; Dörner and Friedrich, unpublished results). However, at this point it is impossible to judge whether this is due to a functional or structural distortion of the complex. The participation of the other conserved aromatic residues on the electron transfer reaction between N3 and N1b (Phe397^{F}), between N1b and N4 (Phe35^{G}) and between N6b and N2 (Phe101^{I}) is experimentally hard to determine, because a substitution of these residues would theoretically have only a mild impact on the electron transfer rates. This decrease is too low to be detected by measuring the overall electron transfer rate.

Another approach towards the determination of electron tunneling in complex I was performed by Hayashi and Stuchebrukhov (2010). They used density functional theory and a semiempirical electronic structure model to determine the electron tunneling pathways and the strength of electronic coupling between the Fe/S clusters. Using this approach, specific amino acids were identified that could serve as electronic wires to connect neighboring Fe/S clusters. In this model, the entrance residue and the exit residue of the wire are always made up of the coordinating cysteine residues. This was motivated by the fact that the Fe/S clusters are oriented in such a way within the protein that the cysteine ligands are pointing towards each other in some cases. One additional amino acid is needed for some transitions and is called the mediator residue. According to the model, these mediator residues are not specific residues as the different side chains are proposed to have similar electronic structures (Hayashi and Stuchebrukhov 2010). Noteworthy, within this model water molecules at the interface between the subunits play an essential role as mediators for electron transfer accelerating the electron transfer reaction rate. To determine whether these water molecules exist, a structure of the peripheral arm of the complex at a much higher resolution is definitively needed.

2.4.2 Role of Cluster N1a

The binuclear Fe/S cluster N1a is not part of the electron transfer chain from the FMN to the putative (ubi)quinone binding site. Its binding motif is present in the sequence of all homologues of NuoE. It is located in 12.3 Å distance to the FMN and 19.4 Å to N3. Thus, it might accept an electron from the reduced FMN, but it is not able to inject the electron into the transfer chain to (ubi)quinone. The midpoint redox potential of N1a differs between −0.38 V in the mitochondrial and −0.25 V in the bacterial complex (Table 2.2). Due to this N1a is easily reduced by NADH in the bacterial but not in the mitochondrial complex.

An antioxidant role for N1a diminishing the amount of reactive oxygen species (ROS) produced by complex I has been proposed (Sazanov 2007). Most investigators agree on the flavin site (Galkin and Brandt 2005; Kussmaul and Hirst 2006), while the (ubi)quinone binding site and the Fe/S clusters N1a and N2 have also been proposed to be responsible for the generation of ROS. Recently, it was proposed that the site where ROS are generated is variable and determined by the reduction state of the individual cofactors (Ohnishi et al. 2010c). Concerning the ROS production at the FMN site, the reduced FMN could pass one electron to cluster N3 resulting in the formation of the flavosemiquinone radical and reduced N3. N1a could accept the second electron from the flavosemiquinone, thus reducing the life time of the fully reduced FMN and the FMN radical state. After N3 is reoxidized by passing the electron into the transfer chain to (ubi)quinone, the electron from N1a might be transferred via the FMN to N3 and further to (ubi)quinone.

The *T. thermophilus* homologue of NuoE is made up of two domains. The N-terminal domain folds in a four-helix bundle and confers interactions with its the neighboring subunits. The C-terminal domain is made up of a thioredoxin-like fold and coordinates N1a (Sazanov and Hinchliffe 2006). This fold is also found in the thioredoxin-like [2Fe2S] ferredoxin (Trx-like-Fd) from *Aquifex aeolicus* (Chatelet et al. 1999; Meyer et al. 2008). However, there are two main differences in the structure between the homologues: First, the Trx-like-Fd is a homodimer and second, the binding motif of the Trx-like-Fd [2Fe2S] cluster is $CysXxx_{10,12}CysXxx_{29,34}CysXxx_3Cys$, while that of N1a is $CysXxx_4CysXxx_{35}CysXxx_3Cys$ (Yano et al. 1994). The additional amino acid residues between the first and the second cysteine residue of the motif build a loop protruding from the surface of the Trx-like-Fd protein and are missing in

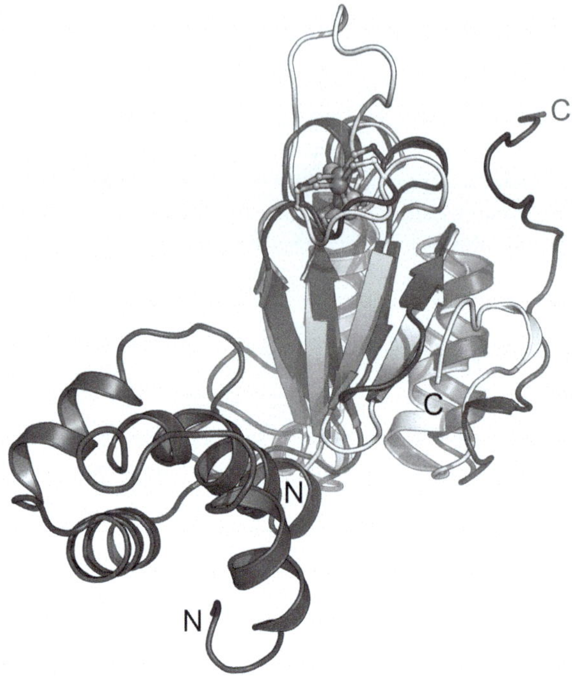

Fig. 2.6 Structural comparison *T. thermophilus* NuoE and *A. aeolicus* Fd×4. NuoE consists of two distinct domains connected by an extended linker region. The C-terminal domain of NuoE (*dark*) is homologous to the thioredoxin-type of [2Fe2S] ferredoxins, as found in Fd×4 from *A. aeolicus* (*light*)

NuoE (Fig. 2.6; Yeh et al. 2002; Meyer et al. 2008). Using the mutagenesis system described above the cysteine residues of the binding motif were individually exchanged to either alanine or serine residues. According to the d-NADH/ferricyanide oxidoreductase activity of the mutant membranes the amount of the complex I Cys92Ala^{E}, C133Ala^{E}, C137Ala^{E} and the Cys97Ser^{E} variants in the membrane was drastically decreased. The Cys97Ala^{E} and Cys133Ser^{E} substitutions had a milder effect on the activity. However, the amount of variants is halved in the mutants compared to the parental strain (Dörner and Friedrich, unpublished results). The membrane proteins of the mutants were extracted with detergent and separated by sucrose gradient centrifugation. The Cys92Ala^{E}, Cys133Ala^{E} and Cys137Ala^{E} variants were not detectable in the fractions of the gradient typically harbouring complex I (Leif et al. 1995; Spehr et al. 1999). The Cys97Ala^{E}, Cys97Ser^{E} and Cys133Ser^{E} mutants showed a small but detectable amount of complex I activity in the typical fractions of the gradient (Dörner and Friedrich, unpublished results). These findings demonstrated that cluster N1a is essential for the stability of the *E. coli* complex I.

The instability of the NuoE cysteine variants was unexpected because extensive mutational studies were performed on the [2Fe2S] cluster of the Trx-like-Fd from *A. aeolicus* and *Clostridium pasteurianum* (Fujinaga et al. 1993; Golinelli et al. 1996, 1998; Yeh et al. 2002; Meyer et al. 2008). In the homologous proteins the individual replacement of three cysteine residues by serine residues led to the formation of a stable protein containing the [2Fe2S] cluster. The structures of two serine variants of the *A. aeolicus* protein were solved and demonstrate the involvement of the serine residues in the coordination of the cluster (Yeh et al. 2002). A variant with just three coordinating cysteine residues was isolated (Meyer et al. 2008). In the *C. pasteurianum* Trx-like-Fd the flexible loop mentioned above is located between the first two cysteine residues of the binding motif. This protruding loop enables the use of the non-coordinating Cys14 as a cluster ligand in the Cys24Ala variant (Golinelli et al. 1996). Moreover, a Cys14Ala/Cys24Ala variant with a cysteine residue inserted at any position between the residues 14 and 26 is capable of coordinating the cluster. Thus, the protruding loop allows the protein environment to adapt to various coordination states of the cluster (Golinelli et al. 1998). This loop has no counterpart in NuoE, which could be an explanation for the instability of the variants created with the *E. coli* complex I (Dörner and Friedrich, unpublished results).

2.4.3 Importance of Cluster N7

The other Fe/S cluster, which is not part of the electron transfer chain to (ubi)quinone, is the tetranuclear cluster N7. It is bound to the C-terminal domain of NuoG in a few bacteria such as *E. coli*, *A. aeolicus* and *T. thermophilus*. The distance to its closest neighbour N4 is 20.5 Å, thus, it seems more than unlikely that N7 participates in the electron transfer reaction to (ubi)quinone (Sazanov and Hinchliffe 2006). It seems to represent an evolutionary relic of a complex I ancestor with a different enzymatic function. Phylogenetic analysis revealed the presence of a common complex I progenitor present in the three domains of life containing the 'minimal' subunits with the exception of NuoE, F, and G (Friedrich et al. 1995; Friedrich and Weiss 1997; Friedrich and Scheide 2000; Friedrich 2001; Sazanov 2007). This ancient progenitor might have worked as a ferredoxin:quinone oxidoreductase. During evolution this oxidoreductase was further equipped with NuoG. As the C-terminus of NuoG is homologous to molybdo-bis(molybdopterin-guanin-dinucleotide) (Mo-bisMGD) enzymes (Rothery et al. 2008; Workun et al. 2008), this addition most likely gave rise to a novel binding site for a different electron donor. Later acquisition of NuoE and NuoF led to the emergence of the present-day complex I that uses NADH as electron donor. Through this acquisition, the presence of the donor substrate binding site, the Mo-bisMGD cofactor and the Fe/S clusters providing the electron pathway from the substrate binding site to the (ubi)quinone site became superfluous and were lost during evolution (Friedrich and Scheide 2000). However, one of the Fe/S clusters connecting the Mo-bisMGD active site to the electron transfer chain leading to the acceptor substrate quinone was retained in some species as cluster N7 of today's complex I. The N-terminal part of NuoG is homologous to FeFe-hydrogenases, which contain Fe/S clusters homologous to N1b, N4 and N5. In addition, FeFe-hydrogenases contain an Fe/S cluster named FS4A, which has no homologue in complex I (Peters et al. 1998). A binding motif for a homologue of cluster FS4A is present in the sequence of *A. aeolicus* NuoG. A first structural investigation of the peripheral arm of *A. aeolicus* complex I shows that such a cluster connecting N7 with the electron pathway to (ubi)quinone reduction site is indeed present in NuoG (Peng et al. 2010). The late acquisition of NuoG and the separate and subsequent acquisition of NuoEF might be reflected in the different chromosomal localization of *nuoG* and *nuoEF* in *A. aeolicus* (Kohlstädt et al. 2008).

The inability of N7 to participate in the physiological electron transfer reaction was shown by EPR spectroscopy (Pohl et al. 2007b). The EPR difference spectrum of complex I reduced either by NADH or by dithionite revealed the signals of a tetranuclear Fe/S cluster, which were attributed to N7. The signals were detectable only at temperatures below 20 K and showed a rhombic symmetry with $g_{x,y,z}$: 1.894, 1.953, and 2.047 (Fig. 2.7). Similar results were obtained from an EPR spectroscopic analysis of the *T. thermophilus* NuoG fragment and the single subunit NuoG from *E. coli*. The EPR spectrum of the cluster detected in the preparations exhibited rhombic symmetry with $g_{x,y,z}$ = 1.91, 1.94, and 2.05 (Nakamaru-Ogiso et al. 2005). The midpoint potential of N7 was determined to −250 ± 40 mV at pH 6.0 (Pohl et al. 2007b) and thus should be easily reduced by NADH if it were connected to the electron transfer chain.

The four cysteine ligands of N7 were identified in the structure of the peripheral arm of the *T. thermophilus* complex. They were individually replaced by alanine residues in *E. coli* complex I. The amount of complex I was significantly reduced in the mutant membranes indicating a disturbed assembly of the complex (Pohl et al. 2007b). The Cys230Ala^F and Cys233Ala^F variants showed a strongly decreased stability, as indicated by the complete loss of complex I activity upon detergent extraction and subsequent sucrose gradient centrifugation. The Cys237Ala^F and Cys265Ala^F variants were stable under these conditions. The preparations of these variants contained N7 in normal amounts, but with shifted g-values indicating a different microenvironment of the cluster (Pohl et al. 2007b). This is most likely due to a change of the ligation pattern of N7. In the Cys265Ala^F variant Asp292^F could replace Cys265^F as ligand. The different coordination of the cluster would induce a sterically allowed rearrangement of the NuoG backbone, explaining the slightly shifted g-values of the N7 signals in the preparation. It is not evident which residue could replace Cys237^F. The possibility that a fourth protein ligand is missing is unlikely because N7 is detected in the reduced preparation while a [3Fe-4S] cluster deriving from a decay of N7 could not be detected in the oxidized preparation. It is likely that Asp260^F replaces Cys237^F by placing a non-protein ligand such as OH^- between the carbonyl oxygen atom of Asp260^F and the Fe atom as described (Cross et al. 2002).

2.4.4 Participation of Fe/S Cluster N2 in Proton Translocation

Cluster N2 is located on subunit NuoB as demonstrated by EPR spectroscopy of site-directed mutants in *E. coli* (Flemming et al. 2003a) and *Y. lipolytica* (Ahlers et al. 2000). It exhibits the most positive midpoint potential of all Fe/S clusters of

Fig. 2.7 EPR spectra of isolated complex I at 13 K and 5 mW microwave power. Aliquots were either reduced by NADH (**a**) or dithionite (**b**); (**c**) shows the dithionite-reduced minus NADH-reduced EPR difference spectrum, (**d**) a simulation of the difference spectrum

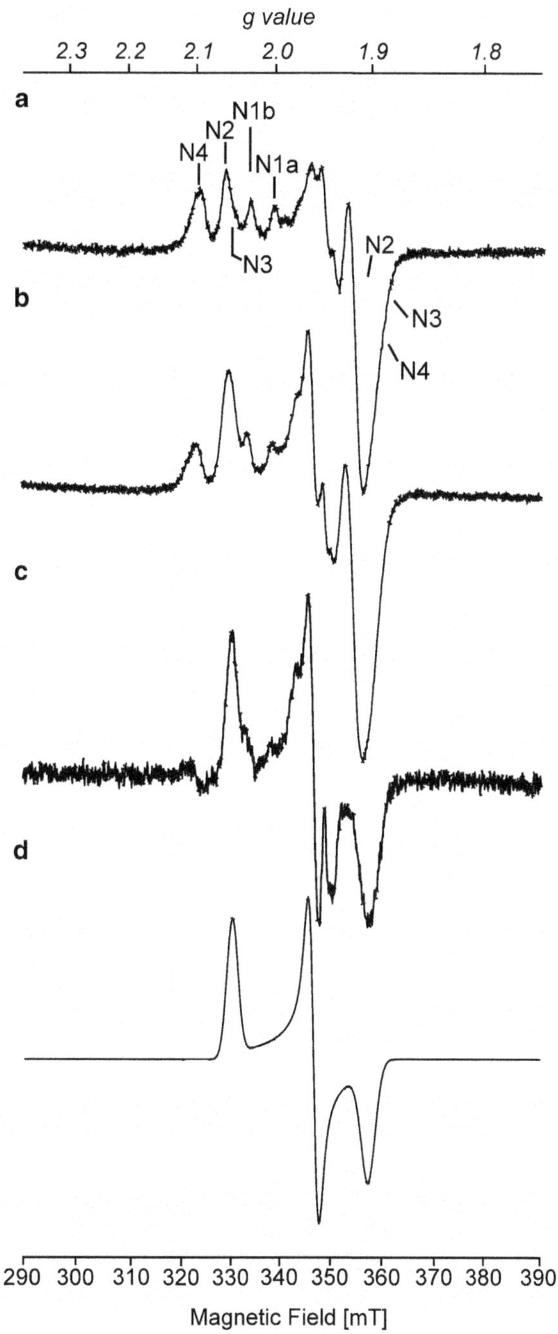

the complex (Table 2.2) and its E_m value is pH-dependent (−60 mV/pH) within the physiological pH range (Ingledew and Ohnishi 1980; Ohnishi 1998; Sled et al. 1993; Leif et al. 1995). N2 is believed to be the electron donor for (ubi)quinone as it is the most distal in the chain of the Fe/S clusters (Sazanov and Hinchliffe 2006). This is substantiated by the finding that there is spin coupling between N2 and the semiquinone species SQN_f (Ohnishi and Salerno 2005). From the strength of the coupling, a distance between N2 and SQN_f of 12 Å was calculated (Ohnishi and Salerno 2005).

The localization of N2 on NuoB raised the question about the nature of its ligands. The first two cysteines of the binding motif are adjacent. It was assumed that for sterical reasons only one of theses cysteines may serve as a ligand for N2 (Ahlers et al. 2000; Duarte et al. 2002; Kashani-Poor et al. 2001). Because the first cysteine residue of this motif is not conserved in the homologous (NiFe) hydrogenase, the second cysteine was considered as a probable ligand (Albracht 1993). It was proposed that either a conserved acidic amino acid on NuoB (Leif et al. 1995) or another amino acid residue located on the neighbouring subunit NuoD that constitutes a functional domain with NuoB (Friedrich et al. 1995; Kashani-Poor et al. 2001) could be the fourth ligand for N2 (Ahlers et al. 2000). However, from a molecular dynamics approach it was proposed that N2 is ligated by the two adjacent cysteine residues on NuoB, because it was found that these residues are capable of binding one [4Fe4S] cluster with a protein backbone conformation compatible to an optimal coordination of the cluster (Gurrath and Friedrich 2004). Finally, the X-ray structure of the peripheral arm of the *T. thermophilus* complex revealed the coordination of N2 by the four cysteine residues on NuoB, including the two immediately adjacent ones (Sazanov and Hinchliffe 2006).

It is still under discussion whether N2 is involved in proton translocation. Due to the pH-dependence of its midpoint potential (Ingledew and Ohnishi 1980), its interaction with SQ_{Nf} (Ohnishi and Salerno 2005) and its expected localization of about 20–25 Å from the membrane surface (Efremov et al. 2010) an involvement was anticipated. In addition it was shown by FTIR spectroscopy, that the reduction of N2 is coupled with a deprotonation of glutamate and possibly aspartate residues (Hellwig et al. 2000) and with the protonation of tyrosine residues (Flemming et al. 2003b). The corresponding amino acid residues were identified by site-directed mutagenesis of NuoB (Flemming et al. 2003b, 2006).

The electrochemically induced FTIR redox difference spectra of complex I are dominated by a large differential feature in the amide I range from 1,700 to 1,600 cm^{-1} (Hellwig et al. 2000). The exact position of these modes depends on the hydrogen bonding strength and is thus related to the type of secondary structure (Goormaghtigh et al. 1994). Based on the distribution of the modes over the amide I range, it was suggested that there is movement of more than one structural element, including β-sheets, loops, and α-helices (Hellwig et al. 2000). This spectral range is dominated by the ν(C=O) vibrations of the polypeptide backbone, mostly contributed by protein reorganizations in response to the redox reaction of the Fe/S clusters. The spectral feature concomitant with the conformational change can be perturbed as a function of pH (Hellwig et al. 2004) or by site-directed mutagenesis. The FTIR spectroscopic detection of conformational changes in the amide I region coupled with the redox reaction of an individual Fe/S cluster is difficult due to the fact that the redox reaction of any Fe/S cluster is associated with a local rearrangement of the protein backbone resulting in signals in the amide I region. This has been reported for the Rieske fragment of the bc_1 complex from *Paracoccus denitrificans* (Ritter et al. 2003) and even for small Fe/S proteins such as rubredoxin or ferrodoxin (Marshall et al. 2006). Two complex I variants were most helpful in identifying the contributions of the mechanistically relevant conformational changes to the amide I region coupled with the redox reaction of N2 (Friedrich and Hellwig 2010). The FTIR redox difference spectra of the enzymatically active Glu67Asp^B variant were compared to those of the enzymatically inactive Glu67Gln^B variant, both containing a similar amount of N2 (Friedrich and Hellwig 2010). The FTIR difference spectra of the inactive Glu67Gln^B variant showed a decreased amplitude of the amide I signal, deriving from several shifts of positive and negative absorptions. The amplitude of the signal in the amide I region was virtually unchanged in the active Glu67Asp^B variant (Friedrich and Hellwig 2010). The signals in question were found to be pH-dependent, implying that the redox reaction of N2 induces conformational changes, which may be part of a proton translocation process.

However, based on the capability of site-directed *Y. lipolytica* mutants for proton translocation it was concluded that N2 is not essential for this process (Zwicker et al. 2006). Mutation of a conserved histidine residue on the NuoD homologue to a methionine residue shifted the midpoint of N2 from −140 mV to −220 mV at pH 7 and completely abolished the redox-Bohr effect. This mutation had only a mild effect on the enyzmatic activity and H^+/e^- stoichiometry measured by the absorbance of the pH indicator neutral red using coupled mitochondria (Zwicker et al. 2006).

A proposal for the participation of N2 in proton translocation was derived from the structure of the peripheral arm of the *T. thermophilus* complex obtained under reduced and oxidized conditions (Berrisford and Sazanov 2009). The coordination of N2 by the two adjacent cysteine residues Cys63^B and Cys64^B is expected to result in a strained conformation of the ligands, which might lead to enhanced conformational flexibility in this region. The electron density between the cluster and the ligands is clearly detectable in the oxidized preparation (Berrisford and Sazanov 2009). Most strikingly, the electron density between Cys64^B and N2 is lost, when the preparation is reduced with NADH. In the dithionite-reduced preparation this density is clearly visible but that between Cys63^B and N2 is lost. This is interpreted by the authors in such a manner that the reduction of N2 leads to the disconnection of one of the adjacent ligating cysteine residues depending on the redox state of the neighbouring Fe/S clusters (Berrisford and Sazanov 2009). From this interpretation, a model for proton translocation coupled with the redox reaction of N2 was proposed involving the conserved Glu67^B and Tyr277^{CD} (Berrisford and Sazanov 2009). It was shown that Glu67^B and Tyr277^{CD} are essential for the activity of complex I (Kashani-Poor et al. 2001; Flemming et al. 2005; Friedrich and Hellwig 2010).

The proposal that N2 in combination with the (ubi)quinone redox chemistry is involved in proton translocation is further supported by the effect of Zn^{2+} on the

E. coli complex I. It was shown that Zn^{2+} is an inhibitor for the mitochondrial complex I (Sharpley and Hirst 2006). The inhibition is most likely due to a block of proton translocation as reported for other enzymes (e.g. Whitehead et al. 2005). The kinetics of Zn^{2+} binding are complex (Sharpley and Hirst 2006). *E. coli* complex I is also inhibited by Zn^{2+} depending on buffer pH, indicating a competition between Zn^{2+} and protons for the same binding sites. Most strikingly, a prior addition of Zn^{2+} to the complex blocks the full reduction of cluster N2, while it has no effect on the reduction of N2 by dithionite (Schulte et al. unpublished results). This indicates that blocking a proton pathway by Zn^{2+} seems to affect the environment of N2 preventing its full reduction. Thus, it is most likely that conformational changes around N2 participate in proton pumping by complex I.

2.5 Quinone Binding

None of the complex I structures published so far contains bound (ubi)quinone, so that its binding site is not yet defined on a molecular level. From labeling the complex with photoreactive quinone-site inhibitors and photoreactive quinones as well as by site-directed mutagenesis it was proposed that subunits NuoB, D, and H are involved in quinone binding, assuming that the quinone-site inhibitors directly act at the quinone-binding site (Earley et al. 1987; Darrouzet et al. 1998; Schuler et al. 1999; Dupuis et al. 2001; Kashani-Poor et al. 2001; Murai et al. 2009; Tocilescu et al. 2010). However, sequence analysis revealed that NuoL, NuoM and NuoN contain a motif for the binding of (ubi)quinones (Fisher and Rich 2000). In addition, NuoM was labeled by a photoreactive quinone derivative (Gong et al. 2003) and NuoL and NuoN were labeled by different quinone-site inhibitors (Nakamaru-Ogiso et al. 2003, 2010). From the structure of the *T. thermophilus* complex the (ubi)quinone-binding site was proposed to correspond to a cavity extending from an interface between NuoB, NuoD and NuoH in the peripheral arm to the membrane arm involving NuoA, K, and N (Sazanov and Hinchliffe 2006; Efremov et al. 2010). The proposed localization of the (ubi)quinone-binding site between NuoB and NuoD fits with site-directed mutagenesis in *E. coli* (Belevich et al. 2007) and *Y. lipolytica* (Zickermann et al. 2003; Tocilescu et al. 2010). The cavity has a diameter of about 30 Å and provides sufficient space for the presence of various inhibitors with partially overlapping binding sites as proposed from inhibitor binding studies (Ohshima et al. 1998; Sekiguchi et al. 2009). Within this cavity the (ubi)quinone headgroup would be within a distance of 14 Å from cluster N2 (Efremov et al. 2010). This value fits nicely the distance between reduced N2 and the semiquinone radical SQ_{Nf} of 12 Å as calculated from their mutual spin-spin interaction (Yano et al. 2005). It was proposed by Ohnishi and collaborators that there is a second (ubi)quinone binding-site in complex I, most likely on NuoN. This site is supposed to harbour the second semiquinone radical named SQ_{Ns} and is located at a distance of more than 30 Å from N2 (Ohnishi et al. 2010a, b).

To localize the (ubi)quinone binding site, $Arg112^{B}$ in close distance to N2 was changed to a cysteine residue in the *E. coli* complex I. The $Arg112Cys^{B}$ variant was isolated and the spin label MTSL (1-oxyl-2,2,5,5,-tetramethyl-Δ3-pyrolline-3-methyl)-methanethiosulfonate was covalently attached to this position (Pohl et al. 2010). Neither the mutation nor the attachment of the label has an effect on the enzymatic activity of the variant. To determine the distance between this position and that of a bound ubiquinone, a decyl-ubiquinone derivative was synthesized with MTSL covalently attached at the C10 position of the alkyl chain (Q_{10}-MTSL). Cyclovoltammetry of Q_{10}-MTSL revealed a single peak at a potential of +196 mV at pH 8.0 coupled to protonation reactions as expected for the quinone moiety. No other peak was detected, indicating that the midpoint potential of the nitroxide group of the label is beyond the physiological range and therefore not expected to interfere with the enzymatic reaction of complex I. Although the activity of complex I with the labeled decyl-ubiquinone was only one fifth of that with decyl-ubiquinone, the reaction was fully sensitive to piericidin A (Pohl et al. 2010). The distance between the label attached to position Arg112^B of the complex and the label attached to Q_{10}-MTSL was determined by measuring the strength of dipolar interactions between two spin labels, which is inversely proportional to the cubed distance (Hubbell et al. 1996; Hustedt et al. 1997; Jeschke 2002). Interspin distances were extracted from cw-EPR spectra of frozen protein samples via simulation (Steinhoff et al. 1997). For this purpose, Q_{10}-MTSL was added in an equimolar ratio to the Arg112^B variant and cw-EPR spectra were recorded at X-Band microwave frequencies at 40 K. Spectral simulations were performed by adjusting two *g* and *A* tensors for the two interacting MTSL radicals, and electron-electron interaction parameters were adjusted to obtain the best possible agreement of simulated and experimental spectra. An isotropic electron-electron interaction tensor of 25 MHz, corresponding to a distance of 13.5 ± 1.5 Å, was obtained (Pohl et al. 2010). This distance is in good agreement with the distance of 12 Å between N2 and SQ_{Nf} as derived from EPR spectroscopy (Yano et al. 2005) and a distance of 14 Å between the proposed (ubi)quinone-binding site and N2 as derived from the structure (Efremov et al. 2010).

2.6 Proton Translocation

The coupling between electron transfer and proton translocation as well as the mechanism of proton translocation are not well understood. Most notably, the membrane arm contains the ATS NuoL, M and N, which derive from a common ancestor and are homologues of monovalent cation/H^+ antiporters (Friedrich and Weiss 1997; Friedrich 2001; Mathiesen and Hagerhall 2003; Moparthi et al. 2011). It is most reasonable to assume that they are involved in proton translocation because it was shown that the *E. coli* subunits are able to functionally substitute their homologues in a *Bacillus subtilis* Na^+/H^+ antiporter (Moparthi et al. 2011). The recently determined structure of complex I from *T. thermophilus* revealed that these subunits are located at the most distal end of the membrane arm (Efremov et al. 2010). NuoL contains an additional C-terminal domain consisting of two TM helices and a long 'horizontal' helix. It was proposed that the helix is used as a piston to transmit the energy released by the redox reaction in the peripheral arm to proton translocation in the membrane arm (Efremov et al. 2010; Ohnishi 2010; Fig. 2.8a).

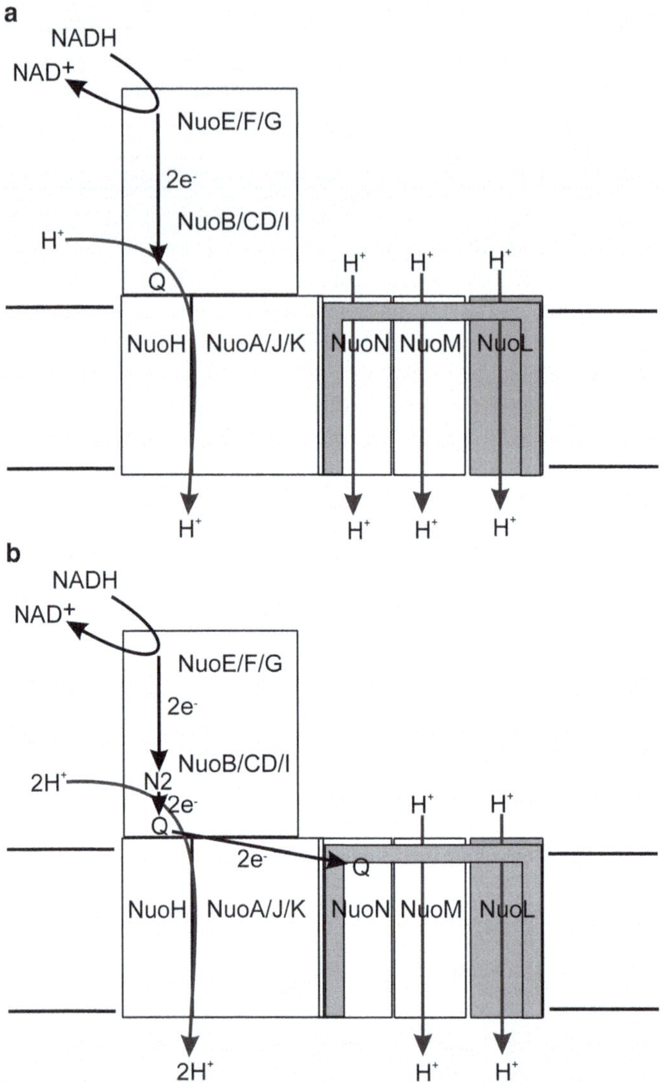

Fig. 2.8 Schematic representation of the models for proton translocation by complex I. The L-shape form of complex with an approximate localization of the subunits named after the Nuo-nomenclature is shown. The redox cofactors in the peripheral arm are omitted for clarity. Subunit NuoL with its additional C-terminal extension (TM helix, 'horizontal' helix and TM helix) are shown in *light gray*. Note that the extension folds back to the proposed (ubi)quinone binding site. The proton pathway is indicated in *dark gray*. For clarity only the translocated protons are shown, not the ones to produce (ubi)quinol. In (**a**) the model by Efremov et al. (2010) is shown. Each ATS translocates one proton, while the fourth proton is translocated by an unknown type of (ubi)quinone redox chemistry. (**b**) shows a model modified from Ohnishi et al. (2010b). Two protons are translocated by the redox chemistry of N2 and Q_{Nf} and the electrons are passed to the second (ubi) quinone bound to NuoN, which is converted to (ubi)quinol. Two additional protons are translocated by NuoM and NuoL. This indirect proton translocation is mediated by the protonation of the second (ubi)quinone

According to the structure of the bacterial complex the proposal was raised that by an unknown mechanism the (ubi)quinone reduction induces a piston-like movement of the amphipathic helix in the membrane arm. Due to this movement, the three ATS are opened and closed depending on the redox state of (ubi)quinone. This movement would lead to a translocation of one proton by each of the subunits. According to the current consensus, four protons are translocated due to the oxidation of one NADH, which is equivalent to the transport of two electrons (Wikstrom 1984; Bogachev et al. 1996; Galkin et al. 2006). Therefore, it was proposed that the fourth proton is translocated at the (ubi)quinone binding site by another yet unknown mechanism (Efremov et al. 2010). On the other hand, it was postulated that only two protons are translocated by the ATS (Ohnishi et al. 2010a, b). Accordingly, NuoN is not involved in proton translocation because mutations of conserved charged residues within the subunit had no effect on proton translocation (Amarneh and Vik 2003). This is further substantiated by the finding that in higher metazoans the three N-terminal helices of NuoN are missing (Birrell and Hirst 2010). On this basis, it was proposed that NuoN rather supports the anchoring of the amphipathic helix of NuoL and promotes the coupling of electron transfer with proton translocation (Birrell and Hirst 2010). Furthermore, NuoN contains a sequence motif proposed to represent a quinone binding motif for respiratory complexes (Fisher and Rich 2000). The assumption that NuoN might be involved in quinone binding is further substantiated by the fact, that it is labeled by a photoaffinity analogue of asimicin, a specific quinone site inhibitor of complex I (Nakamaru-Ogiso et al. 2010).

The role of the additional C-terminal domain of NuoL for proton translocation was investigated by producing truncated versions of NuoL in *E. coli* (Steimle et al. 2011). The variants either lacked the C-terminal TM helix (W592^{L}Stop), lacked this helix plus approximately half of the amphipathic helix (Y544^{L}Stop) or lacked the entire subunit NuoL (ΔNuoL). The variants were purified to homogeniety by affinity chromatography. All preparations were stable and contained the expected protein composition as derived from SDS-PAGE in combination with mass spectrometry. EPR spectroscopy revealed the presence of all Fe/S clusters in the variants (Steimle et al. 2011). The presence of fully assembled and structurally intact NuoL variants in *E. coli* was not unexpected as it was shown that NuoL can be selectively removed from the complex by changing the detergent without decreasing the stability of the residual complex (Baranova et al. 2007).

The preparations were reconstituted in proteoliposomes and assayed for electron transfer and proton translocation (Steimle et al. 2011). The NuoL variants exhibited 80–85% of the NADH:decyl-ubiquinone oxidoreductase activity of the fully assembled complex. They were inhibited to 90% by the complex I specific inhibitor piericidin A. Thus, the loss of NuoL or the loss of the C-terminal domain of NuoL had only a very mild effect on the electron transfer activity. This might have been expected due to the distal location of NuoL in the membrane arm. However, as the long horizontal helix of NuoL folds back to the interface between the two arms, the C-terminal TM helix is at a 40–50 Å distance from the proposed (ubi)quinone binding site.

Proton translocation by the variants was measured by the quench of the ACMA fluorescence. Under the given conditions the fully assembled complex exhibited a fluorescence quench of 40%. Proteoliposomes containing the preparation of the ΔNuoL variant quenched the ACMA signal by 18%, those containing the $Y544^{L}$Stop and the $W592^{L}$Stop variants quenched the fluorescence by approximately 15%. All signals were fully sensitive to CCCP and piericidin A demonstrating that they derive from a proton gradient generated by the redox reaction of complex I (Steimle et al. 2011). Assuming that the parental complex has a $H^{+}/2e^{-}$ stoichiometry of four, the data indicated that the ΔNuoL variant exhibits a $H^{+}/2e^{-}$ stoichiometry of about two. The Stop-variants exhibited a slightly lower stoichiometry (Steimle et al. 2011). The difference could be due to a disordered or flexible arrangement of the amphipathic helix no longer connected to the membrane arm by the C-terminal TM helix of NuoL. The disordered amphipathic helix might allow a random opening and closing of the ATS not coupled with the electron transfer reaction. This would lead to a decrease of the proton gradient across the membrane.

If the mechanism proposed from structural data is correct, the ΔNuoL variant should be able to translocate one $H^{+}/2e^{-}$ (Efremov et al. 2010). In contrast, the data implicate that NuoL is essential for the translocation of two protons. Thus, the three ATS are most likely involved in the translocation of two $H^{+}/2e^{-}$. The two residual protons required to fulfill the experimentally determined stoichiometry are probably translocated by some kind of (ubi)quinone redox chemistry, involving the participation of two different semiquinones as proposed (Ohnishi et al. 2010a, b). It is possible that cluster N2 is also engaged in this reaction (Fig. 2.8b).

The data demonstrate that there are at least two coupling sites in complex I. One site involving the (ubi)quinone(s) and probably N2 directly couples proton translocation with electron transfer, while the other site involving NuoL and M is indirectly coupled with the electron transfer reaction. The C-terminal domain of NuoL is essential for the transmission of the energy from the quinone binding site to NuoL and M. Both sites contribute with a stoichiometry of 2 $H^{+}/2e^{-}$ to the total stoichiometry of 4 $H^{+}/2e^{-}$ (Fig. 2.8b). The mechanism of the complex as a chimera of a directly and indirectly driven proton pump was proposed from the evolution of the complex (Friedrich 2001).

Acknowledgment The work in the authors groups was supported by the Deutsche Forschungsgemeinschaft (DFG; to TF and OE), by the Volkswagen Foundation (to TF and PH) and the Agence Nationale de Recherche (ANR; to PH). We thank Linda Williams for her help in correcting the manuscript.

References

Ahlers PM, Zwicker K, Kerscher S, Brandt U (2000) Function of conserved acidic residues in the PSST homologue of complex I (NADH:ubiquinone oxidoreductase) from *Yarrowia lipolytica*. J Biol Chem 275:23577–23582

Albracht SPJ (1993) Intimate relationships of the large and the small subunits of all nickel hydrogenases with two nuclear-encoded subunits of mitochondrial NADH:ubiquinone oxidoreductase. Biochim Biophys Acta 1144:221–224

Amarneh B, Vik SB (2003) Mutagenesis of subunit N of the *Escherichia coli* complex I. Identification of the initiation codon and the sensitivity of mutants to decylubiquinone. Biochemistry 42:4800–4808

Auriol C, Bestel-Corre G, Claude JB, Soucaille P, Meynial-Salles I (2011) Stress-induced evolution of *Escherichia coli* points to original concepts in respiratory cofactor selectivity. Proc Natl Acad Sci USA 108:1278–1283

Baranova EA, Holt PJ, Sazanov LA (2007) Single particle analysis confirms distal location of subunits NuoL and NuoM in *Escherichia coli* complex I. J Mol Biol 366:140–154

Belevich G, Euro L, Wikström M, Verkhovskaya M (2007) Role of the conserved arginine 274 and histidine 224 and 228 residues in the NuoCD subunit of complex I from *Escherichia coli*. Biochemistry 46:526–533

Berrisford JM, Sazanov LA (2009) Structural basis for the mechanism of respiratory complex I. J Biol Chem 284:29773–29783

Birrell JA, Hirst J (2010) Truncation of subunit ND2 disrupts the threefold symmetry of the antiporter-like subunits in complex I from higher metazoans. FEBS Lett 584:4247–4252

Bogachev AV, Murtazina RA, Skulachev VP (1996) H^+/e^- stoichiometry for NADH dehydrogenase I and dimethyl sulfoxide reductase in anaerobically grown *Escherichia coli* cells. J Bacteriol 178:6233–6237

Brandt U (2006) Energy converting NADH:quinone oxidoreductase (complex I). Annu Rev Biochem 75:69–92

Bungert S, Krafft B, Schlesinger R, Friedrich T (1999) One-step purification of the NADH dehydrogenase fragment of the *Escherichia coli* by means of Strep-tag affinity chromatography. FEBS Lett 460:207–211

Chatelet C, Gaillard J, Pétillot Y, Louwagie M, Meyer J (1999) A [2Fe-2S] protein from the hyperthermophilic bacterium *Aquifex aeolicus*. Biochem Biophys Res Commun 261:885–889

Clason T, Ruiz T, Schägger H, Peng G, Zickermann V, Brandt U, Michel H, Radermacher M (2010) The structure of eukaryotic and prokaryotic complex I. J Struct Biol 169:81–88

Cross M, Xiao Z, Maes EM, Czernuszewicz RS, Drew SC, Pilbrow JR, George GN, Wedd AG (2002) Removal of a cysteine ligand from rubredoxin: assembly of Fe(2)S(2) and Fe(S-Cys)(3)(OH) centres. J Biol Inorg Chem 7:781–790

Darrouzet E, Issartel JP, Lunardi J, Dupuis A (1998) The 49-kDa subunit of NADH-ubiquinone oxidoreductase (complex I) is involved in the binding of piericidin and rotenone, two quinone-related inhibitors. FEBS Lett 431:34–38

Datsenko KA, Wanner BL (2000) One-step inactivation of chromosomal genes in *Escherichia coli* K-12 using PCR products. Proc Natl Acad Sci USA 97:6640–6645

de Jongh HH, Goormaghtigh E, Ruysschaert JM (1997) Amide-proton exchange of water-soluble proteins of different structural classes studied at the submolecular level by infrared spectroscopy. Biochemistry 36:13603–13610

Djavadi-Ohaniance L, Hatefi Y (1975) Oxidation of NADPH by submitochondrial particles from beef heart in complete absence of transhydrogenase activity from NADPH to NAD. J Biol Chem 250:9397–9403

Duarte M, Pópulo H, Videira A, Friedrich T, Schulte U (2002) Disruption of iron-sulphur cluster N2 from NADH: ubiquinone oxidoreductase by site-directed mutagenesis. Biochem J 364:833–839

Dupuis A, Prieur I, Lunardi J (2001) Toward a characterization of the connecting module of complex I. J Bioenerg Biomembr 33:159–168

Earley FGP, Patel SD, Ragan CI, Attardi G (1987) Photolabelling of a mitochondrially encoded subunit of NADH dehydrogenase with [3H]dihydrorotenone. FEBS Lett 219:108–112

Efremov RG, Baradaran R, Sazanov LA (2010) The architecture of respiratory complex I. Nature 465:441–445

Ernster L, Hoberman HD, Howard RL, King TE, Lee CP, Mackler B, Sottocasa G (1965) Stereospecificity of certain soluble and particulate preparations of mitochondrial reduced nicotinamide-adenine dinucleotide dehydrogenase from beef heart. Nature 207:940–941

Fearnley IM, Walker JE (1992) Conservation of sequences of subunits of mitochondrial complex I and their relationships with other proteins. Biochim Biophys Acta 1140:105–134

Fisher N, Rich PR (2000) A motif for quinone binding sites in respiratory and photosynthetic systems. J Mol Biol 296:1153–1162

Flemming D, Schlitt A, Spehr V, Bischof T, Friedrich T (2003a) Iron-sulfur cluster N2 of the *Escherichia coli* NADH:ubiquinone oxidoreductase (complex I) is located on subunit NuoB. J Biol Chem 278:47602–47609

Flemming D, Hellwig P, Friedrich T (2003b) Involvement of tyrosines 114 and 139 of subunit NuoB in the proton pathway around cluster N2 in *Escherichia coli* NADH:ubiquinone oxidoreductase. J Biol Chem 278:3055–3062

Flemming D, Hellwig P, Lepper S, Kloer DP, Friedrich T (2006) Catalytic importance of acidic amino acids on subunit NuoB of the *Escherichia coli* NADH:ubiquinone oxidoreductase (complex I). J Biol Chem 281:24781–24789

Flemming D, Stolpe S, Schneider D, Hellwig P, Friedrich T (2005) A possible role for iron-sulfur cluster N2 in proton translocation by the NADH:ubiquinone oxidoreductase (complex I). J Mol Microbiol Biotechnol 10:208–222

Friedrich T (1998) The NADH:ubiquinone oxidoreductase (complex I) from *Escherichia coli*. Biochim Biophys Acta 1364:134–146

Friedrich T (2001) Complex I: a chimaera of a redox and conformation driven proton-pump? J Bioenerg Biomembr 33:169–179

Friedrich T, Hellwig P (2010) Redox-induced conformational changes within the *Escherichia coli* NADH ubiquinone oxidoreductase (complex I): an analysis by mutagenesis and FTIR spectroscopy. Biochim Biophys Acta 1797:659–663

Friedrich T, Pohl T (13 Aug 2007, posting date) Chapter 3.2.4, NADH as Donor. In: Böck A, Curtis III R, Kaper JB, Neidhardt FC, Nyström T, Slauch JM, Squires CL (eds) EcoSal – Escherichia coli and Salmonella: cellular and molecular biology. ASM Press, Washington, DC. http://www.ecosal.org

Friedrich T, Scheide D (2000) The respiratory complex I of bacteria, archaea and eukarya and its module common with membrane-bound multisubunit hydrogenases. FEBS Lett 479:1–5

Friedrich T, Weiss H (1997) Modular evolution of the respiratory NADH:ubiquinone oxidoreductase and the origin of its modules. J Theor Biol 187:529–541

Friedrich T, Steinmüller K, Weiss H (1995) The proton-pumping respiratory complex I of bacteria and mitochondria and its homologue in chloroplasts. FEBS Lett 367:107–111

Friedrich T, Stolpe S, Schneider D, Barquera B, Hellwig P (2005) Ion translocation by the *Escherichia coli* NADH:ubiquinone oxidoreductase (complex I). Biochem Soc Trans 33:836–839

Fujinaga J, Gaillard J, Meyer J (1993) Mutated forms of a [2Fe-2S] ferredoxine with serine ligands to the iron-sulfur-cluster. Biochem Biophys Res Commun 194:104–111

Galkin A, Brandt U (2005) Superoxide radical formation by pure complex I (NADH:ubiquinone oxidoreductase) from *Yarrowia lipolytica*. J Biol Chem 280:30129–30135

Galkin A, Dröse S, Brandt U (2006) The proton pumping stoichiometry of purified mitochondrial complex I reconstituted in proteoliposomes. Biochim Biophys Acta 1757:1575–1581

Garczarek F, Gerwert K (2006) Functional waters in intraprotein proton transfer monitored by FTIR difference spectroscopy. Nature 439:109–112

Golinelli MP, Akin LA, Crouse BR, Johnson MK, Meyer J (1996) Cysteine ligand swapping on a deletable loop of the [2Fe-2S] ferredoxin from *Clostridium pasteurianum*. Biochemistry 35:8995–9002

Golinelli MP, Chatelet C, Duin EC, Johnson MK, Meyer J (1998) Extensive ligand rearrangements around the [2Fe-2S] cluster of *Clostridium pasteurianum* ferredoxin. Biochemistry 37:10429–10437

Gong X, Xie T, Yu L, Hesterberg M, Scheide D, Friedrich T, Yu CA (2003) The ubiquinone-binding site in NADH:ubiquinone oxidoreductase from *Escherichia coli*. J Biol Chem 278:25731–25737

Goormaghtigh E, Cabiaux V, Ruysschaert JM (1994) Determination of soluble and membrane protein structure by Fourier transform infrared spectroscopy. III. Secondary structures. Subcell Biochem 23:405–450

Gregory RB, Lumry R (1985) Hydrogen-exchange evidence for distinct structural classes in globular proteins. Biopolymers 24:301–326
Grigorieff N (1999) Structure of the respiratory NADH:ubiquinone oxidoreductase (complex I). Curr Opin Struct Biol 9:476–483
Guénebaut V, Schlitt A, Weiss H, Leonard K, Friedrich T (1998) Consistent structure between bacterial and mitochondrial NADH:ubiquinone oxidoreductase (complex I). J Mol Biol 276:105–112
Gunner MR, Saleh MA, Cross E, ud-Doula A, Wise M (2000) Backbone dipoles generate positive potentials in all proteins: Origins and implications of the effect. Biophys J 78:1126–1144
Gurrath M, Friedrich T (2004) Adjacent cysteines are capable of ligating the same tetranuclear iron-sulfur cluster. Proteins: Struct Funct Bioinform 56:556–563
Guzman L, Belin D, Carson MJ, Beckwith J (1995) Tight regulation, modulation, and high-level expression by vectors containing the arabinose pBAD promoter. J Bacteriol 177:4121–4130
Hamilton CM, Aldea M, Washburn BK, Babitzke P, Kushner SR (1989) New method for generating deletions and gene replacements in *Escherichia coli*. J Bacteriol 171:4617–4622
Haris PI, Chapman D (1995) The conformational analysis of peptides using Fourier transform IR spectroscopy. Biopolymers 37:251–263
Hatefi Y, Hanstein WG (1973) Interactions of reduced and oxidized triphosphopyridine nucleotides with the electron-transport system of bovine heart mitochondria. Biochemistry 12:3515–3522
Hayashi T, Stuchebrukhov AA (2010) Electron tunneling in respiratory complex I. Proc Natl Acad Sci USA 107:19157–19162
Hellwig P, Scheide D, Bungert S, Mäntele W, Friedrich T (2000) FTIR spectroscopic characterization of NADH:ubiquinone oxidoreductase (complex I) from *Escherichia coli*: oxidation of FeS cluster N2 is coupled with the protonation of an aspartate or glutamate side chain. Biochemistry 39:10884–10891
Hellwig P, Stolpe S, Friedrich T (2004) Fourier transform infrared spectroscopic study on the conformational reorganization in *Escherichia coli* complex I due to redox-driven proton translocation. Biopolymers 74:69–72
Hielscher R (2009) The role of lipids and nucleotides in the catalytic mechanism of proteins from the respiratory chain: an electrochemical and infrared spectroscopic approach, PhD thesis, University of Strasbourg
Hielscher R, Wenz T, Stolpe S, Hunte C, Friedrich T, Hellwig P (2006) Monitoring redox-dependent contribution of lipids in Fourier transform infrared difference spectra of complex I from *Escherichia coli*. Biopolymers 82:291–294
Hielscher R, Friedrich T, Hellwig P (2011) Far and mid infrared spectroscopic analysis on the substrate induced structural dynamics of the respiratory complex I. Chemphyschem 12:217–224
Hinchliffe P, Carroll J, Sazanov LA (2006) Identification of a novel subunit of respiratory complex I from *Thermus thermophilus*. Biochemistry 45:4413–4420
Hirst J (2010) Towards the molecular mechanism of respiratory complex I. Biochem J 425: 327–339
Hirst J, King MS, Pryde KR (2008) The production of reactive oxygen species by complex I. Biochem Soc Trans 36:976–980
Hubbell WL, Mchaourab HS, Altenbach C, Lietzow MA (1996) Watching proteins move using site-directed spin labeling. Structure 4:779–783
Hunte C, Zickermann V, Brandt U (2010) Functional modules and structural basis of conformational coupling in mitochondrial complex I. Science 329:448–451
Hustedt EJ, Smirnov AI, Laub FL, Cobb CE, Beth AH (1997) Molecular distances from dipolar coupled spin-labels: the global analysis of multifrequency CW-EPR data. Biophys J 74:1861–1877
Hvidt A, Linderstrøm-Lang K (1954) Exchange of hydrogen atoms in insulin with deuterium atoms in aqueous solutions. Biochim Biophys Acta 14:574–575
Hvidt A, Linderstrøm-Lang K (1955) The kinetics of the deuterium exchange of insulin with D2O; an amendment. Biochim Biophys Acta 16:168–169

Hvidt A, Nieslen SO (1966) Hydrogen exchange in proteins. Adv Protein Chem 21:287–386

Ibarra RU, Edwards JS, Palsson BO (2002) *Escherichia coli* K-12 undergoes adaptive evolution to achieve in silico predicted optimal growth. Nature 420:186–189

Ingledew WJ, Ohnishi T (1980) An analysis of some thermodynamic properties of iron-sulphur centres in site I of mitochondria. Biochem J 186:111–117

Jeschke G (2002) Distance measurements in the nanometer range by pulse EPR. Chemphyschem 3:927–932

Kashani-Poor N, Zwicker K, Kerscher S, Brandt U (2001) A central functional role for the 49-kDa subunit within the catalytic core of mitochondrial complex I. J Biol Chem 276:24082–24087

Kikuno R, Miyata T (1985) Sequence homologies among mitochondrial DNA-coded URF2, URF4 and URF5. FEBS Lett 189:85–88

Kim KS, Fuchs JA, Woodward CK (1993) Hydrogen exchange identifies native-state motional domains important in protein folding. Biochemistry 32:9600–9608

Knox DG, Rosenberg A (1980) Fluctuations of protein structure as expressed in the distribution of hydrogen exchange rate constants. Biopolymers 19:1049–1068

Kohlstädt M, Dörner K, Labatzke R, Koç C, Hielscher R, Schiltz E, Einsle O, Hellwig P, Friedrich T (2008) Heterologous production, isolation, characterization and crystallization of a soluble fragment of the NADH:ubiquinone oxidoreductase (complex I) from *Aquifex aeolicus*. Biochemistry 47:13036–13045

Krimm S, Bandekar J (1986) Vibrational spectroscopy and conformation of peptides, polypeptides, and proteins. J Adv Protein Chem 38:181–364

Kussmaul L, Hirst J (2006) The mechanism of superoxide production by NADH:ubiquinone oxidoreductase (complex I) from bovine heart mitochondria. Proc Natl Acad Sci USA 103:7607–7612

Leif H, Sled VD, Ohnishi T, Weiss H, Friedrich T (1995) Isolation and characterization of the proton-translocating NADH:ubiquinone oxidoreductase from *Escherichia coli*. Eur J Biochem 230:538–548

Lesk AM (1995) NAD-binding domains of dehydrogenases. Curr Opin Struct Biol 5:775–783

Mamedova AA, Holt PJ, Carroll J, Sazanov LA (2004) Substrate-induced conformational change in bacterial complex I. J Biol Chem 279:23830–23836

Marshall D, Fisher N, Grigic L, Zickermann V, Brandt U, Shannon RJ, Hirst J, Lawrence R, Rich PR (2006) ATR-FTIR redox difference spectroscopy of *Yarrowia lipolytica* and bovine complex I. Biochemistry 45:5458–5467

Mathiesen C, Hagerhall C (2003) The 'antiporter module' of respiratory chain complex I includes the MrpC/NuoK subunit – a revision of the modular evolution scheme. FEBS Lett 549:7–13

Meyer J, Andrade SL, Einsle O (2008) Thioredoxin-like [2Fe-2S] ferredoxin. In: Messerschmidt A (ed) Handbook of metalloproteins. Wiley, New York

Moparthi VK, Kumar B, Mathiesen C, Hägerhäll C (2011) Homologous protein subunits from Escherichia coli NADH:quinone oxidoreductase can functionally replace MrpA and MrpD in *Bacillus subtilis*. Biochim Biophys Acta 1807:427–436

Morina K, Schulte M, Hubrich F, Dörner K, Steimle S, Stolpe S, Friedrich T (2011) Engineering the respiratory complex I to an energy-converting NADPH:ubiquinone oxidoreductase. J Biol Chem 286:34627–34634

Moser CC, Farid TA, Chobot SE, Dutton PL (2006) Electron tunneling chains of mitochondria. Biochim Biophys Acta 1757:1096–1109

Murai M, Sekiguchi K, Nishioka T, Miyoshi H (2009) Characterization of the inhibitor binding site in mitochondrial NADH-ubiquinone oxidoreductase by photoaffinity labeling using a quinazoline-type inhibitor. Biochemistry 48:688–698

Nakamaru-Ogiso E, Sakamoto K, Matsuno-Yagi A, Miyoshi H, Yagi T (2003) The ND5 subunit was labeled by a photoaffinity analogue of fenpyroximate in bovine mitochondrial complex I. Biochemistry 42:746–754

Nakamaru-Ogiso E, Yano T, Yagi T, Ohnishi T (2005) Characterization of the iron-sulfur cluster N7 (N1c) in the subunit NuoG of the proton-translocating NADH-quinone oxidoreductase from *Escherichia coli*. J Biol Chem 280:301–307

Nakamaru-Ogiso E, Han H, Matsuno-Yagi A, Keinan E, Sinha SC, Yagi T, Ohnishi T (2010) The ND2 subunit is labeled by a photoaffinity analogue of asimicin, a potent complex I inhibitor. FEBS Lett 584:883–888

Oden KL, DeVeaux LC, Vibat CR, Cronan JE Jr, Gennis RB (1990) Genomic replacement in *Escherichia coli* K-12 using covalently closed circular plasmid DNA. Gene 96:29–36

Ohnishi T (1998) Iron-sulfur clusters/semiquinones in complex I. Biochim Biophys Acta 1364:186–206

Ohnishi T (2010) Piston drives a proton pump. Nature 465:428–429

Ohnishi T, Salerno JC (2005) Conformation-driven and semiquinone-gated proton-pump mechanism in the NADH-ubiquinone oxidoreductase (complex I). FEBS Lett 579:4555–4561

Ohnishi ST, Salerno JC, Ohnishi T (2010a) Possible roles of two quinone molecules in direct and indirect proton pumps of bovine heart NADH-quinone oxidoreductase (complex I). Biochim Biophys Acta 1797:1891–1893

Ohnishi T, Nakamaru-Ogiso E, Ohnishi ST (2010b) A new hypothesis on the simultaneous direct and indirect proton pump mechanisms in NADH-quinone oxidoreductase (complex I). FEBS Lett 584:4131–4137

Ohnishi ST, Shinzawa-Itoh K, Ohta K, Yoshikawa S, Ohnishi T (2010c) New insights into the superoxide generation sites in bovine heart NADH-ubiquinone oxidoreductase (complex I): the significance of protein-associated ubiquinone and the dynamic shifting of generation sites between semiflavin and semiquinone radicals. Biochim Biophys Acta 1797:1901–1909

Ohshima M, Miyoshi H, Sakamoto K, Takegami K, Iwata J, Kuwabara K, Iwamura H, Yagi T (1998) Characterization of the ubiquinone reduction site of mitochondrial complex I using bulky synthetic ubiquinones. Biochemistry 37:6436–6445

Peng G, Fritzsch G, Zickermann V, Schägger H, Mentele R, Lottspeich F, Bostina M, Radermacher M, Huber R, Stetter KO, Michel H (2003) Isolation, characterization and electron microscopic single particle analysis of the NADH:ubiquinone oxidoreductase (complex I) from the hyperthermophilic eubacterium *Aquifex aeolicus*. Biochemistry 42:3032–3039

Peng G, Ermler U, Clason T, Bornemann S, Hedderich T, Ruiz T, Meyer B, Radermacher M, Karras M, Michel H (2010) The structure of complex I from the hyperthermophilic eubacterium *Aquifex aeolicus*. Biochim Biophys Acta 1797(Supplement):22

Peters JW, Lanzilotta WN, Lemon BJ, Seefeldt LC (1998) X-ray crystal structure of the Fe-only hydrogenase (CpI) from *Clostridium pasteurianum* to 1.8 angstrom resolution. Science 282:1853–1858

Pohl T, Uhlmann M, Kaufenstein M, Friedrich T (2007a) Lambda Red-mediated mutagenesis and efficient large scale affinity purification of the *Escherichia coli* NADH:ubiquinone oxidoreductase (complex I). Biochemistry 46:10694–10702

Pohl T, Bauer T, Dörner K, Stolpe S, Sell P, Zocher G, Friedrich T (2007b) Iron-sulfur cluster N7 of the NADH:ubiquinone oxidoreductase (complex I) is essential for stability but not involved in electron transfer. Biochemistry 46:6588–6596

Pohl T, Walter J, Stolpe S, Defeu Soufo HJ, Graumann PL, Friedrich T (2007c) Effects of the deletion of the *Escherichia coli* frataxin homologue CyaY on the respiratory NADH:ubiquinone oxidoreductase. BMC Biochem 8:13

Pohl T, Schneider D, Hielscher R, Stolpe S, Dörner K, Kohlstädt M, Böttcher B, Hellwig P, Friedrich T (2008) Nucleotide-induced conformational changes in the *Escherichia coli* NADH:ubiquinone oxidoreductase (complex I). Biochem Soc Trans 36:971–975

Pohl T, Spatzal T, Aksoyoglu M, Schleicher E, Rostas AM, Lay H, Glessner U, Boudon C, Hellwig P, Weber S, Friedrich T (2010) Spin labeling of the *Escherichia coli* NADH:ubiquinone oxidoreductase (complex I). Biochim Biophys Acta 1797:1894–1900

Ragan CI, Racker E (1973) Resolution and reconstitution of the mitochondrial electron transport system. IV. The reconstitution of rotenone-sensitive reduced nicotinamide adenine dinucleotide-ubiquinone reductase from reduced nicotinamide adenine dinucleotide dehydrogenase and phospholipids. J Biol Chem 248:6876–6884

Ried JL, Collmer A (1987) An *nptI-sacB-sacR* cartridge for constructing directed, unmarked mutations in gram-negative bacteria by marker exchange-eviction mutagenesis. Gene 57:239–246

Ritter M, Anderka O, Ludwig B, Mäntele W, Hellwig P (2003) Electrochemical and FTIR spectroscopic characterization of the cytochrome *bc1* complex from *Paracoccus denitrificans*: evidence for protonation reactions coupled to quinone binding. Biochemistry 42:12391–12399

Rothery RA, Workun GJ, Weiner JH (2008) The prokaryotic complex iron-sulfur molybdoenzyme family. Biochim Biophys Acta 1778:1897–1929

Sazanov LA (2007) Respiratory complex I: mechanistic and structural insights provided by the crystal structure of the hydrophilic domain. Biochemistry 46:2275–2288

Sazanov LA, Hinchliffe P (2006) Structure of the hydrophilic domain of respiratory complex I from *Thermus thermophilus*. Science 311:1430–1436

Schneider D, Pohl T, Walter J, Dörner K, Kohlstädt M, Berger A, Spehr V, Friedrich T (2008) Assembly of the *Escherichia coli* complex I. Biochim Biophys Acta 1777:735–739

Schuler F, Yano T, DiBernardo S, Yagi T, Yankovskaya V, Singer T, Casida JE (1999) NADH-quinone oxidoreductase: PSST subunit couples electron transfer from iron-sulfur cluster N2 to quinone. Proc Natl Acad Sci USA 96:4149–4153

Screpanti E, Hunte C (2007) Discontinuous membrane helices in transport proteins and their correlation with function. J Struct Biol 159:261–267

Sekiguchi K, Murai M, Miyoshi H (2009) Exploring the binding site of acetogenin in the ND1 subunit of bovine mitochondrial complex I. Biochim Biophys Acta 1787:1106–1111

Sharpley MS, Hirst J (2006) The inhibition of mitochondrial complex I (NADH:ubiquinone oxidoreductase) by Zn^{2+}. J Biol Chem 281:34803–34809

Sharpley MS, Shannon RJ, Draghi F, Hirst J (2006) Interactions between phospholipids and NADH:ubiquinone oxidoreductase (complex I) from bovine mitochondria. Biochemistry 45:241–248

Sled VD, Friedrich T, Leif H, Weiss H, Meinhardt SW, Fukumori Y, Calhoun MW, Gennis RB, Ohnishi T (1993) Bacterial NADH-quinone oxidoreductases: iron-sulfur clusters and related problems. J Bioenerg Biomembr 25:347–356

Sled VD, Rudnitzky NI, Hatefi Y, Ohnishi T (1994) Thermodynamic analysis of flavin in mitochondrial NADH:ubiquinone oxidoreductase (complex I). Biochemistry 33:10069–10075

Smith MA, Finel M, Korolik V, Mendz GL (2000) Characteristics of the aerobic respiratory chains of the microaerophiles *Campylobacter jejuni* and *Helicobacter pylori*. Arch Microbiol 174:1–10

Spehr V, Schlitt A, Scheide D, Guénebaut V, Friedrich T (1999) Overexpression of the *Escherichia coli nuo*-operon and isolation of the overproduced NADH:ubiquinone oxidoreductase (complex I). Biochemistry 38:16261–16267

Steimle S, Bajzath C, Dörner K, Schulte M, Bothe V, Friedrich T (2011) Role of subunit NuoL for proton translocation by respiratory complex I. Biochemistry 50:3386–3393

Steinhoff HJ, Radzwill N, Thevis W, Lenz V, Brandenburg D, Antson A, Dodson G, Wollmer A (1997) Determination of interspin distances between spin labels attached to insulin: comparison of electron paramagnetic resonance data with the X-ray structure. Biophys J 73:3287–3298

Stolpe S, Friedrich T (2004) The *Escherichia coli* NADH:ubiquinone oxidoreductase (complex I) is a primary proton pump but may be capable of secondary sodium antiport. J Biol Chem 279:18377–18383

Tocilescu MA, Zickermann V, Zwicker K, Brandt U (2010) Quinone binding and reduction by respiratory complex I. Biochim Biophys Acta 1797:1883–1890

Uhlmann M, Friedrich T (2005) EPR signals assigned to Fe/S cluster N1c of the *Escherichia coli* NADH:ubiquinone oxidoreductase (complex I) derive from cluster N1a. Biochemistry 44:1653–1658

Unden G, Bongaerts J (1997) Alternative respiratory pathways of *Escherichia coli*: energetics and transcriptional regulation in response to electron acceptors. Biochim Biophys Acta 1320:217–234

van Hellemond JJ, van der Klei A, van Weelden SW, Tielens AG (2003) Biochemical and evolutionary aspects of anaerobically functioning mitochondria. Philos Trans R Soc Lond B Biol Sci 358:205–213

Verkhovskaya ML, Belevich N, Euro L, Wikström M, Verkhovsky MI (2008) Real-time electron transfer in respiratory complex I. Proc Natl Acad Sci USA 105:3763–3767

Vigano C, Smeyers M, Raussens V, Scherilinckx F, Ruysschaert JM, Goormaghtigh E (2004) Hydrogen-deuterium exchange in membrane proteins monitored by IR spectroscopy: a new tool to resolve protein structure and dynamics. Biopolymers 74:19–26

Vinogradov AD (1998) Catalytic properties of the mitochondrial NADH-ubiquinone oxidoreductase (complex I) and the pseudo-reversible active/inactive enzyme transition. Biochim Biophys Acta 1364:169–185

Vogel R, Siebert F (2000) Vibrational spectroscopy as a tool for probing protein function. Curr Opin Chem Biol 4:518–523

Walker JE (1992) The NADH:ubiquinone oxidoreductase (complex I) of respiratory chains. Quart Rev Biophys 25:253–324

Wallace BJ, Young IG (1977) Role of quinones in electron transport to oxygen and nitrate in *Escherichia coli*. Studies with a ubiA- menA- double quinone mutant. Biochim Biophys Acta 461:84–100

Weerakoon DR, Olson JW (2008) The *Campylobacter jejuni* NADH:ubiquinone oxidoreductase (complex I) utilizes flavodoxin rather than NADH. J Bacteriol 190:915–925

Weidner U, Geier S, Ptock A, Friedrich T, Leif H, Weiss H (1993) The gene locus of the proton-translocating NADH: ubiquinone oxidoreductase in *Escherichia coli*. J Mol Biol 233:109–122

Weiss H, Friedrich T, Hofhaus G, Preis D (1991) The respiratory-chain NADH dehydrogenase (complex I) of mitochondria. Eur J Biochem 197:563–576

Whitehead SJ, Rossington KE, Hafiz A, Cotton NP, Jackson JB (2005) Zinc ions selectively inhibit steps associated with binding and release of NADP(H) during turnover of proton-translocating transhydrogenase. FEBS Lett 579:2863–2867

Wikström M (1984) Two protons are pumped from the mitochondrial matrix per electron transferred between NADH and ubiquinone. FEBS Lett 169:300–304

Wittekindt C, Schwarz M, Friedrich T, Koslowski T (2009) Aromatic amino acids as stepping stones in charge transfer in respiratory complex I: an unusual mechanism deduced from atomistic theory and bioinformatics. J Am Chem Soc 131:8134–8140

Workun GJ, Moquin K, Rothery RA, Weiner JH (2008) Evolutionary persistence of the molybdopyranopterin-containing sulfite oxidase protein fold. Microbiol Mol Biol Rev 72:228–248

Yagi T, Matsuno-Yagi A (2003) The proton-translocating NADH-quinone oxidoreductase in the respiratory chain: the secret unlocked. Biochemistry 42:2266–2274

Yano T, Sled VD, Ohnishi T, Yagi T (1994) Expression of the 25-kilodalton iron-sulfur subunit of the energy-transducing NADH-ubiquinone oxidoreductase of *Paracoccus denitrificans*. Biochemistry 33:494–499

Yano T, Chu SS, Sled VD, Ohnishi T, Yagi T (1997) The proton-translocating NADH-quinone oxidoreductase (NDH-1) of thermophilic bacterium *Thermus thermophilus* HB-8. Complete DNA sequence of the gene cluster and thermostable properties of the expressed NQO2 subunit. J Biol Chem 272:4201–4211

Yano T, Magnitsky S, Ohnishi T (2000) Characterization of the complex I-associated ubisemiquinone species: toward the understanding of their functional roles in the electron/proton transfer reaction. Biochim Biophys Acta 1459:299–304

Yano T, Dunham WR, Ohnishi T (2005) Characterization of the delta μH^+-sensitive ubisemiquinone species (SQ_{Nf}) and the interaction with cluster N2: new insight into the energy-coupled electron transfer in complex I. Biochemistry 44:1744–1754

Yeh AP, Ambroggio XI, Andrade SLA, Einsle O, Chatelet C, Meyer J, Rees DC (2002) High resolution crystal structures of the wild type and Cys-55 → Ser and Cys-59 → Ser variants of the thioredoxin-like [2Fe-2S] ferredoxin from *Aquifex aeolicus*. J Biol Chem 277:34499–34507

Zickermann V, Bostina M, Hunte C, Ruiz T, Radermacher M, Brandt U (2003) Functional implications from an unexpected position of the 49-kDa subunit of NADH:ubiquinone oxidoreductase. J Biol Chem 278:29072–29078

Zscherp C, Barth A (2001) Reaction-induced infrared difference spectroscopy for the study of protein reaction mechanisms. Biochemistry 40:1875–1883

Zwicker K, Galkin A, Dröse S, Grgic L, Kerscher S, Brandt U (2006) The Redox-Bohr group associated with iron-sulfur cluster N2 of complex I. J Biol Chem 281:23013–23017

Chapter 3
Iron–Sulfur Clusters in Complex I

Eiko Nakamaru-Ogiso

Abstract Multiple iron-sulfur clusters are found in a variety of metalloproteins such as ferredoxins, hydrogenases, and nitrogenases as well as in complex I (NADH:ubiquinone oxidoreductase) and complex II (succinate:ubiquinone oxidoreductase) of the mitochondrial electron transport chain. Complex I contains two binuclear iron-sulfur clusters, N1a and N1b, and at least six tetranuclear iron-sulfur clusters, N2, N3, N4, N5, N6a, and N6b. An additional cluster N7 is found in some bacteria. The electron transfer pathway in complex I was finally shown in the crystal structure of the hydrophilic portion of *Thermus thermophilus* complex I. Electrons from NADH are transferred over approximately 95 Å from FMN (flavin mononucleotide) to cluster N2 in the hydrophilic domain of complex I. In addition, the binuclear cluster N1a and the tetranuclear cluster N7 are found not to participate in this pathway. However, not all iron-sulfur clusters have been identified by electron paramagnetic resonance (EPR), which has been the most informative technique for the study of iron-sulfur clusters in complex I for the past four decades. In this chapter, current knowledge on characteristics and roles of each iron-sulfur cluster is overviewed.

Keywords Bovine heart • Complex I • *E. coli* • Electron transfer • EPR • Flavin • g values • Iron-sulfur cluster • NADH • Redox potential

---

E. Nakamaru-Ogiso (✉)
Johnson Research Foundation, Department of Biochemistry and Biophysics,
Perelman School of Medicine, University of Pennsylvania,
Philadelphia, PA 19104, USA
e-mail: eikoo@mail.med.upenn.edu

L. Sazanov (ed.), *A Structural Perspective on Respiratory Complex I: Structure and Function of NADH:ubiquinone oxidoreductase*, DOI 10.1007/978-94-007-4138-6_3,

3.1 Introduction

Iron-sulfur (Fe/S) clusters play an important role in oxidation-reduction reactions. They participate in many redox and electron transfer reactions, especially in photosynthetic and respiratory electron transfer. Fe/S clusters are usually acid-labile sulfide-linked di-, tri-, or tetrairon centers in variable oxidation states. Usually, they are ligated to four cysteinyl sulfurs from the polypeptide chain. However, there are several cases in which one or more cysteine ligands are replaced by non-S-containing ligands such as aspartate and histidine residues.

Multiple Fe/S clusters are found in a variety of metalloproteins such as ferredoxins, hydrogenases, and nitrogenases as well as in complex I (NADH:ubiquinone oxidoreductase) and complex II (succinate:ubiquinone oxidoreductase) of the mitochondrial electron transport chain.

The complex I reaction involves the transfer of electrons more than 95 Å, from the NADH binding site to the ubiquinone binding site in the hydrophobic membrane domain of complex I (Sazanov 2007). Complex I contains several cofactors: flavin mononucleotide (FMN); two binuclear ([2Fe-2S]) Fe/S clusters, N1a and N1b; and at least six tetranuclear ([4Fe-4S]) Fe/S clusters, N2, N3, N4, N5, N6a, and N6b. An additional [4Fe-4S] cluster N7 is found in some bacteria.

In 2006, a major breakthrough in the study of complex I was accomplished by Sazanov and Hinchliffe. They determined the crystal structure of the hydrophilic portion of *Thermus thermophilus* complex I at 3.3 Å resolution. The FMN moiety and nine Fe/S clusters (*T. thermophilus* complex I contains an additional [4Fe-4S] cluster N7) were clearly shown in the X-ray crystal structure, providing the best available model to date for the electron transfer pathway in mitochondrial complex I (Fig. 3.1a, Ref. Sazanov and Hinchliffe 2006).

3.2 Location of Redox Centers in Individual Hydrophilic Subunits

Complex I has an unusual L-shaped structure consisting of a membrane part in addition to a promontory (or peripheral) part protruding into the mitochondrial matrix (or bacterial cytoplasm). The mammalian mitochondrial enzyme consists of 45 different subunits (Carroll et al. 2006), while bacterial complex I is composed of at least 14 subunits usually encoded in the same operon (Yagi et al. 1998). All of the 14 subunits have homologues in mammalian complex I. Thus, those 14 'core subunits' are sufficient for energy transduction and therefore regarded as a minimal structure of complex I. The core subunits comprise two distinct groups. The seven predominantly hydrophilic subunits are encoded in the nucleus and bear all of the Fe/S clusters and FMN. The remaining seven subunits are highly hydrophobic proteins folded into a total of 63 α-helices in the membrane (Efremov et al. 2010); in eukaryotes, they are encoded by the mitochondrial genome (Yagi and Matsuno-Yagi 2003).

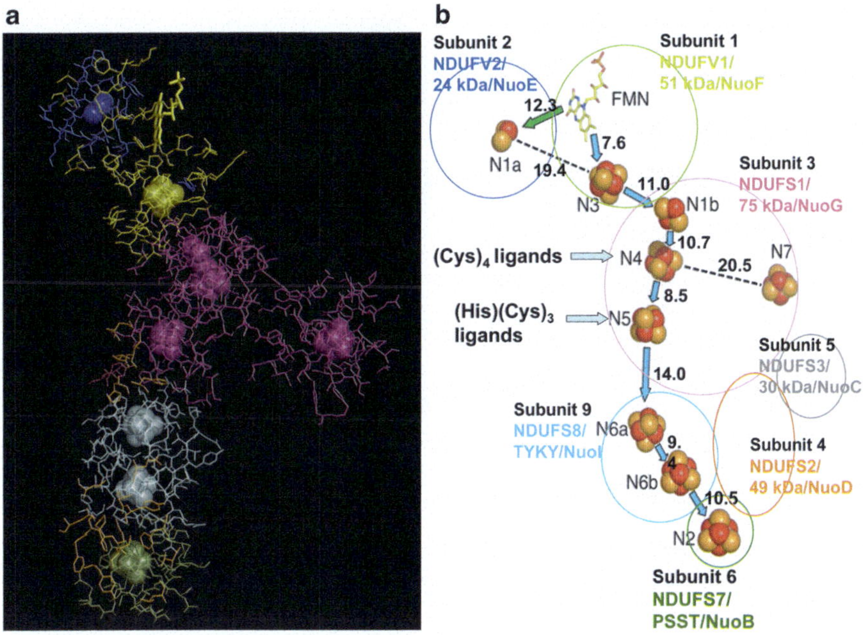

Fig. 3.1 Arrangement of redox centers in the hydrophilic domain of *T. thermophilus* complex I. (**a**), Three dimensional representation of redox centers surrounded by amino acid residues within 8 Å of each Fe/S cluster based on the X-ray structure (3IAS). (**b**), Two dimensional representation of redox centers with subunit boundaries modified from the figure published previously (Sazanov and Hinchliffe 2006). Edge-to edge distances are in Å. The conventional Fe/S cluster names are listed based on EPR spectroscopy and recent identification studies. Cluster N1a is in subunit 2 (NDUFV2/24 kDa/NuoE); cluster N3 and FMN in subunit 1 (NDUFV1/51 kDa/NuoF); clusters N1b, N4, N5, and N7 in subunit 3 (NDUFS1/75 kDa/NuoG); clusters N6a and N6b in subunit 9 (NDUFS8/TYKY/NuoI); and cluster N2 in subunit 6 (NDUFS7/PSST/NuoB)

Figure 3.1 and Table 3.1 summarize the location of each Fe/S cluster in the hydrophilic subunits. The nomenclature of *Homo sapiens*/*Bos taurus*/*Escherichia coli* complexes I is used throughout this chapter, regardless of the species referred to. Before the X-ray crystal structure of complex I was determined, all clusters had been predicted correctly by sequence analyses, overexpression of putative cluster-containing subunits, and EPR analyses (Braun et al. 1998; Flemming et al. 2003; Friedrich 1998; Maly et al. 2006; Nakamaru-Ogiso et al. 2002, 2005; Uhlmann and Friedrich 2005; Velazquez et al. 2005; Yano et al. 1994, 1995, 1999, 2003). Since EPR signals of Fe/S clusters in complex I were detected more than 30 years ago, long before DNA sequences of complex I subunits were determined, Fe/S clusters have been named mostly based on EPR analyses (Table 3.1, Ohnishi 1975).

EPR spectroscopy has been the most informative technique for the study of Fe/S clusters in complex I. EPR analyses have also provided useful information on the spin–spin interactions between neighboring redox centers. However, both sensitivity and spectral resolution of the EPR technique are generally much lower than

Table 3.1 The location of each Fe/S cluster bound by the hydrophilic subunits in complex I

Hydrophilic core subunits			Fe/S clusters		Possible assignments of EPR signals to clusters proposed by		g-values (g_z, g_y, g_x)	
Homo sapiens	*Bos taurus*	*Escherichia coli*		Proposed name	Ohnishi[a]	Hirst[b]	Bovine heart[c]	*E. coli*
NDUFV1	51 kDa	NuoF	[4Fe-4S]	4Fe[FV1/51/F]	N3	N3	2.04, 1.93, 1.87	2.04,1.92, 1.88
NDUFV2	24 kDa	NuoE	[2Fe-2S]	2Fe[FV2/24/E]	N1a	N1a	2.03, 1.95, 1.91[d]	2.00, 1.95, 1.92
							2.00, 1.94, 1.92[e]	
NDUFS1	75 kDa	NuoG	[2Fe-2S]	2Fe[FS1/75/G]	N1b	N1b	2.02, 1.94, 1.92	2.03, 1.94, 1.94
			[4Fe-4S]	4Fe[FS1/75/G]C	N4	N5	2.10, 1.94, 1.88	2.09, 1.93, 1.89
			[4Fe-4S][f]	4Fe[FS1/75/G]H	N5		2.07, 1.93, 1.90	?, ?, 1.90[g]
			[4Fe-4S]		N7	N7	Not present	2.05, 1.94, 1.91
NDUFS2	49 kDa	NuoD						
NDUFS3	30 kDa	NuoC						
NDUFS7	PSST	NuoB	[4Fe-4S]	4Fe[FS7/PS/B]	N2	N2	2.05, 1.92, 1.92	2.05, 1.91, 1.91
NDUFS8	TYKY	NuoI	[4Fe-4S]	4Fe[FS8/TY/I]1	N6a	N4	?, ?, ?	2.09, ?, 1.89?[h]
			[4Fe-4S]	4Fe[FS8/TY/I]2	N6b	N4	?, ?, ?	2.05, ?, 1.94[i]

[a] Ref. Ohnishi (1998)
[b] Ref. Yakovlev et al. (2007)
[c] The g-values were given based on Ohnishi's assignment (Sled et al. 1993)
[d] Undefined signals at extreme low Eh (Ingledew and Ohnishi 1980)
[e] Detected in NADH dehydrogenase subcomplex (Reda et al. 2008)
[f] Coordinated with one histidine and three cysteines
[g] Detected in *E. coli* NuoCDEFG subcomplex at 3 K (Nakamaru-Ogiso et al. 2008)
[h] Detected in the P71A in NuoI (N6b) mutant of *E. coli* complex I (Shinha et al. unpublished data)
[i] Detected in the P110A in NuoI (N6a) mutant of *E. coli* complex I (Shinha et al. unpublished data)

those of spectrophotometric techniques. In addition, EPR spectral properties of each Fe/S cluster, such as principal g values, line widths, and spin relaxation rates can be very different and affected by the microenvironment around the Fe/S cluster. It may especially be affected in a delicate multi-component membrane protein like complex I. It is likely that EPR signals from clusters residing in the overexpressed single subunit may be altered from those in the intact complex I system. Therefore, EPR identification of each Fe/S cluster has been a very difficult task.

Among the eight Fe/S clusters present in mitochondrial complex I, only five have been distinguished in EPR spectra of isolated complex I from bovine heart mitochondria: N1b, N2, N3, N4, and N5 in order of increasing spin relaxation rate (Table 3.1, Fig. 3.2a). Clusters N1a and N1b are [2Fe-2S] clusters. Since the spin relaxation of [2Fe-2S] clusters is much slower than that of [4Fe-4S] clusters, EPR signals from [2Fe-2S] clusters can be detected at liquid nitrogen temperatures (77 K), while EPR signals from [4Fe-4S] clusters are usually detected only at low temperatures (<20 K) (Fig. 3.2). As for *E. coli* complex I, which contains nine Fe/S clusters, only six clusters N1a, N1b, N2, N3, N4, and N7 are unambiguously identified (Table 3.1, Fig 3.2b). However, since clusters N6a and N6b have yet to be identified on EPR spectra of complex I, possibly due to signal overlapping, assignments of EPR signals from clusters N4 and N5 to structurally defined Fe/S clusters currently remain a matter of debate (Table 3.1, Ohnishi and Nakamaru-Ogiso 2008; Yakovlev et al. 2007). To avoid unnecessary confusion, Ohnishi's assignments, which are historically and broadly accepted, are adopted in this chapter.

3.3 Electron Transfer Pathway

Based on the structural information, the main electron pathway has been defined as FMN-N3-N1b-N4-N5-N6a-N6b-N2, (Fig. 3.1b). Surprisingly, it was found that clusters N1a and N7 are too distant from the path to be directly involved (Sazanov and Hinchliffe 2006). The midpoint redox potential at pH 7 ($E_{m,7}$) of NADH is −320 mV, while that of ubiquinone (UQ), an electron acceptor of complex I, is 110 mV. The energy gap (ΔE_m) between NADH and UQ is ~430 mV, which provides a considerable amount of free energy for ATP production. The $E_{m,7}$ values of individual clusters are all similar, with the exception of cluster N2, which is the last center in the electron pathway before UQ (Table 3.2). The $E_{m,7}$ of cluster N2 is ~150 mV for bovine heart complex I. Since clusters N3 to N6b are all believed to be isopotential (~250 mV), it is speculated that electrons do not go directly from NADH to the UQ molecule. They frequently jump back and forth between neighboring redox centers. However, redox titrations rely on equilibrium potentials: the enzyme is in a resting state that may not be catalytically relevant, and equilibrium potentials in the absence of membrane potentials may not reflect the free-energy changes during turnover.

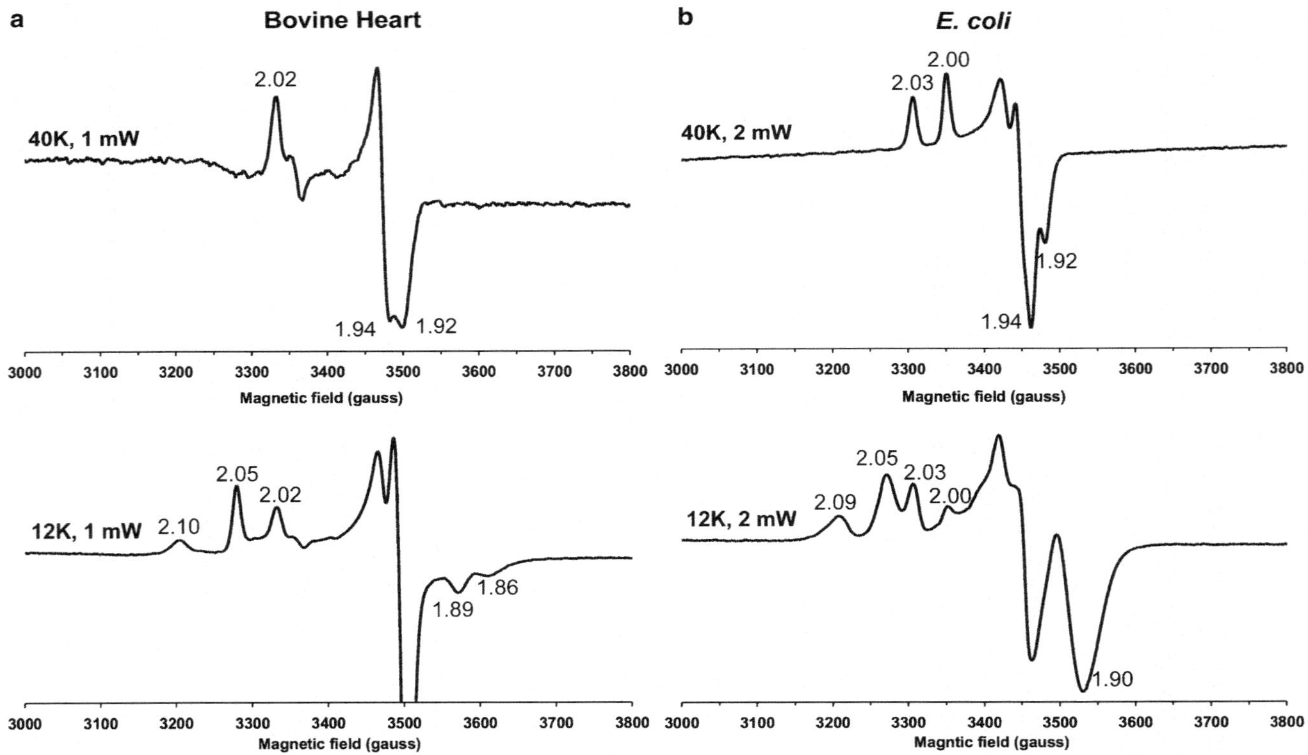

Fig. 3.2 EPR spectra at 40 K (*top*) and 12 K (*bottom*) of Fe/S clusters in isolated bovine heart complex I reduced with NADH (*left*) and *E. coli* complex I reduced with NADH and dithionite (*right*). EPR operating conditions: microwave frequency, 9.44 GH_Z; modulation amplitude, 6 G; microwave power, 1 or 2 mW; modulation frequency, 100 kHz; time constant, 82 ms. The g_z and g_x values are indicated

Table 3.2 Single-electron redox center midpoint potentials in complex I[a]

Redox center	$E_{m,7}$ (mV)	
	Bovine heart mitochondria	E. coli cytoplasmic membranes
NADH/NAD+	−0.32	−0.32 (−0.35)[b]
$FMNH_2$/FMNH•	−0.293[c]	N.A.[d]
FMNH•/FMN	−0.389[c]	N.A.
QH2/SQ•	−0.063[e]	N.A.
SQ•/Q	−0.045[e]	N.A.
N1a	< −0.37	−0.33 (−0.38)
N1b	−0.25	−0.23 (−0.265)
N2	−0.15	−0.22 (~ −0.25)
N3	−0.25	−0.27 (−0.315)
N4	−0.25	−0.27 (−0.33)
N5	−0.26	? (−0.33)
N6a	−0.25?[f]	−0.27?[f]
N6b	−0.25?[f]	−0.27?[f]
N7	Not present	? (−0.315)

[a]Refs. Leif et al. (1995), Sled et al. (1993)
[b]Values in parenthesis were determined for isolated *E. coli* complex I after activation with phospholipids (Euro et al. 2008)
[c]Ref. Sled et al. (1994)
[d]*N.A.* not available
[e]Ref. Ohnishi et al. (2005)
[f]Estimated values: clusters N1b, N3, N4, N5, N6a, and N6b are thought to be almost equipotential (Ohnishi 1998)

3.4 Characteristics and Roles of Individual Clusters

3.4.1 Cluster N1a

The [2Fe-2S] cluster N1a is conserved in complex I across all species, indicating its functional importance. However, it has been found to be distal from the main redox chain. The $E_{m,7}$ values of cluster N1a have been reported to be dependent on pH, but voltammetric studies have shown that the pH dependence has resulted from pH-linked changes in protein charge due to nonspecific protonation events, as opposed to the coupling of a specific ionizable residue (Zu et al. 2002). In addition, the low $E_{m,7}$ of cluster N1a ($E_{m,7}$<−0.37 mV) is out of line with the potentials of the other iron-sulfur clusters in complex I. It has been speculated that cluster N1a may play a role in preventing excessive generation of reactive oxygen species (ROS) by complex I (Sazanov 2007).

FMN accepts two electrons simultaneously from NADH and transfers one electron at a time to one-electron carrier Fe/S clusters. It is postulated that cluster N1a can still be suitable to accept one electron from flavosemiquinone, which has a significantly lower $E_{m,7}$ value (~ −390 mV). Thus, two electrons from FMN can

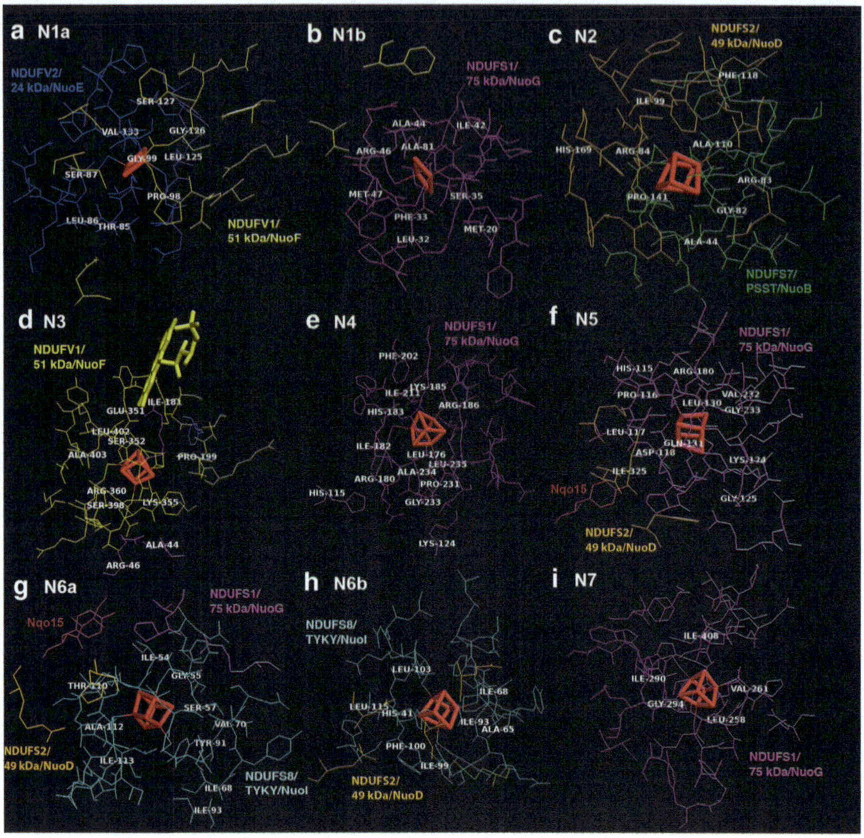

Fig. 3.3 Local environment of each Fe/S cluster in complex I. Amino acid residues shown in the figure are located within 8 Å of Fe/S clusters, and residues within 5 Å of Fe/S clusters are labeled. Amino acid residues from different subunits are depicted in different colors as used in Fig. 3.1: subunit 1 (NDUFV1/51 kDa/NuoF), *yellow*; subunit 2 (NDUFV2/24 kDa/NuoE), *blue*; subunit 3 (NDUFS1/75 kDa/NuoG), *pink*; subunit 4 (NDUFS2/49 kDa/NuoD), *orange*; subunit 5 (NDUF S3/30 kDa/NuoC), *gray*; subunit 6 (NDUFS7/PSST/NuoB), *green*; subunit 9 (NDUFS8/TYKY/NuoI), *pale green*; subunit 15, *darker pink*

be donated nearly simultaneously to clusters N3 (7.6 Å) and N1a (11.3 Å), preventing accumulation of the unstable flavosemiquinone intermediate. However, cluster N1a is never detected by EPR in complex I from any source except *E. coli*, in which the $E_{m,7}$ of cluster N1a is relatively high (−0.33 mV). While Ohnishi observed cluster N1a signals in pigeon heart mitochondria (Ohnishi 1975), in which cluster N1a was separated from cluster N1b by potentiometric titration, a recent report suggests that the EPR signal from cluster N1a has not been observed in bovine complex I at any potential down to −1 V (Reda et al. 2008). Therefore, the role of N1a remains unclear and could be complex. Cluster N1a is close to the interface of the NADH binding subunit, NDUFV1/51 kDa/NuoF (Fig. 3.3a). It is surrounded by relatively more hydrophobic amino acid residues than other clusters, which may explain its low midpoint potential. Interestingly, cluster N1a in the flavoprotein subcomplex containing the 51 and 24 kDa subunits of bovine heart mitochondrial complex I (Reda et al. 2008) was not reduced by NADH, whereas cluster N1a in a mutant NuoEF subcomplex of *E. coli* complex I was reduced with NADH, even though it lacks cluster N3 (Velazquez et al. 2005). It was shown that *E. coli* complex I produces approximately 20% superoxide and 80% hydrogen peroxide, while bovine complex I produces 95% superoxide (Esterhazy et al. 2008). The specific role of cluster N1a might be different between *E. coli* and mitochondrial complex I.

3.4.2 Cluster N1b

Binuclear cluster N1b is reducible with NADH but has a slightly lower spin concentration stoichiometry of 0.8/FMN (Ohnishi et al. 1981). The EPR spectra from cluster N1b can be observed over a wide temperature range from <10 to 130 K (Nakamaru-Ogiso et al. 2008). Recently, it was reported that cluster N1b exhibits lower potential than clusters N3, N4, and N5 (Reda et al. 2008). Moreover, the EPR signals from cluster N1b at 40 K is 3.3 times greater at −1 V than those at −0.4 V, indicating electrostatic interactions between neighboring cluster(s). In addition, cluster N1b is hardly reduced with dithionite (Kowal et al. 1986), suggesting its inaccessibility from outside and/or its lower midpoint redox potential. Cluster N1b contains ${}_3Cys^{45}$ (the prefix indicates the subunit number, see Fig. 3.1), a cysteine ligand oriented toward the subunit boundary with NDUFV1/51 kDa/NuoF through neighboring ${}_3Ala^{44}$, ${}_3Arg^{46}$, and ${}_2Thr^{357}$ (Fig. 3.3b).

3.4.3 Cluster N2

The terminal cluster N2 generally exhibits a more positive midpoint potential, receives the electrons from isopotential Fe/S clusters (Table 3.2), and transfers electrons to quinone (Ohnishi 1998). The redox midpoint potential for cluster N2 depends on the host organism from which complex I has been isolated. Except for values of −300 mV in *T. thermophilus* membranes (Meinhardt et al. 1990) and < −450 mV in isolated *Rhodothermus marinus* complex I (Fernandes et al. 2006), the $E_{m,7}$ for cluster N2 varies from ~ −150 mV in most organisms, including bovine, *Neurospora crassa*, *Paracoccus denitrificans,* and *E. coli* (Sled et al. 1993; Wang et al. 1991), to −20 mV in pigeon heart submitochondrial particles (Ohnishi 1975). This feature of higher redox midpoint potential makes cluster N2 the electron sink of complex I. After electrons from NADH were injected into the electron transfer pathway, the reduction of cluster N2 was observed within 90 μs (Verkhovskaya et al. 2008). In addition, the midpoint potential of the cluster is pH-dependent, becoming more positive with decreasing pH (Sled et al. 1993). Therefore, cluster N2 is considered to play a key role in the coupling of redox reactions to the translocation of protons across the membrane.

For over a decade, the location of cluster N2 was under debate. Careful mutational studies (Flemming et al. 2003) and the crystal structure of the hydrophilic part of complex I undoubtedly has assigned cluster N2 to the NDUFS7/PSST/NuoB subunit (Sazanov and Hinchliffe 2006). This subunit contains a conserved binding motif for a [4Fe-4S] cluster with two adjacent cysteines. It was previously questioned whether they could be ligands of the same cluster due to a possible steric hindrance (Gurrath and Friedrich 2004). Currently, there are only few examples of a motif with two consecutive cysteines binding a Fe/S cluster: a [2Fe-2S] cluster in the *E. coli* FhuF protein, a siderophore-iron reductase (Muller et al. 1998); and a [4Fe-4S] cluster in adenosine 5′-phosphosulfate reductase from *Pseudomonas aeruginosa* (Chartron et al. 2006) and *Mycobacterium tuberculosis* (Bhave et al. 2011). This unique and unusual coordination of cluster N2 could allow direct protonation of the cluster upon reduction, leading to its pH dependence (the redox-Bohr effect) (Zwicker et al. 2006).

Based on site directed mutagenesis studies and redox titrations monitored by EPR spectroscopy, the conserved $_{4}His^{169}$ in the NDUFS2/49 kDa/NuoD subunit was identified as the major redox-Bohr group modulating the redox midpoint potential of cluster N2 (Grgic et al. 2004; Zwicker et al. 2006). In fact, $_{4}His^{169}$ is hydrogen-bonded to cluster N2 (Fig. 3.3c), and the mutation of histidine to methionine abolished its pH dependence and shifted the midpoint potential from −140 to −220 mV at neutral pH. However, this mutation had no significant effect on electron transfer and proton pumping activities of complex I. This indicates that the release of the energy that drives proton transfer across the membrane is linked to the redox chemistry of the quinone, not to that of cluster N2. Oxidation of cluster N2 by ubiquinone is a rate limiting step. The redox-driven coupling should be provided by the interaction of cluster N2 with the semiquinone species detected under steady state conditions (Magnitsky et al. 2002; Ohnishi et al. 2005; Yano et al. 2005).

The mutation of the conserved $_{4}Arg^{84}$ in the NDUFS2/49 kDa/NuoD subunit led to the drastic reduction of cluster N2 signals (Grgic et al. 2004). Since the $_{4}Arg^{84}$ residue is only 5 Å away from cluster N2, the R84A replacement could lower the E_m of this cluster. In fact, reduction by dithionite of the corresponding R84A mutant of the *E. coli* complex I resulted in a large increase in the EPR signals of cluster N2 (Belevich et al. 2007). However, in the corresponding R84A mutant of *Yarrowia lipolytica*, no N2 signal appeared with NADH plus dithionite (Grgic et al. 2004; Kashani-Poor et al. 2001).

The highly conserved $_{4}Tyr^{87}$ in the NDUFS2/49 kDa/NuoD subunit is within 8 Å of cluster N2 (Fig. 3.3c) and faces the quinone binding site. It is found to be essential for activity. It could participate in both electron and proton transfer. It has been speculated that proton transfer toward the UQ-binding site may be controlled through tandem cysteines depending on the redox state of clusters N2 and N6b (Berrisford and Sazanov 2009; Sazanov 2007). The conserved ionizable residues $_{4}Tyr^{87}$ and $_{6}Glu^{49}$ in the NDUFS7/PSST/NuoB subunit have been suggested to be a proton delivery pathway (Flemming et al. 2006; Grgic et al. 2004; Kashani-Poor et al. 2001). Alternatively, it has been suggested that the hydroxyl group of $_{4}Tyr^{87}$ might be involved in the binding of the quinone head group via a hydrogen bond between the aromatic hydroxyl and the UQ carbonyl (Tocilescu et al. 2010), while the protonation of $_{6}$Glu^{49} might be coupled with a conformational change (Flemming et al. 2006).

Cluster N2 displays paramagnetic interactions with a semiquinone, called SQ_{Nf}. This fast relaxing semiquinone species is formed during catalytic turnover, is only found in the presence of $\Delta\mu H^+$ and is sensitive to the complex I inhibitors rotenone and piericidin A (Yano et al. 2005). Based on a study of the dipolar interactions of the two paramagnetic species, the distance between cluster N2 and SQ_{Nf} is 12 Å (Yano et al. 2005). The second semiquinone species, SQ_{NS}, has slow spin relaxation and is not sensitive to the membrane potential (Ohnishi et al. 2005). Recently, SQ_{NS} was proposed to reside in the membrane subunit ND2 (Ohnishi et al. 2010). The tight coupling between electron transfer and proton translocation is also illustrated by the full reversibility of the complex I reaction.

3.4.4 Cluster N3

The [4Fe-4S] cluster N3 is located near the C terminus of NDUFV1/51 kDa/NuoF (Fig. 3.1a). Cluster N3 is within ~8 Å of FMN, and unlike the distal cluster N1a, it resides in the main redox pathway. It has been reported that the spin relaxation of flavosemiquinone in complex I is much faster than that of flavodoxin (which contains no transition metal ions in the vicinity of the flavin) due to strong dipole-dipole interaction with cluster N3 (Sled et al. 1994). Thus, cluster N3 receives an electron from FMN and feeds single electrons into the electric wiring of Fe/S clusters. It has been noted that the binding motif for cluster N3, "CxxCxxCx_{40}C" is strikingly conserved not only in complex I but also in a variety of hydrogenases and dehydrogenases (Velazquez et al. 2005). Interestingly, cluster N3 is coordinated within a four-helix bundle, which is a common fold but has not been previously observed to bind Fe-S clusters (Sazanov and Hinchliffe 2006). Cluster N3 is close to the subunit boundary of NDUFV2/24 kDa/NuoE via $_{2}$Phe^{66} and NDUFS1/75 kDa/NuoG via $_{3}$Arg^{46} (Fig. 3.3d).

The disease-causing mutation of Thr434 to methionine in *C. elegans* NDUFV1/51 kDa/NuoF gene, which corresponds to Ser398 in *T. thermophilus*, demonstrates hallmark features of complex I dysfunction such as lactic acidosis and decreased NADH-dependent mitochondrial respiration (Grad and Lemire 2004). Since $_{1}$Ser^{398} is located at 4.1 Å from cluster N3 (Fig. 3.3d), it is likely that the mutation destabilizes cluster N3 or modifies the redox potential.

3.4.5 Cluster N4

The [4Fe-4S] cluster N4 is coordinated in a 2×[4Fe-4S] ferredoxin-like fold (Sazanov and Hinchliffe 2006). In [FeFe]-hydrogenases, this fold contains an additional tetranuclear cluster FS4A, which is normally absent in complex I.

However, in sequence data of the NDUFS1/75 kDa/NuoG subunit from *Aquifex aeolicus*, *Campylobacter jejuni*, and *Helicobacter pylori*, this cysteine binding motif for FS4A is conserved. Therefore, it is highly likely that complex I in these organisms contains an extra Fe/S cluster, which could be referred to cluster N8, assuming it exists at the protein level. It is expected that cluster N7 is connected to cluster N4 in the main electron transfer chain through cluster N8.

As seen in Table 3.1, the principal g values of cluster N4 in *E. coli* complex I are $g_{z,y,x}$ = 2.09, ~1.93, 1.89 (Leif et al. 1995), and are similar to the $g_{z,y,x}$ = 2.10, ~1.94, 1.89 of cluster N4 in the bovine heart complex I (Ohnishi 1998). However, the g values of cluster N4 in singly overexpressed *P. denitrificans*, *T. thermophilus*, and *E. coli* NDUFS1/75 kDa/NuoG subunits are $g_{z,y,x}$ = 2.07, 1.95, 1.88–1.89 (Nakamaru-Ogiso et al. 2002, 2005; Yano et al. 1995). Ohnishi et al. interpreted these shifted g values as reflecting some modification, because cluster N4 is especially sensitive to alteration of the protein environment around the cluster while other clusters are not (Ohnishi 1998). However, since these $g_{z,y,x}$ values are actually similar to the g values of cluster N5 in purified complex I from the bovine heart and *Y. lipolytica*, the assignment of EPR signals in isolated complex I ($g_{z,y,x}$ = 2.10, ~1.94, 1.89) to cluster N4 is still under debate.

Cysteine ligands of N1b ($_3$Cys^{34}) and N4 ($_3$Cys^{184}) are well aligned within the same NDUFS1/75 kDa/NuoG subunit (Hayashi and Stuchebrukhov 2010). The main electron tunneling pathway of N1b → N4 is along these cysteines with a 1.9 Å through-space jump, the shortest in the pathway. However, the edge-to-edge distance from cluster N3 to N4 is 13.8 Å, which is longer than the distance from cluster N1b to N4 (10.7 Å), but it is still within 14 Å of the maximum distance for physiological electron transfer (Page et al. 1999). Cluster N4 appears to be a convergent relay point in the main electron transfer pathway. The change of conserved leucine at position 235 to valine in human complex I leads to infant death and a large decrease in enzyme activity (Martin et al. 2005). Since $_3$Leu^{235} is within hydrogen bonding distance of cluster N4 (Fig. 3.3e), it is likely that the mutation destabilizes cluster N4 or changes the redox potential.

3.4.6 Cluster N5

The [4Fe-4S] cluster N5 is uniquely coordinated by three cysteines and one histidine. Similar to the [FeFe]-hydrogenase, the N^{ε} atom of histidine is a ligand for iron from cluster N5, in contrast to other known examples where N^{δ} is the coordinating ligand (Peters et al. 1998). The EPR properties of cluster N5 are unusual due to its histidine ligation, as it exhibits very fast spin relaxation and is presumed to be in mixed S = 1/2 and S = 3/2 ground states (Ohnishi 1998). Cluster FS0 in nitrate reductase A (which is also ligated by three cysteines and one histidine (Bertero et al. 2003)), has a fast spin relaxation and an unusual high-spin ground state based on the EPR analysis (Rothery et al. 2004). However, for the detection of the S = 3/2 state

EPR spectrum in the g=~5 region, a much higher enzyme concentration was required than that which was used for EPR analysis of the S=1/2 spin state. No report on the S=3/2 state EPR spectrum in complex I is available. In addition, even detecting the S=1/2 state EPR spectrum is very difficult, as it requires an extremely low temperature. Thus, cluster N5 signals were reported only in a limited number of complex I preparations such as the bovine heart and *Y. lipolytica* complex I. Cluster N5 signals have yet to be detected in isolated *E. coli* complex I.

Currently, the assignment of the EPR signals from clusters N4 and N5 and the identity of the two Fe/S clusters are controversial as mentioned above. The underlying problems are originated in the large shift of g values of cluster N4 (from $g_{z,y,x}$=2.07, 1.95, 1.88–1.89 in overexpressed subunit to $g_{z,y,x}$=2.10, ~1.94, 1.89 in complex I) (Table 3.1) and data from mutational experiments. Exchange of the histidine ligand with alanine in *Y. lipolytica* complex I did not change the EPR signature of cluster N5 (Waletko et al. 2005), while the same mutation in *P. denitrificans* NDUFS1/75 kDa/NuoG subunit caused severe reduction in the incorporation of all Fe/S clusters to the subunit and another histidine to cysteine mutation led to the alteration of cluster N5 signals (Yano et al. 2003). A recent site-directed mutagenesis study on *E. coli* complex I suggests that the unique EPR properties of cluster N5 can be attributed to its unique coordination of His(Cys)3 at soluble subcomplex of complex I levels, even though the histidine mutation in *E. coli* complex I completely destabilized assembly and dissociated the NADH dehydrogenase subcomplex from the membrane (Nakamaru-Ogiso et al. 2008).

Since clusters N5 and N6a have the largest edge to edge distance (14 Å), the transfer path N5 → N6a is expected to be the major kinetic bottleneck of the electron transfer pathway. Cluster N5 is close to the boundaries of the 49 kDa and the *T. thermophilic* specific Nqo15 subunits through hydrophobic residues $_4$Phe^{328} and $_{15}$Tyr^{42}, respectively (Fig. 3.3f). Several polar residues are within hydrogen-bonding distances of the cluster (Fig. 3.3f and Sazanov and Hinchliffe 2006). Such an environment in the protein interior as well as the unique histidine ligation appear to indicate that cluster N5 is not simply a "stepping stone" in the electron transfer pathway, but that it may also have an additional role in the mechanism of complex I.

3.4.7 Clusters N6a and N6b

The [4Fe-4S] clusters N6a and N6b, which follow cluster N5 in the redox chain, are localized in the NDUFS8/TYKY/NuoI subunit. The fold is similar to that of the 2×[4Fe-4S] bacterial ferredoxin family (Peters et al. 1998). Clusters N6a and N6b have the stereotypical coordination of two Fe/S cluster structures with eight highly conserved cysteine residues ($_9$Cys^{53}, $_9$Cys^{56}, $_9$Cys^{59}, $_9$Cys^{108}) (Fig. 3.3g)— ($_9$Cys^{63}, $_9$Cys^{98}, $_9$Cys^{101}, $_9$Cys^{104}) (Fig. 3.3h) by cross bridging two cubanes, which greatly contributes to the overall protein structure. The clusters are linked by short α helices, and appear to provide a "connecting chain" of two clusters between cluster N5 and the terminal cluster N2 (Fig. 3.1).

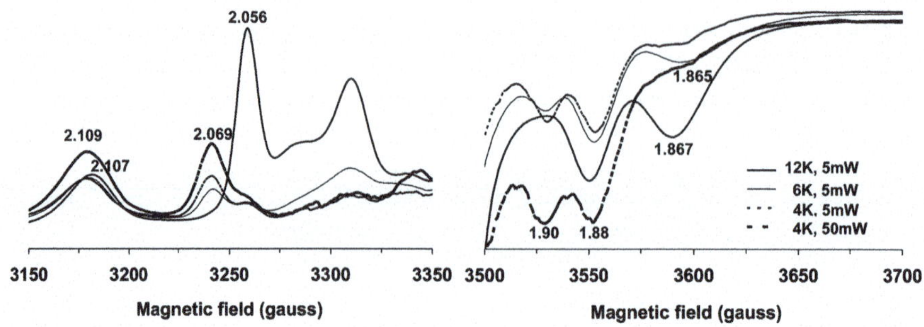

Fig. 3.4 EPR spectra of isolated bovine heart complex I reduced with NADH under anaerobic conditions. The g_z and g_x values are indicated

In 1999, Yano et al. succeeded in the first EPR demonstration of the presence of two [4Fe-4S] clusters in the overexpressed *P. denitrificans* NDUFS8/TYKY/NuoI subunit (Yano et al. 1999). Multiple signals were observed at g=2.08, 2.05, 1.92, 1.89 (at 14 K and 10 mW), and an outside-pair $g_{z,x}=2.08$ and 1.89 showed much faster spin relaxation ($P_{1/2}=342$ mW) than the inside-pair $g_{z,x}=2.05$ and 1.92 signals ($P_{1/2}=8$ mW) at 14 K.

In 2001, broad near axial-type N6 signals ($g_{z,y,x}=2.09$, ~1.91, 1.89) were detected from the connecting fragment of *E. coli* complex I, which consists of NuoBCDI subunits, and thus, do not contain cluster N4, but rather cluster N2 (Rasmussen et al. 2001). In this case, the other cluster N6 signals seemed to overlap with cluster N2 signals.

The recent mutational analyses by Yagi's group suggest that the outside-pair species of $g_{z,x}=2.08$, ~1.89 signals are probably arising from cluster N6a and the inside-pair species of $g_{z,x}=2.05$ and 1.92 are from cluster N6b (Shinha et al. unpublished results). The definitive assignment is still awaited. The outside N6 signals ($g_{z,y,x}=2.09$, 1.91, 1.89) were previously proposed to arise from the cluster N6b based on the R84A mutant study of the *E. coli* NDUFS2/49 kDa/NuoD subunit (Belevich et al. 2007). However, considering the complete loss of g=2.05 signals, which could be attributed not only to cluster N2 but also to cluster N6b, and their locations within a 5 and 8 Å distance of $_4Arg^{84}$ (Fig. 3.3f, g), it seems reasonable that cluster N6a was partially affected by the R84A mutation as well as nearby clusters N2 and N6b.

As early as in 1975, Ohnishi reported EPR signals ($g_{z,y,x}=2.11$, 2.06, 2.03, 1.93, 1.90, 1.88) of complex I from pigeon heart mitochondria, which were detectable only at extremely low temperatures (Ohnishi 1975). She suggested the presence of two different [4Fe-4S] clusters. By applying our current knowledge of EPR spectra of mitochondrial complex I, EPR signals of $g_{z,x}=2.06$ and 1.90 can be attributed to cluster N5, and the other EPR signals of $g_{z,x}=2.11$ and 1.88 can probably be attributed to cluster N4. Cluster N4 signals in the NuoBCDEFG subcomplex of *E. coli* complex I, which does not contain clusters N6a and N6b, showed very similar fast relaxation properties (Nakamaru-Ogiso et al. 2008). In fact, as shown in Fig. 3.4, the g=2.07 and 1.90 regions of cluster N5 signals, apparent heterogeneity at different temperatures and power dependence can clearly be recognized in the g=2.10 and

1.88 regions of EPR signals from purified bovine heart complex I. Although clusters N6a and N6b are historically believed to be EPR silent signals, this data in Fig. 3.4 strongly suggest that there are overlapping signals: cluster N6a and N6b do exist in the spectra. However, only very limited information is available for clusters N6a and N6b. Since the g values of clusters N2 ($g_{z,y,x}$=2.05, 1.92, 1.92) and N6b ($g_{z,y,x}$=2.05, 1.92?, 1.92?), N4 ($g_{z,y,x}$=2.10, 1.94, 1.88) and N6a ($g_{z,y,x}$=2.09, ~1.94, 1.89) are very similar, further discrimination between those two cluster pairs is necessary for their definite assignment.

3.4.8 Cluster N7

The [4Fe-4S] cluster N7 is not conserved, and is found only in *T. thermophilus, E. coli* and some other bacteria (Nakamaru-Ogiso et al. 2002). Cluster N7 is regarded as an evolutionary remnant of a molybdopterin-containing enzyme, in which the corresponding cluster participates in electron transfer. In complex I, cluster N7 is too far away from the main redox chain, and thus, it confers structural stability on the fold of the Nqo3 subunit (Pohl et al. 2007), although most prokaryotes and all eukaryotes seemingly have lost cluster N7. In *E. coli* NDUFS1/75 kDa/NuoG subunit, mutation of cysteine ligands for cluster N7 led to nearly complete loss of ability to incorporate all other Fe/S clusters in this subunit (Nakamaru-Ogiso et al. 2005). Cluster N7 resides in a mostly hydrophobic environment (Fig. 3.3i).

As mentioned above (Sect. 3.4.5), microaerophiles *C. jejuni* and *H. pylori* both contain an extra cysteine motif for an additional Fe/S cluster predicted to be cluster N8, which could connect cluster N7 to the main redox chain (Sazanov 2007). Since these organisms lack the genes corresponding to the NADH binding subunit NDUFV1/51 kDa/NuoF and the NDUFV2/24 kDa/NuoE subunit in their complex I operon, it is conceivable that complex I in these organisms accepts electrons not from NADH but instead from an unknown substrate and that cluster N7 takes part in the main electron transfer pathway (Finel 1998; Sazanov 2007).

The NDUFS1/75 kDa subunit of respiratory complex I was identified as a critical caspase 3 substrate in mitochondria (Ricci et al. 2004). The corresponding cluster N7 binding site in mammalian complex I was found to be the cleavage site by caspase 3. Cells expressing a noncleavable mutant of the NDUFS1/75 kDa subunit sustained $\Delta\Psi m$ and ATP levels during apoptosis and ROS production in response to apoptotic stimuli had dampened. Therefore, it is believed that the caspase cleavage of NDUFS1 is required for several mitochondrial changes associated with apoptosis.

3.5 Conclusion

The reason for which complex I contains so many Fe/S clusters particularly remains enigmatic. One plausible answer might be that various building blocks from other enzymes were picked up in an evolutionary process and the electron transfer chain happened to be an approximately 95 Å long. It will be very exciting to find out how the long-range electron transfer pathway is linked to vectorial proton translocation in the future.

References

Belevich G, Euro L, Wikstrom M, Verkhovskaya M (2007) Role of the conserved arginine 274 and histidine 224 and 228 residues in the NuoCD subunit of complex I from Escherichia coli. Biochemistry 46:526–533

Berrisford JM, Sazanov LA (2009) Structural basis for the mechanism of respiratory complex I. J Biol Chem 284:29773–29783

Bertero MG, Rothery RA, Palak M, Hou C, Lim D, Blasco F, Weiner JH, Strynadka NC (2003) Insights into the respiratory electron transfer pathway from the structure of nitrate reductase A. Nat Struct Biol 10:681–687

Bhave DP, Hong JA, Lee M, Jiang W, Krebs C, Carroll KS (2011) Spectroscopic studies on the [4Fe-4S] cluster in adenosine 5′-phosphosulfate reductase from Mycobacterium tuberculosis. J Biol Chem 286:1216–1226

Braun M, Bungert S, Friedrich T (1998) Characterization of the overproduced NADH dehydrogenase fragment of the NADH:ubiquinone oxidoreductase (complex I) from Escherichia coli. Biochemistry 37:1861–1867

Carroll J, Fearnley IM, Skehel JM, Shannon RJ, Hirst J, Walker JE (2006) Bovine complex I is a complex of 45 different subunits. J Biol Chem 281:32724–32727

Chartron J, Carroll KS, Shiau C, Gao H, Leary JA, Bertozzi CR, Stout CD (2006) Substrate recognition, protein dynamics, and iron-sulfur cluster in Pseudomonas aeruginosa adenosine 5′-phosphosulfate reductase. J Mol Biol 364:152–169

Efremov RG, Baradaran R, Sazanov LA (2010) The architecture of respiratory complex I. Nature 465:441–445

Esterhazy D, King MS, Yakovlev G, Hirst J (2008) Production of reactive oxygen species by complex I (NADH:ubiquinone oxidoreductase) from Escherichia coli and comparison to the enzyme from mitochondria. Biochemistry 47:3964–3971

Euro L, Bloch DA, Wikstrom M, Verkhovsky MI, Verkhovskaya M (2008) Electrostatic interactions between FeS clusters in NADH:ubiquinone oxidoreductase (complex I) from Escherichia coli. Biochemistry 47:3185–3193

Fernandes AS, Sousa FL, Teixeira M, Pereira MM (2006) Electron paramagnetic resonance studies of the iron-sulfur centers from complex I of Rhodothermus marinus. Biochemistry 45:1002–1008

Finel M (1998) Does NADH play a central role in energy metabolism in Helicobacter pylori? Trends Biochem Sci 23:412–413

Flemming D, Schlitt A, Spehr V, Bischof T, Friedrich T (2003) Iron-sulfur cluster N2 of the Escherichia coli NADH:ubiquinone oxidoreductase (complex I) is located on subunit NuoB. J Biol Chem 278:47602–47609

Flemming D, Hellwig P, Lepper S, Kloer DP, Friedrich T (2006) Catalytic importance of acidic amino acids on subunit NuoB of the Escherichia coli NADH:ubiquinone oxidoreductase (complex I). J Biol Chem 281:24781–24789

Friedrich T (1998) The NADH:ubiquinone oxidoreductase (complex I) from Escherichia coli. Biochim Biophys Acta 1364:134–146

Grad LI, Lemire BD (2004) Mitochondrial complex I mutations in Caenorhabditis elegans produce cytochrome c oxidase deficiency, oxidative stress and vitamin-responsive lactic acidosis. Hum Mol Genet 13:303–314

Grgic L, Zwicker K, Kashani-Poor N, Kerscher S, Brandt U (2004) Functional significance of conserved histidines and arginines in the 49-kDa subunit of mitochondrial complex I. J Biol Chem 279:21193–21199

Gurrath M, Friedrich T (2004) Adjacent cysteines are capable of ligating the same tetranuclear iron-sulfur cluster. Proteins 56:556–563

Hayashi T, Stuchebrukhov AA (2010) Electron tunneling in respiratory complex I. Proc Natl Acad Sci USA 107:19157–19162

Ingledew WJ, Ohnishi T (1980) An analysis of some thermodynamic properties of iron-sulphur centres in site I of mitochondria. Biochem J 186:111–117

Kashani-Poor N, Zwicker K, Kerscher S, Brandt U (2001) A central functional role for the 49-kDa subunit within the catalytic core of mitochondrial complex I. J Biol Chem 276:24082–24087

Kowal AT, Morningstar JE, Johnson MK, Ramsay RR, Singer TP (1986) Spectroscopic characterization of the number and type of iron-sulfur clusters in NADH:ubiquinone oxidoreductase. J Biol Chem 261:9239–9245

Leif H, Sled VD, Ohnishi T, Weiss H, Friedrich T (1995) Isolation and characterization of the proton-translocating NADH: ubiquinone oxidoreductase from Escherichia coli. Eur J Biochem 230:538–548

Magnitsky S, Toulokhonova L, Yano T, Sled VD, Hagerhall C, Grivennikova VG, Burbaev DS, Vinogradov AD, Ohnishi T (2002) EPR characterization of ubisemiquinones and iron-sulfur cluster N2, central components of the energy coupling in the NADH-ubiquinone oxidoreductase (complex I) in situ. J Bioenerg Biomembr 34:193–208

Maly T, Grgic L, Zwicker K, Zickermann V, Brandt U, Prisner T (2006) Cluster N1 of complex I from Yarrowia lipolytica studied by pulsed EPR spectroscopy. J Biol Inorg Chem 11:343–350

Martin MA, Blazquez A, Gutierrez-Solana LG, Fernandez-Moreira D, Briones P, Andreu AL, Garesse R, Campos Y, Arenas J (2005) Leigh syndrome associated with mitochondrial complex I deficiency due to a novel mutation in the NDUFS1 gene. Arch Neurol 62:659–661

Meinhardt SW, Wang DC, Hon-nami K, Yagi T, Oshima T, Ohnishi T (1990) Studies on the NADH-menaquinone oxidoreductase segment of the respiratory chain in Thermus thermophilus HB-8. J Biol Chem 265:1360–1368

Muller K, Matzanke BF, Schunemann V, Trautwein AX, Hantke K (1998) FhuF, an iron-regulated protein of Escherichia coli with a new type of [2Fe-2S] center. Eur J Biochem 258:1001–1008

Nakamaru-Ogiso E, Yano T, Ohnishi T, Yagi T (2002) Characterization of the iron-sulfur cluster coordinated by a cysteine cluster motif (CXXCXXXCX27C) in the Nqo3 subunit in the proton-translocating NADH-quinone oxidoreductase (NDH-1) of Thermus thermophilus HB-8. J Biol Chem 277:1680–1688

Nakamaru-Ogiso E, Yano T, Yagi T, Ohnishi T (2005) Characterization of the iron-sulfur cluster N7 (N1c) in the subunit NuoG of the proton-translocating NADH-quinone oxidoreductase from Escherichia coli. J Biol Chem 280:301–307

Nakamaru-Ogiso E, Matsuno-Yagi A, Yoshikawa S, Yagi T, Ohnishi T (2008) Iron-sulfur cluster N5 is coordinated by an HXXXCXXCXXXXXC motif in the NuoG subunit of Escherichia coli NADH:quinone oxidoreductase (complex I). J Biol Chem 283:25979–25987

Ohnishi T (1975) Thermodynamic and EPR characterization of iron-sulfur centers in the NADH-ubiquinone segment of the mitochondrial respiratory chain in pigeon heart. Biochim Biophys Acta 387:475–490

Ohnishi T (1998) Iron-sulfur clusters/semiquinones in complex I. Biochim Biophys Acta 1364:186–206

Ohnishi T, Nakamaru-Ogiso E (2008) Were there any "misassignments" among iron-sulfur clusters N4, N5 and N6b in NADH-quinone oxidoreductase (complex I)? Biochim Biophys Acta 1777:703–710

Ohnishi T, Blum H, Galante YM, Hatefi Y (1981) Iron-sulfur N-1 clusters studied in NADH-ubiquinone oxidoreductase and in soluble NADH dehydrogenase. J Biol Chem 256:9216–9220

Ohnishi T, Johnson JE Jr, Yano T, Lobrutto R, Widger WR (2005) Thermodynamic and EPR studies of slowly relaxing ubisemiquinone species in the isolated bovine heart complex I. FEBS Lett 579:500–506

Ohnishi T, Nakamaru-Ogiso E, Ohnishi ST (2010) A new hypothesis on the simultaneous direct and indirect proton pump mechanisms in NADH-quinone oxidoreductase (complex I). FEBS Lett 584:4131–4137

Page CC, Moser CC, Chen X, Dutton PL (1999) Natural engineering principles of electron tunnelling in biological oxidation-reduction. Nature 402:47–52

Peters JW, Lanzilotta WN, Lemon BJ, Seefeldt LC (1998) X-ray crystal structure of the Fe-only hydrogenase (CpI) from Clostridium pasteurianum to 1.8 angstrom resolution. Science 282:1853–1858

Pohl T, Bauer T, Dorner K, Stolpe S, Sell P, Zocher G, Friedrich T (2007) Iron-sulfur cluster N7 of the NADH:ubiquinone oxidoreductase (complex I) is essential for stability but not involved in electron transfer. Biochemistry 46:6588–6596

Rasmussen T, Scheide D, Brors B, Kintscher L, Weiss H, Friedrich T (2001) Identification of two tetranuclear FeS clusters on the ferredoxin-type subunit of NADH:ubiquinone oxidoreductase (complex I). Biochemistry 40:6124–6131

Reda T, Barker CD, Hirst J (2008) Reduction of the iron-sulfur clusters in mitochondrial NADH:ubiquinone oxidoreductase (complex I) by EuII-DTPA, a very low potential reductant. Biochemistry 47:8885–8893

Ricci JE, Munoz-Pinedo C, Fitzgerald P, Bailly-Maitre B, Perkins GA, Yadava N, Scheffler IE, Ellisman MH, Green DR (2004) Disruption of mitochondrial function during apoptosis is mediated by caspase cleavage of the p75 subunit of complex I of the electron transport chain. Cell 117:773–786

Rothery RA, Bertero MG, Cammack R, Palak M, Blasco F, Strynadka NC, Weiner JH (2004) The catalytic subunit of Escherichia coli nitrate reductase A contains a novel [4Fe-4S] cluster with a high-spin ground state. Biochemistry 43:5324–5333

Sazanov LA (2007) Respiratory complex I: mechanistic and structural insights provided by the crystal structure of the hydrophilic domain. Biochemistry 46:2275–2288

Sazanov LA, Hinchliffe P (2006) Structure of the hydrophilic domain of respiratory complex I from Thermus thermophilus. Science 311:1430–1436

Sled VD, Friedrich T, Leif H, Weiss H, Meinhardt SW, Fukumori Y, Calhoun MW, Gennis RB, Ohnishi T (1993) Bacterial NADH-quinone oxidoreductases: iron-sulfur clusters and related problems. J Bioenerg Biomembr 25:347–356

Sled VD, Rudnitzky NI, Hatefi Y, Ohnishi T (1994) Thermodynamic analysis of flavin in mitochondrial NADH:ubiquinone oxidoreductase (complex I). Biochemistry 33:10069–10075

Tocilescu MA, Fendel U, Zwicker K, Drose S, Kerscher S, Brandt U (2010) The role of a conserved tyrosine in the 49-kDa subunit of complex I for ubiquinone binding and reduction. Biochim Biophys Acta 1797:625–632

Uhlmann M, Friedrich T (2005) EPR signals assigned to Fe/S cluster N1c of the Escherichia coli NADH:ubiquinone oxidoreductase (complex I) derive from cluster N1a. Biochemistry 44:1653–1658

Velazquez I, Nakamaru-Ogiso E, Yano T, Ohnishi T, Yagi T (2005) Amino acid residues associated with cluster N3 in the NuoF subunit of the proton-translocating NADH-quinone oxidoreductase from Escherichia coli. FEBS Lett 579:3164–3168

Verkhovskaya ML, Belevich N, Euro L, Wikstrom M, Verkhovsky MI (2008) Real-time electron transfer in respiratory complex I. Proc Natl Acad Sci USA 105:3763–3767

Waletko A, Zwicker K, Abdrakhmanova A, Zickermann V, Brandt U, Kerscher S (2005) Histidine 129 in the 75-kDa subunit of mitochondrial complex I from Yarrowia lipolytica is not a ligand for [Fe4S4] cluster N5 but is required for catalytic activity. J Biol Chem 280:5622–5625

Wang DC, Meinhardt SW, Sackmann U, Weiss H, Ohnishi T (1991) The iron-sulfur clusters in the two related forms of mitochondrial NADH: ubiquinone oxidoreductase made by Neurospora crassa. Eur J Biochem 197:257–264

Yagi T, Matsuno-Yagi A (2003) The proton-translocating NADH-quinone oxidoreductase in the respiratory chain: the secret unlocked. Biochemistry 42:2266–2274

Yagi T, Yano T, Di Bernardo S, Matsuno-Yagi A (1998) Procaryotic complex I (NDH-1), an overview. Biochim Biophys Acta 1364:125–133

Yakovlev G, Reda T, Hirst J (2007) Reevaluating the relationship between EPR spectra and enzyme structure for the iron sulfur clusters in NADH:quinone oxidoreductase. Proc Natl Acad Sci USA 104:12720–12725

Yano T, Sled VD, Ohnishi T, Yagi T (1994) Expression of the 25-kilodalton iron-sulfur subunit of the energy-transducing NADH-ubiquinone oxidoreductase of Paracoccus denitrificans. Biochemistry 33:494–499

Yano T, Yagi T, Sled VD, Ohnishi T (1995) Expression and characterization of the 66-kilodalton (NQO3) iron-sulfur subunit of the proton-translocating NADH-quinone oxidoreductase of Paracoccus denitrificans. J Biol Chem 270:18264–18270

Yano T, Magnitsky S, Sled VD, Ohnishi T, Yagi T (1999) Characterization of the putative 2×[4Fe-4S]-binding NQO9 subunit of the proton-translocating NADH-quinone oxidoreductase (NDH-1) of Paracoccus denitrificans. Expression, reconstitution, and EPR characterization. J Biol Chem 274:28598–28605

Yano T, Sklar J, Nakamaru-Ogiso E, Takahashi Y, Yagi T, Ohnishi T (2003) Characterization of cluster N5 as a fast-relaxing [4Fe-4S] cluster in the Nqo3 subunit of the proton-translocating NADH-ubiquinone oxidoreductase from Paracoccus denitrificans. J Biol Chem 278:15514–15522

Yano T, Dunham WR, Ohnishi T (2005) Characterization of the delta muH+-sensitive ubisemiquinone species (SQ(Nf)) and the interaction with cluster N2: new insight into the energy-coupled electron transfer in complex I. Biochemistry 44:1744–1754

Zu Y, Di Bernardo S, Yagi T, Hirst J (2002) Redox properties of the [2Fe-2S] center in the 24 kDa (NQO2) subunit of NADH:ubiquinone oxidoreductase (complex I). Biochemistry 41:10056–10069

Zwicker K, Galkin A, Drose S, Grgic L, Kerscher S, Brandt U (2006) The Redox-Bohr group associated with iron-sulfur cluster N2 of complex I. J Biol Chem 281:23013–23017

Chapter 4
Current Topics of the Inhibitors of Mitochondrial Complex I

Hideto Miyoshi

Abstract Acetogenins isolated from *Annonaceous* plants are very potent inhibitors of bovine heart mitochondrial complex I and unique in chemical structure among a wide variety of inhibitors of the enzyme. The structure-activity studies of numerous acetogenins revealed that the inhibitors exhibit potent inhibition only when the two pharmacophores (i.e. the γ-lactone and the hydroxylated THF moieties) are directly linked by the alkyl spacer and cooperatively bind to the two putative binding sites. Regardless of marked changes in the conformation and/or the length of the spacer region, the spacer dynamically regulates the cooperative binding of the two pharmacophores to the sites. Through structural modification of acetogenins, we developed new types of inhibitors termed Δlac-acetogenins and piperazine-type inhibitors, whose inhibitory effects on complex I significantly differ from those of traditional inhibitors. In particular, examination of the inhibition mechanism of Δlac-acetogenins provided valuable insights into the terminal electron transfer step of complex I including a mechanism of the generation of superoxide in the presence of inhibitor. Using a photoaffinity labeling technique, the binding sites of the new inhibitors (Δlac-acetogenin and piperazine) as well as traditional inhibitors (acetogenin and quinazoline) were identified at sub-subunit level. The photoaffinity labeling studies clearly indicated that the 49 kDa and ND1 subunits construct a large inhibitor binding pocket in bovine complex I.

Keywords Complex I • Inhibitor • Acetogenin • Photoaffinity labeling • Structure-activity relationship

---

H. Miyoshi (✉)
Division of Applied Life Sciences, Graduate School of Agriculture,
Kyoto University, Kyoto 606-8502, Japan
e-mail: miyoshi@kais.kyoto-u.ac.jp

L. Sazanov (ed.), *A Structural Perspective on Respiratory Complex I: Structure and Function of NADH:ubiquinone oxidoreductase*, DOI 10.1007/978-94-007-4138-6_4,

4.1 Introduction

There are a variety of inhibitors of complex I (Friedrich et al. 1994; Degli Esposti 1998). With the exception of a few inhibitors that inhibit electron input into complex I such as rhein (Kean et al. 1971) and diphenyleneiodonium (Majander et al. 1994), all inhibitors act at the terminal electron transfer step of this enzyme (Friedrich et al. 1994). Although these inhibitors are generally believed to act at the ubiquinone reduction site, there have been conflicting reports (Heinrich and Werner 1992; Gong et al. 2003). Radio- and fluorescent-ligand binding (Okun et al. 1999) and photoaffinity labeling (Schuler et al. 1999; Nakamaru-Ogiso et al. 2003) studies indicated that a certain marker inhibitor is displaced by numerous competitors, i.e. other complex I inhibitors. These findings are consistent with the model of inhibitor binding sites proposed by Okun et al. (1999), wherein a wide variety of inhibitors share a common large binding domain with partially overlapping sites.

It should, however, be realized that in all these papers (Schuler et al. 1999; Okun et al. 1999; Nakamaru-Ogiso et al. 2003), the authors demonstrated that the binding of a certain marker inhibitor to the enzyme is suppressed in the presence of an excess amount of competitors. Under these experimental conditions, one cannot rule out the possibility that even though the binding sites of the marker inhibitor and competitors are different, the binding of an excess amount of competitors induced structural change of the binding site of the marker (i.e. allosteric effect), which resulted in suppression of the marker binding (Yagi and Matsuno-Yagi 2003). In addition, some studies suggested that diverse complex I inhibitors can be grouped into two (Friedrich et al. 1994) or three (Degli Esposti 1998; Okun et al. 1999) classes, though criterion of the classification seems to be not general and precise. It remains, therefore, to be learned how the binding positions of chemically diverse inhibitors are related to each other.

On the other hand, taking into consideration that complex I undergoes dynamic conformational change (Belogrudov and Hatefi 1994; Yamaguchi et al. 1998; Mamedova et al. 2004; Berrisford et al. 2008), one cannot exclude the possibility that there are further diverse chemicals that disturb the function of the membrane domain differently depending on their structural nature. In fact, we recently found several new types of inhibitors whose inhibition mechanism is different from that of traditional inhibitors, as described in the Section 4.4.1. This paper reviews recent advances in our understanding of the action mechanism of some inhibitors of bovine complex I.

4.2 Recent Structure-Activity Relationship (SAR) Studies of Complex I Inhibitors

4.2.1 *SAR Studies of Traditional Inhibitors*

To fully understand the action mechanism of biologically active chemicals, identification of the critical structural factors required for eliciting the activity is valuable. To this end, the structure-activity relationship (SAR) studies using systematically selected analogues is necessary. The same is true for numerous inhibitors of complex I. A number of papers on the SAR studies for complex I inhibitors have been published. The purpose of this section is not to describe each SAR study, instead the author just lists recent representative studies from which the readers will be able to access earlier literature: iromycins (Surup et al. 2008), nafuredins (Shiomi et al. 2005), piericidins (Schnermann et al. 2006), polyene amides (Yoshida et al. 2007; Feutrill et al. 2008), pyridabens (Purohit et al. 2008), quinazolines (Yamashita et al. 2004), rotenoids (Nicolaou et al. 2000; Granados-Covarrubias and Maldonado 2009), and verticipyrones (Leiris et al. 2010). Regarding the SAR studies of capsaicins and pyridiniums (i.e. MPP^+ analogues), the author was not able to find out a new literature after the publication of our earlier works (capsaicins; Satoh et al. 1996, pyridiniums; Miyoshi 1998; Miyoshi et al. 1998a).

4.2.2 SAR Studies of Acetogenins

Since the isolation of the first acetogenin, uvaricin, more than 400 members of the family have been isolated from *Annonaceous* plants (Zeng et al. 1996; Alali et al. 1999; Bermejo et al. 2005). Acetogenins have very potent and diverse biological effects such as antimalarial, pesticidal, antifeedant, and probably most important, antitumor activities (Alali et al. 1999; Bermejo et al. 2005). The inhibitory effect of acetogenins on mitochondrial complex I is of particular importance since their diverse biological activities are thought to be attributable to this effect. Some acetogenins, such as bullatacin (Fig. 4.1) and rolliniastatin-1, are the most potent inhibitors of bovine heart mitochondrial complex I identified to date (Degli Esposti et al. 1994; Friedrich et al. 1994; Miyoshi 2001).

Among a wide variety of complex I inhibitors, natural acetogenins are very unique in chemical structure. Acetogenins are extremely flexible molecules since they are biogenetical derivatives of long-chain fatty acids (C32 or C34). There are few structural similarities between acetogenins and ordinary complex I inhibitors such as piericidins, rotenones and several synthetic agrochemicals like fenpyroximate. However, an electron paramagnetic resonance (EPR) spectroscopic study on the redox state of iron-sulfur clusters showed that acetogenins act at the terminal electron transfer step of complex I (Friedrich et al. 1994), as is the case for other traditional inhibitors. Thus, considering the unusual structural characteristics as well as the very strong inhibitory effect of acetogenins, a detailed analysis of the inhibitory action of these inhibitors would provide valuable insights into the terminal electron transfer step of complex I. Toward this end, identification of the crucial structural factors of acetogenins responsible for the potent inhibition is necessary.

The chemical structure of most natural acetogenins is characterized by four segments (Fig. 4.1): (i) an α,β-unsaturated γ-lactone ring, (ii) one to three tetrahydrofuran (THF) ring(s) with one or two flanking OH group(s), (iii) a long alkyl tail, and (iv) an alkyl spacer linking the two pharmacophores (i.e. γ-lactone and THF moieties). Based on the SAR studies carried out by ourselves and other research groups using systematically selected natural and synthetic acetogenins, the roles of

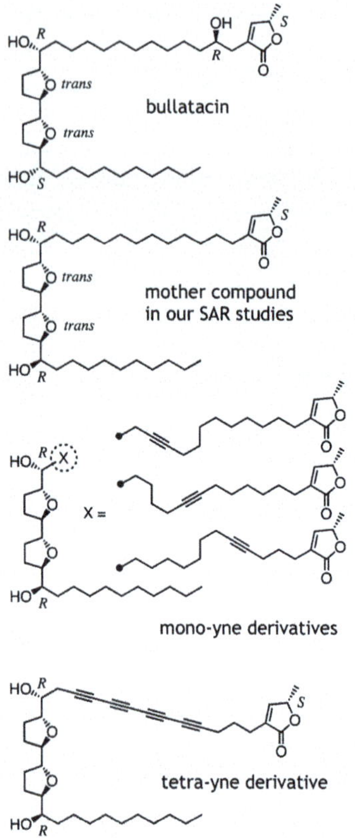

Fig. 4.1 Structures of natural acetogenin (bullatacin) and synthetic derivatives. The IC_{50} values of bullatacin and mother compound, which was used as a standard compound in the SAR studies conducted in the author's laboratory (Abe et al. 2004, 2005, 2008), are 0.90 and 0.92 nM, respectively. The IC_{50} values of mono-yne and tetra-yne derivatives, in which the total number of carbon atoms of the spacer was fixed at 13 as in the mother compound, are 0.90 ~ 1.1 nM and 1.7 nM, respectively

each segment in the inhibitory action with bovine heart mitochondrial complex I have been elucidated as follows: (i) a natural α,β-unsaturated γ-lactone ring itself is not crucial for the activity and is substitutable with other structures (Takada et al. 2000; Hoppen et al. 2000; Tormo et al. 2001, 2003; Yabunaka et al. 2003), (ii) neither the number of THF rings nor the stereochemistry around the hydroxylated THF ring moiety is an essential factor (Degli Esposti et al. 1994; Miyoshi et al. 1998b; Makabe et al. 2004a, b), and the presence of either of two OH groups adjacent to the THF ring(s) sufficiently sustains the potent activity (Abe et al. 2004), (iii) a long alkyl tail is preferable, but not essential, for the activity since even a methyl derivative elicited strong inhibition at the nanomolar level (Motoyama et al. 2002), and (iv) the presence of polar functional group(s) such as hydroxy and carbonyl groups in the space is not crucial for the activity (Degli Esposti et al. 1994; Miyoshi et al. 1998b; Tormo et al. 1999a, b). Thus the crucial structural factors of acetogenins are entirely ambiguous, suggesting that complex I recognizes each of the multiple functional groups of the inhibitors in a fairly loose way. In this sense, the manner of binding of acetogenins might not follow the so-called "key and keyhole" relation. However, it should be noted that neither the hydroxylated THF nor γ-lactone ring moiety synthesized separately has an inhibitory effect by itself and that there is no synergistic enhancement of the inhibitory effect between the two pharmacophoric moieties (Kuwabara et al. 2000; Hamada et al. 2004). Thus acetogenins act as strong inhibitors only when the THF and γ-lactone ring moieties are directly linked by an alkyl spacer.

Accordingly, clarification of the function of the spacer moiety is crucial to elucidate the inhibitory action of acetogenins in detail. We therefore performed SAR studies of a series of synthetic bis-THF acetogenins in which the spacer region was systematically modified, and provided the following findings (Abe et al. 2005, 2008). First, the optimal length of the spacer required for strong inhibition is about 13 carbon atoms, which corresponds in length to the spacers of the most potent natural acetogenins such as bullatacin. Second, the decrease in inhibitory potency caused by elongating the spacer from 13 carbons is much more drastic than that caused by shortening it. Concerning this point, it should be noted that even the derivatives possessing a short spacer (7 ~ 10 carbon atoms) elicit potent inhibitory effects at the nanomolar level. Third, neither the local flexibility in a specific region of the spacer nor the flexibility of the whole spacer moiety is critical for the inhibition. For instance, mono-yne derivatives and a tetra-yne derivative (Fig. 4.1), in which the spacer's flexibility was drastically reduced by hardening the spacer covering ten carbon atoms into a rod-like shape, retain a strong activity at the nanomolar level.

Taken together, these findings reveal that acetogenins exhibit potent inhibition of bovine complex I only when the two pharmacophoric moieties are directly linked by the alkyl spacer and cooperatively bind to the two putative binding sites. Regardless of remarkable changes in the conformation and/or the length of the spacer, the spacer region dynamically regulates the cooperative binding of the two pharmacophores to the sites. Our studies strongly suggest that there is no receptor wall for the spacer region (Abe et al. 2008). This is probably due to the fact that the ND1 subunit, to which acetogenins bind (see Chap. 3), is located at the end of the membrane domain, as revealed by the crystal structure of the entire complex I from *Thermus thermophilus* (Efremov et al. 2010). Motivated by unique structural features and attractive biological activities of acetogenins, a wide variety of acetogenin derivatives including biochemical probes have been synthesized. Kojima and Tanaka (2009) concisely reviewed the design, synthesis, and biological evaluation of such synthetic acetogenin derivatives.

4.3 Identification of the Binding Site of Inhibitor by Photoaffinity Labeling

4.3.1 Design Synthesis of Photoaffinity Labeling Probes

Photoaffinity labeling technique is a powerful means of studying the interaction of biologically active chemicals with their target macromolecules (Brunner 1993; Dormán and Prestwich 2000). The technique allows for the identification of the target protein

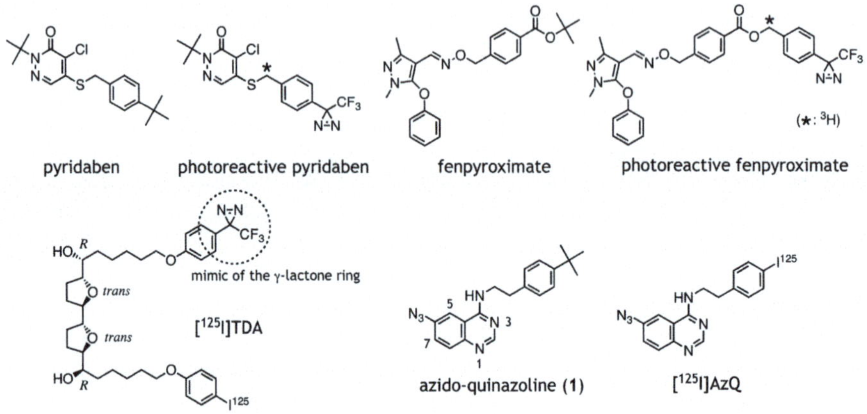

Fig. 4.2 Structures of pyridaben, fenpyroximate, azido-quinazoline (**1**), and their photoreactive derivatives used in previous photoaffinity labeling studies. $[^{125}I]$TDA is a photoreactive derivative of natural-type acetogenin. In $[^{125}I]$TDA molecule, the (trifluoromethyl)phenyldiazirinyl ring acts as a mimic of the γ-lactone ring of natural acetogenin as well as a photoreactive functional group. $[^{125}I]$AzQ, ^{125}I-labeled 6-*az*ido-4-(4-iodophenetylamino)*q*uinazoline; $[^{125}I]$TDA, ^{125}I-labeled (*t*rifluoromethyl)phenyl*d*iazirinyl *a*cetogenin

and also the binding domain within the target. In general, a photoaffinity probe is synthesized by attaching a photolabile cross-linking group, such as aryldiazirine and arylazido, and an indicator tag, such as radioisotope and biotin, to the ligand (Dormán and Prestwich 2000). To avoid a significant decrease in an affinity of the ligand, the photolabile group is generally introduced at a position remote from a pharmacophoric moiety and inherently unessential for exhibiting high affinity.

Actually, the same holds true for the photoaffinity inhibitor probes used in researching complex I. In earlier photoaffinity labeling studies using pyridaben (Schuler et al. 1999) and fenpyroximate (Nakamaru-Ogiso et al. 2003), a photolabile aryldiazirine group was attached to the moiety where it is remote from the pharmacophoric heterocyclic ring and merely enhances the hydrophobicity of the molecule (Fig. 4.2). However this strategy in the compound design has fundamental problems: (i) the region cross-linked by the photolabile group may not be responsible for critical interaction with the inhibitor and hence for important functionality of the enzyme and (ii) the unessential moiety of the inhibitor would be too flexible to fix the position of a photolabile group in the enzyme. In fact, some phenomena that would be associated with these problems can be observed in early photoaffinity labeling studies. For instance, a pyridaben derivative predominantly binds to the PSST subunit, but with significant nonspecific binding to the ND1 subunit (Schuler et al. 1999). A fenpyroximate derivative binds to the ND5 subunit (Nakamaru-Ogiso et al. 2003). This subunit (Nqo12 in *T. thermophilus*) is however located far from the hydrophilic domain, i.e. at the distal end of the membrane domain, except for a long amphipathic α-helix HL (Efremov et al.

2010). It would therefore be ideal if some photoreactive group could substitute for a pharmacophoric core structure with no loss of function. Although this idea seems unrealistic at first sight, our group synthesized such photoreactive inhibitor probes. In this chapter, the author will introduce the photoaffinity labeling studies performed in our laboratory.

4.3.2 Binding Site of Acetogenin

Taking into consideration the unusual structural characteristics as well as the very strong inhibitory effect of acetogenins, exploration of the inhibition manner of acetogenins including identification of the binding site would provide valuable insights into the terminal electron transfer step of complex I. We synthesized a photoreactive acetogenin {[^{125}I](*t*rifluoromethyl)phenyl*d*iazirinyl *a*cetogenin, [^{125}I]TDA, Fig. 4.2}, in which an aryldiazirine group serves both as a photoreactive group and a substitute for the γ-lactone ring that is a common pharmacophore of numerous natural acetogenins, and carried out photoaffinity labeling experiment using bovine heart submitochondrial particles (SMP) (Murai et al. 2007). When SMP were UV-irradiated in the presence of [^{125}I]TDA, radioactivity was predominantly incorporated into a ~30 kDa protein of complex I. Immnoprecipitation of the [^{125}I]TDA-labeled complex I with anti-bovine ND1 antibody unequivocally indicated that the labeled protein is the membrane subunit ND1. A variety of complex I inhibitors such as piericidin A and rotenone efficiently suppressed the specific binding of [^{125}I]TDA to ND1, indicating that they may share a common binding domain.

To pinpoint the binding site of [^{125}I]TDA in the ND1 subunit, we carried out limited proteolysis of the subunit cross-linked by [^{125}I]TDA using various proteases (lysylendopeptidase, endoprotease Asp-N, V8-protease or trypsin) and carefully analyzed the fragmentation patterns (Sekiguchi et al. 2009). The results show that the cross-linked residue is located within the region of the fourth to fifth transmembrane helices (Val144-Glu192) of the subunit. This study has revealed, for the first time, the inhibitor binding site in complex I at the sub-subunit level. The SAR studies for numerous acetogenins strongly suggested that the γ-lactone ring buries into the hydrophobic environment of the enzyme with the support of the hydrophobic alkyl spacer (Abe et al. 2005, 2008). Therefore it is reasonable to consider that the amino acid residue cross-linked by the phenyldiazirine ring of [^{125}I]TDA resides in the hydrophobic fourth or fifth transmembrane helix.

We tried to identify the specific amino acid residue(s) cross-linked by [^{125}I]TDA by MALDI-TOF MS of tryptic digests of the ND1 subunit. However the sequence coverage was low (15%) and the region covering the fourth to fifth transmembrane helices was not detected (Sekiguchi et al. 2009). Similar results were reported in earlier MS study of the ND1 subunit (Carroll et al. 2003, 2006). This is probably due to large hydrophobicity of the subunit.

The question of whether the binding sites for ubiquinone and different inhibitors in complex I are identical remains to be answered. It is however realized that competition experiment between strong inhibitors and exogenous short-chain ubiquinones would be difficult since the binding affinity of the two ligands differs too much (about three orders of magnitude) to examine efficient competitive behavior in a limited concentration range. We recently showed that an excess amount of short-chain ubiquinones such as ubiquinone-2 (Q_2) suppress the cross-linking of ND1 subunit by [^{125}I]TDA in a concentration-dependent way (Sekiguchi et al. 2009), providing direct evidence that acetogenin and ubiquinone competitively bind to the enzyme. Considering this result along with the competition test conducted in earlier photoaffinity labeling studies (Schuler et al. 1999; Gong et al. 2003; Nakamaru-Ogiso et al. 2003), we propose that not all inhibitors acting at the terminal electron transfer step of complex I necessarily bind to the ubiquinone binding site itself. From mutagenesis studies of yeast *Yarrowia lipolytica* complex I, Brandt and colleagues demonstrated that although the mutations that exhibit inhibitor resistance (to quinazoline and rotenone) and reduce the ubiquinone reductase activity tend to be comparable, specific binding sites for ubiquinone and different inhibitors are not identical (Tocilescu et al. 2007; Fendel et al. 2008).

4.3.3 Binding Site of Quinazoline

Quinazoline-type chemicals are strong inhibitors of bovine heart complex I and highly fluorescent (Okun et al. 1999; Ino et al. 2003). We focused our attention on quinazoline-type inhibitors as test compounds for two reasons. First, on the basis of competition experiments taking advantage of a partial quenching of the fluorescence of quinazolines upon specific binding to complex I, Okun et al. (1999) and Ino et al. (2003) suggested that a variety of complex I inhibitors share a common binding domain with the quinazolines. Second, an azido-derivative of quinazoline (compound **1**, Fig. 4.2), which possesses a photolabile azido group on the pharmacophoric quinazoline ring, maintains a strong inhibitory effect on bovine complex I (Yamashita et al. 2004) and hence is suitable as a candidate photoaffinity probe, as described in the Sect. 4.3.1.

We synthesized a ^{125}I-labeled azido-quinazoline [6-*az*ido-4-(4-iodophenetylamino) *q*uinazoline, [^{125}I]AzQ, Fig. 4.2] and carried out photoaffinity labeling with bovine SMP and isolated complex I (Murai et al. 2009; Murai et al. 2011). Our study revealed that [^{125}I]AzQ binds to the 49 kDa and ND1 subunits with a frequency of ~4:1. The combination of different proteolysis of the cross-linked 49 kDa subunit indicated the binding site of [^{125}I]AzQ to be located within the region Asp41-Arg63 (23 amino acids). The labeled site in ND1 is between Asp199 and Lys262, mostly likely in the third matrix loop that connects the fifth and sixth transmembrane helices.

Our findings strongly suggest that the 49 kDa and ND1 subunits are in contact with each other and that the arylazido group of [^{125}I]AzQ is located between them, consistent with the crystal structure of the entire complex I from *T. thermophilus* (Efremov et al. 2010). Although the large cytoplasmic loops of the Nqo8 subunit (ND1 in bovine) were not resolved, our results reveal that the third matrix-side loop interacts with the N-terminal region of the 49 kDa subunit.

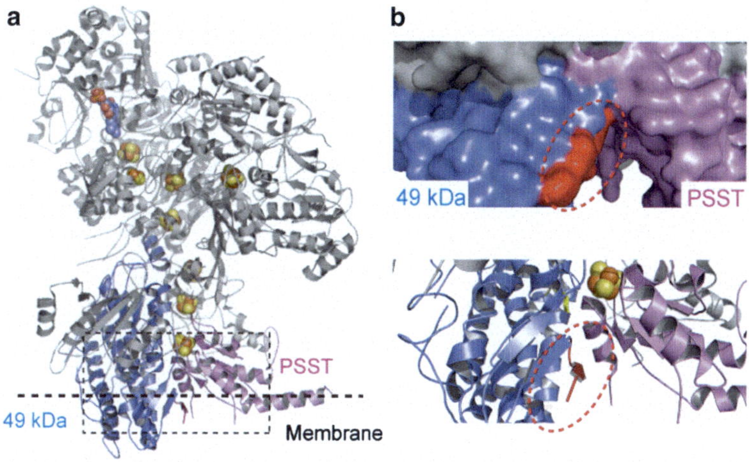

Fig. 4.3 The binding site of [^{125}I] AzQ in complex I. (**a**) The crystal structure of the peripheral arm of complex I from *T. thermophilus* (Sazanov and Hinchliffe 2006): Nqo4 (49 kDa in *blue*) and Nqo6 (PSST in *magenta*). (**b**) Close-up of the Nqo4 and Nqo6 subunits. As some N-terminal residues in the sequence corresponding to the labeled region Asp41-Arg63 in bovine were not modeled in the crystal structure, only the resolved region Pro35-Arg42, which corresponds to Pro56-Arg63 in bovine, is shown in *red*. Tyr87 (Tyr144 in *Y. lipolytica*) is shown in *yellow*

On the basis of an amino acid sequence alignment of the 49 kDa subunit from various organisms, we consider the functionality of the cross-linked region (i.e., Asp41-Arg63) (Murai et al. 2009). In the crystal structure of the Nqo4 subunit (49 kDa) in *T. thermophilus* (Sazanov and Hinchliffe 2006; Efremov et al. 2010), some N-terminal residues corresponding to this region were not modeled. Nevertheless, taking into consideration the overall architecture of complex I in *T. thermophilus*, this region must comprise a large pocket for the quinone/inhibitor binding, as illustrated in Fig. 4.3. Tocilescu et al. (2010) showed that mutations of Tyr144 (Y144F and Y144W) of the 49 kDa subunit in *Y. lipolytica* (Tyr87 in *T. thermophilus*) result in drastic resistance, by about three orders of magnitude, to a quinazoline-type inhibitor (DQA). According to the crystal structure of the subunit (Sazanov and Hinchliffe 2006), this residue is far from the cross-linked region by [^{125}I]AzQ and close to iron-sulfur cluster N2, only ~7 angstrom away from the cluster (Fig. 4.3b).

4.4 New Type of Inhibitors of Bovine Complex I

4.4.1 Δlac-Acetogenins

In the course of structural modification of natural acetogenins, we deleted the γ-lactone ring that is a structural feature common to a large number of natural acetogenins, from the mother skeleton of the natural products and named the

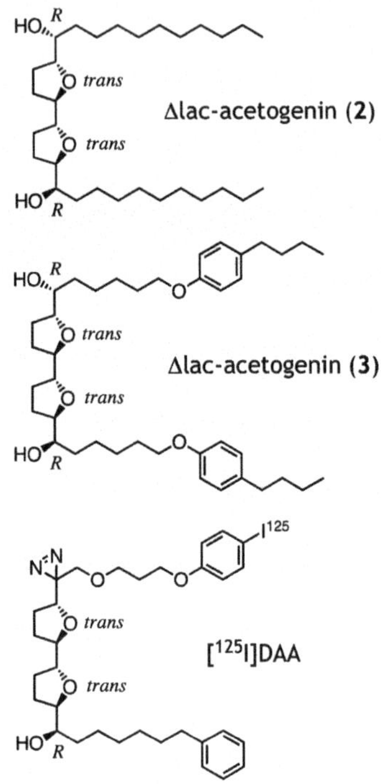

Fig. 4.4 Structures of representative Δlac-acetogenins (**2** and **3**) and a photoreactive derivative [^{125}I]DAA. [^{125}I]DAA, ^{125}I-labeled *d*i*a*zirinylated Δlac-*a*cetogenin

resultant derivatives "Δlac-acetogenins" (Hamada et al. 2004). Unexpectedly, Δlac-acetogenins (e.g. compounds **2** and **3** in Fig. 4.4) retained a strong inhibitory effect on bovine complex I at nanomolar level. An EPR spectroscopic study on the redox state of iron-sulfur clusters indicated that the inhibition site of Δlac-acetogenins is downstream of the iron-sulfur cluster N2, as is the case for other traditional complex I inhibitors (Hamada et al. 2004). However, detailed biochemical characterization of the inhibition mechanism of Δlac-acetogenins with bovine complex I revealed that they are a new type of complex I inhibitors, as summarized below. First, the most characteristic feature of Δlac-acetogenins is that they induce superoxide generation from complex I, regardless of whether endogenous or exogenous ubiquinone is used, at much lower rates than do traditional inhibitors (Ichimaru et al. 2005; Murai et al. 2006). Second, the inhibitory effect of Δlac-acetogenins on ubiquinol-NAD^+ oxidoreductase activity (reverse electron transfer) is significantly weaker than that on NADH oxidase activity (forward electron transfer), i.e., they are direction-specific inhibitors (Murai et al. 2006). Third, traditional inhibitors such as rotenone and piericidin A prevent the specific photocross-linking of the ND1 subunit by [^{125}I]TDA, but Δlac-acetogenins do not (Murai et al. 2007). Thus, the inhibition manner of Δlac-acetogenins is significantly different from that of traditional inhibitors. In view of the proposed dynamic function of the membrane domain (Belogrudov and Hatefi 1994; Yamaguchi et al. 1998; Mamedova et al. 2004; Berrisford et al. 2008), it is not strange that there are diverse chemicals that disturb the function of the membrane domain differently depending on their structural nature.

Among unique inhibitory effects of Δlac-acetogenins mentioned above, the first point, i.e. the production of much less superoxide, makes it interesting to consider the major site of superoxide generation from complex I in the presence of inhibitor. As described later, the binding site of Δlac-acetogenins appeared to reside in the membrane subunit ND1 (Kakutani et al. 2010), which is located downstream of all the redox cofactors (FMN and iron-sulfur clusters) residing in the hydrophilic domain. When the reduction of ubiquinone in complex I is completely blocked by traditional inhibitors or Δlac-acetogenin, the extent to which electrons accumulate in the cofactors (FMN and iron-sulfur clusters) located in the hydrophilic arm should be comparable (Lambert and Brand 2004a). Therefore, the remarkable difference in superoxide generation would not be attributable to a difference in the electron leakage from the cofactors to molecular oxygen. It is likely that interference of the reduction of ubiquinone by Δlac-acetogenin causes less electron leakage from ubisemiquinone to molecular oxygen than traditional inhibitors do (Kakutani et al. 2010). Genova et al. (2001) claimed that the ubisemiquinone radical is not a candidate for a direct electron donor to molecular oxygen in the presence of an ordinary inhibitor because rotenone remarkably diminished the formation of ubisemiquinone radical as revealed by EPR spectroscopy (Burbaev et al. 1989). However, we cannot exclude the possibility that a destabilized ubisemiquinone and/or the residual ubisemiquinone serves as an electron donor to molecular oxygen (Lambert and Brand 2004a, b; Ohnishi et al. 2005).

To identify the binding site of Δlac-acetogenins, we recently synthesized a photoreactive probe that has a small photolabile diazirine ring attached to the bis-THF ring moiety ([^{125}I]diazinylated Δlac-acetogenin, [^{125}I]DAA, Fig. 4.4) and carried out photoaffinity labeling (Kakutani et al. 2010). Characterization of the inhibitory effects of DAA on bovine complex I revealed unique features specific to, though not completely the same as those of, the original Δlac-acetogenin. Analysis of the photocross-linked protein revealed that [^{125}I]DAA binds to the ND1 subunit with high specificity. Examination of fragmentation patterns generated by digestion of the [^{125}I]DAA-labeled ND-1 using Lys-C, Asp-N or trypsin strongly suggested that the cross-linked residues are located at two different sites in the third matrix-side loop connecting the fifth and sixth transmembrane helices: site A in the Ala135-Phe198 region and site B in the Asp199-Lys262 region (Fig. 4.5). We proposed that sites A and B is the site that accommodates the bis-THF ring moiety of natural acetogenins and Δlac-acetogenins, respectively (Kakutani et al. 2010). The cytoplasmic loops of Nqo8 (ND1) were not resolved in the crystal structure of complex I from *T. thermophilus* (Efremov et al. 2010). Given that an excess of Q_2 selectively blocked the binding of [^{125}I]DAA to site B, this site may participate in the binding or the access path of ubiquinone to the "quinone cavity" which may be formed by the PSST and 49 kDa subunits (Sazanov 2007).

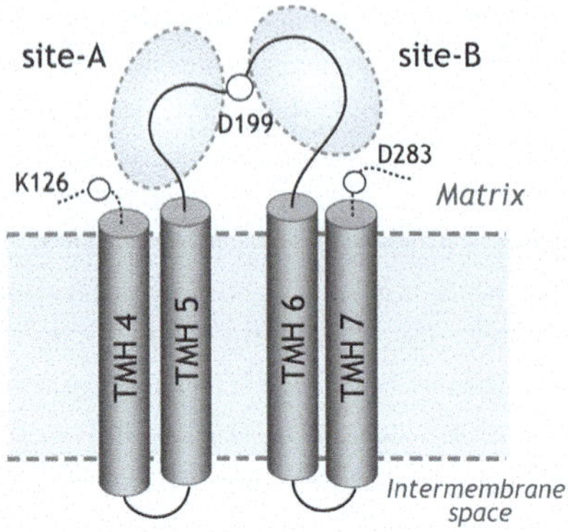

Fig. 4.5 A hypothetical spatial arrangement of the two binding sites (A and B) in the membrane topology model of the ND1 subunit

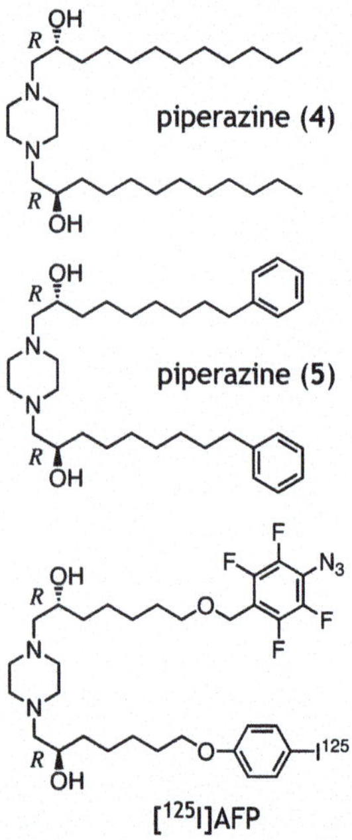

Fig. 4.6 Structures of representative piperazines (**4** and **5**) and a photoreactive [^{125}I]AFP. [^{125}I]AFP, ^{125}I-labeled 4-*a*zido-2,3,5,6-tetra*f*luorobenzyloxyalkyl *p*iperazine

4.4.2 Piperazine-Type Inhibitors

On the basis of the structural similarity between the hydroxylated bis-THF ring and piperazine, we synthesized a series of piperazine derivatives as new analogues of Δlac-acetogenins (e.g., compounds **4** and **5**, Fig. 4.6). The inhibition manner of the piperazine series, however, turned out to be significantly different from that of the original Δlac-acetogenins (Ichimaru et al. 2008). In particular, the inhibitory effect of Δlac-acetogenins on the reverse electron transfer in complex I (ubiquinol-NAD$^+$ oxidoreductase activity) in the presence of the proton-motive force is unique.

Compared with the original Δlac-acetogenins as well as traditional inhibitors, piperazines elicit a much weaker inhibitory effect, in terms of the IC_{50} value (pmol inhibitor/mg of protein), on the reverse electron transfer than the forward electron transfer. For instance, approximately 50-fold higher concentrations of **5** were needed to inhibit the reverse electron transfer. The piperazines are, to our knowledge, the most definitive "direction-specific" inhibitors among diverse complex I inhibitors. Consequently, we suggest that the structure of the binding site of piperazines in the forward electron transfer state differ from that in the reverse electron transfer state probably because the two reactions have different rate-limiting steps. A postulate of two different electron transfer pathways is an unlikely explanation for the direction-specific effect of the piperazines.

A photoaffinity labeling study using a photoreactive piperazine (4-azido-2,3,5,6-tetrafluorobenzyloxyalkyl piperazine, [^{125}I]AFP, Fig. 4.6) revealed that this inhibitor binds to the 49 kDa subunit and an unidentified subunit, not PSST, TYKY and ND1, with a frequency of ~1:3 (Ichimaru et al. 2008). While the piperazine does not bind to the ND1 subunit, it prevents the specific binding of [^{125}I]TDA to the ND1 (Ichimaru et al. 2008). A variety of traditional complex I inhibitors as well as Δlac-acetogenins suppressed the specific binding of [^{125}I]AFP to the two subunits. The apparent competitive behavior observed in some pairs of inhibitors, each of which seems to bind to different sites, may be due to structural changes at the binding site (i.e. an allosteric effect), rather than occupying the same site (Ichimaru et al. 2008).

4.4.3 Photoresponsive Inhibitor: Light-Control of Complex I Activity

The SAR studies of a number of Δlac-acetogenins indicated that their inhibitory effect is remarkably influenced by the physicochemical properties of the alkyl side chains (Hamada et al. 2004; Ichimaru et al. 2005, 2006). For instance, the *para-n*-butyl derivative (**6**, Fig. 4.7) has a very potent activity, but the *ortho-n*-butyl derivative (**7**) elicits almost no inhibition. This means that the molecular shape of the side chains has a remarkable influence on the inhibitory potency of this series of inhibitors. On the basis of this finding, we speculated that if the shape of the side chains can be changed by photoirradiation, the inhibitory effect could be switched on and off in a reversible manner. Accordingly, such a photoresponsive inhibitor may allow rapid, remote and noninvasive control of complex I activity.

We synthesized Δlac-acetogenin (**8**) possessing an azobenzene unit in the side chains not only because azobenzene has been widely used as an efficient photoresponsive switch since the pioneering work by Shinkai et al. (1980), but also because the change in molecular shape of the substituted azobenzene owning to *trans-cis* isomerization seems to be similar to that seen between compounds **6** and **7** (Fig. 4.5). ^{1}H-NMR, HPLC and UV-visible absorption analyses showed that the azobenzene unit in **8** is rapidly and reversibly *trans-cis* isomerized by photoirradiation in organic solvents (Fujita et al. 2006). The inhibitory effect of (*trans*, *trans*)-**8** on complex I

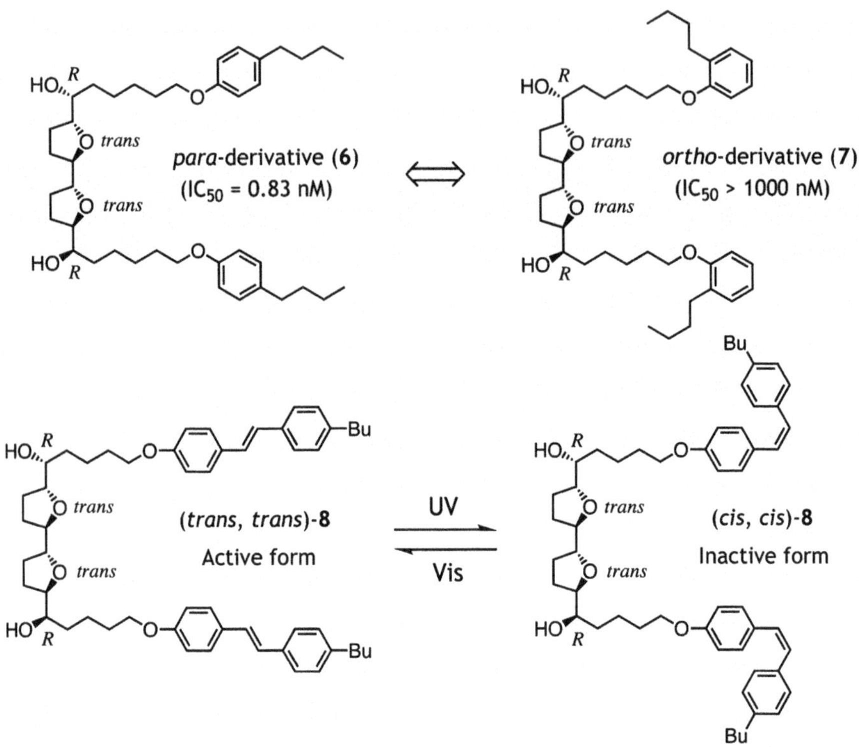

Fig. 4.7 Structures of representative Δlac-acetogenins (**6** and **7**) and a photoresponsive inhibitor (**8**)

activity in SMP was more potent than that of (*cis*, *cis*)-**8**. When **8** was applied at the nanomolar level to complex I, the inhibitory effect was reversibly reduced and enhanced by alternating irradiation by UV and visible light, respectively, though complete on-off switching was not achieved (Fujita et al. 2006).

Thus the concept that complex I activity can be controlled by a photoresponsive inhibitor was in principle verified. Our study therefore gives a positive clue to the light-control of complex I activity. The light-control in combination with other techniques would provide a new experimental methodology to the complex I researches.

References

Abe M, Kenmochi A, Ichimaru N, Hamada T, Nishioka T, Miyoshi H (2004) Essential structural features of acetogenins: role of hydroxy groups adjacent to the bis-THF rings. Bioorg Med Chem Lett 14:779–782

Abe M, Murai M, Ichimaru N, Kenmochi A, Yoshida T, Kubo A, Kimura Y, Moroda A, Makabe H, Nishioka T, Miyoshi H (2005) Dynamic function of the alkyl spacer of acetogenins in their inhibitory action with mitochondrial complex I (NADH-ubiquinone oxidoreductase). Biochemistry 44:14898–14906

Abe M, Kubo A, Yamamoto S, Hatoh Y, Murai M, Hattori Y, Makabe H, Nishioka T, Miyoshi H (2008) Dynamic function of the spacer region of acetogenins in the inhibition of bovine mitochondrial NADH-ubiquinone oxidoreductase (complex I). Biochemistry 47:6260–6266

Alali FQ, Liu XX, McLaughlin JL (1999) Annonaceous acetogenins: recent progress. J Nat Prod 62:504–540

Belogrudov GI, Hatefi Y (1994) Catalytic sector of complex I (NADH-ubiquinone oxidoreductase): subunit stoichiometry and substrate-induced conformation changes. Biochemistry 33:4571–4576

Bermejo A, Figadere B, Zafra-Polo MC, Barrachina I, Estornell E, Cortes D (2005) Acetogenins from Annonaceae: recent progress in isolation, synthesis and mechanisms of action. Nat Prod Rep 22:269–303

Berrisford JM, Thompson CJ, Sazanov LA (2008) Chemical and NADH-induced, ROS-dependent, cross-linking between subunits of complex I from Escherichia coli and Thermus thermophilus. Biochemistry 47:10262–10270

Brunner J (1993) New photolabeling and crosslinking methods. Annu Rev Biochem 62:483–514

Burbaev DS, Moroz IA, Kotlyar AB, Sled VD, Vinogradov AD (1989) Ubisemiquinone in the NADH-ubiquinone reductase region of the mitochondrial respiratory chain. FEBS Lett 254:47–51

Carroll J, Fearnley IE, Shannon RJ, Hirst J, Walker JE (2003) Analysis of the subunit composition of complex I from bovine heart mitochondria. Mol Cell Proteomics 2:117–126

Carroll J, Fearnley IM, Walker JE (2006) Definition of the mitochondrial proteome by measurement of molecular masses of membrane proteins. Proc Natl Acad Sci USA 103:16170–16175

Degli Esposti M (1998) Inhibitors of NADH-ubiquinone reductase: an overview. Biochim Biophys Acta 1364:222–235

Degli Esposti M, Ghelli A, Ratta M, Cortes D, Estornell E (1994) Natural substances (acetogenins) from the family *Annonaceae* are powerful inhibitors of mitochondrial NADH dehydrogenase (complex I). Biochem J 301:161–167

Dormán G, Prestwich GD (2000) Using photolabile ligands in drug discovery and development. Trends Biotechnol 18:64–76

Efremov RG, Baradaran R, Sazanov LA (2010) The architecture of respiratory complex I. Nature 465:441–445

Fendel U, Tocilescu MA, Kerscher S, Brandt U (2008) Exploring the inhibitor binding pocket of respiratory complex I. Biochim Biophys Acta 1777:660–665

Feutrill JT, Lilly MJ, White JM, Rizzacasa MA (2008) Asymmetric total synthesis of the myxobacteria metabolites crocacins A-D. Tetrahedron 64:4880–4895

Friedrich T, Van Heek P, Leif H, Ohnishi T, Forche E, Kunze B, Jansen R, Trowitzsch-Kienast W, Höfle G, Reichenbach H, Weiss H (1994) Two binding sites of inhibitors in NADH:ubiquinone oxidoreductase (complex I). Eur J Biochem 219:691–698

Fujita D, Murai M, Nishioka T, Miyoshi H (2006) Light control of mitochondrial complex I activity by a photoresponsive inhibitor. Biochemistry 45:6581–6586

Genova ML, Ventura B, Giuliano G, Bovine C, Formiggini G, Castelli GP, Lenaz G (2001) The site of production of superoxide radical in mitochondrial complex I is not a bound ubisemiquinone but presumably iron-sulfur cluster N2. FEBS Lett 505:364–368

Gong X, Xie T, Yu L, Hesterberg M, Scheide D, Friedrich T, Yu CA (2003) The ubiquinone-binding site in NADH:ubiquinone oxidoreductase from *Escherichia coli*. J Biol Chem 278: 25731–25737

Granados-Covarrubias EH, Maldonado LA (2009) Protected cyanohydrins in the synthesis of rotenoids: (±)-munduserone and (±)-*cis*-12a-hydroxymunduserone. J Org Chem 74:5097–5099

Hamada T, Ichimaru N, Abe M, Fujita D, Kenmochi A, Nishioka T, Zwicker K, Brandt U, Miyoshi H (2004) Synthesis and inhibitory action of novel acetogenin mimics with bovine heart mitochondrial complex I. Biochemistry 43:3651–3658

Heinrich H, Werner S (1992) Identification of the ubiquinone-binding site of NADH-ubiquinone oxidoreductase (complex I) from *Neurospora crassa*. Biochemistry 31:11413–11419

Hoppen S, Emde U, Friedrich T, Grubert L, Koert U (2000) Natural-product hybrid: design, synthesis, and biological evaluation of quinone-annonaceous acetogenins. Angew Chem Int Ed 39:2099–2102

Ichimaru N, Murai M, Abe M, Hamada T, Yamada Y, Makino S, Nishioka T, Makabe H, Makino A, Kobayashi T, Miyoshi H (2005) Synthesis and inhibition mechanism of Δlac-acetogenins: a novel type of inhibitor of bovine heart mitochondrial complex I. Biochemistry 44:816–825

Ichimaru N, Abe M, Murai M, Senoh M, Nishioka T, Miyoshi H (2006) Function of the alkyl side chains of Δlac-acetogenins in the inhibitory effect on mitochondrial complex I (NADH-ubiquinone oxidoreductase). Bioorg Med Chem Lett 16:3555–3558

Ichimaru N, Murai M, Kakutani N, Kako J, Ishihara A, Nakagawa Y, Nishioka T, Yagi T, Miyoshi H (2008) Synthesis and characterization of new piperazine-type inhibitors for mitochondrial NADH-ubiquinone oxidoreductase (complex I). Biochemistry 47:10816–10826

Ino T, Nishioka T, Miyoshi H (2003) Characterization of inhibitor binding sites of mitochondrial complex I using fluorescent inhibitor. Biochim Biophys Acta 1605:15–20

Kakutani N, Murai M, Sakiyama N, Miyoshi H (2010) Exploring the binding site of Δlac-acetogenin in bovine heart mitochondrial NADH-ubiquinone oxidoreductase. Biochemistry 49:4794–4803

Kean EA, Gutman M, Singer TP (1971) Studies on the respiratory chain-linked nicotinamide adenine dinucleotide dehydrogenase. XXII. Rhein, a competitive inhibitor of the dehydrogenase. J Biol Chem 246:2346–2353

Kojima N, Tanaka T (2009) Medicinal chemistry of annonaceous acetogenins: design, synthesis, and biological evaluation of novel analogues. Molecules 14:3621–3661

Kuwabara K, Takada M, Iwata J, Tatsumoto K, Sakamoto K, Iwamura H, Miyoshi H (2000) Design syntheses and mitochondrial complex I inhibitory activity of novel acetogenin mimics. Eur J Biochem 267:2538–2546

Lambert AJ, Brand MD (2004a) Inhibitors of the quinone-binding site allow rapid superoxide production from mitochondrial NADH-ubiquinone oxidoreductase (complex I). J Biol Chem 279:39414–39420

Lambert AJ, Brand MD (2004b) Superoxide production by NADH:ubiquinone oxidoreductase (complex I) depends on the pH gradient across the mitochondrial inner membrane. Biochem J 382:511–517

Leiris SJ, Khdour OM, Segerman ZJ, Tsosie KS, Chapuis JC, Hecht SM (2010) Synthesis and evaluation of verticipyrone analogues as mitochondrial complex I inhibitors. Bioorg Med Chem 18:3481–3493

Majander A, Finel M, Wikström M (1994) Diphenyleneiodonium inhibits reduction of iron-sulfur clusters in the mitochondrial NADH-ubiquinone oxidoreductase (complex I). J Biol Chem 269:21037–21042

Makabe H, Hattori Y, Kimura Y, Konno H, Abe M, Miyoshi H, Tanaka A, Oritani T (2004a) Total synthesis of *cis*-solamin and its inhibitory action with bovine heart mitochondrial complex I. Tetrahedron 60:10651–10657

Makabe H, Miyawaki A, Takahashi R, Hattori Y, Konno H, Abe M, Miyoshi H (2004b) Synthesis of two possible diastereomers of reticulatain-1. Tetrahedron Lett 45:973–977

Mamedova AA, Holt PJ, Carroll J, Sazanov LA (2004) Substrate-induced conformational change in bacterial complex I. J Biol Chem 279:23830–23836

Miyoshi H (1998) Structure-activity relationships of some complex I inhibitors. Biochim Biophys Acta 1364:236–244

Miyoshi H (2001) Probing the ubiquinone reduction site in bovine mitochondrial complex I using a series of synthetic ubiquinones and inhibitors. J Bioenerg Biomembr 33:223–231

Miyoshi H, Iwata J, Sakamoto K, Furukawa H, Takada M, Iwamura H, Watanabe T, Kodama Y (1998a) Specificity of pyridinium inhibitors of the ubiquinone reduction sites in mitochondrial complex I. J Biol Chem 273:17368–17374

Miyoshi H, Ohshima M, Shimada H, Akagi T, Iwamura H, McLaughlin JL (1998b) Essential structural factors of annonaceous acetogenins as potent inhibitors of mitochondrial complex I. Biochim Biophys Acta 1365:443–452

Motoyama T, Yabunaka H, Miyoshi H (2002) Essential structural factors of acetogenins, potent inhibitors of mitochondrial complex I. Bioorg Med Chem Lett 12:2089–2092

Murai M, Ichimaru N, Abe M, Nishioka T, Miyoshi H (2006) Mode of inhibitory action of Δlac-acetogenins, a new class of inhibitors of bovine heart mitochondrial complex I. Biochemistry 45:9778–9787

Murai M, Ishihara A, Nishioka T, Yagi T, Miyoshi H (2007) The ND1 subunit constructs the inhibitor binding domain in bovine heart mitochondrial complex I. Biochemistry 46: 6409–6416

Murai M, Sekiguchi K, Nishioka T, Miyoshi H (2009) Characterization of the inhibitor binding site in mitochondrial NADH-ubiquinone oxidoreductase by photoaffinity labeling using a quinazoline-type inhibitor. Biochemistry 48:688–698

Murai M, Mashimo Y, Hirst J, Miyoshi H (2011) Exploring interactions between the 49 kDa and ND1 subunits in mitochondrial NADH-ubiquinone oxidoreductase (complex I) by photoaffinity labeling. Biochemistry 50:6901–6908

Nakamaru-Ogiso E, Sakamoto K, Matsuno-Yagi A, Miyoshi H, Yagi T (2003) The ND5 subunit was labeled by a photoaffinity analogue of fenpyroximate in bovine mitochondrial complex I. Biochemistry 42:746–754

Nicolaou KC, Pfefferkorn JA, Schuler F, Roecker AJ, Cao GQ, Casida JE (2000) Combinatorial synthesis of novel and potent inhibitors of NADH:ubiquinone oxidoreductase. Chem Biol 7:979–992

Ohnishi ST, Ohnishi T, Muranaka S, Fujita H, Kimura H, Uemura K, Yoshida K, Utsumi K (2005) A possible site of superoxide generation in the complex I segment of rat heart mitochondria. J Bioenerg Biomembr 37:1–15

Okun JG, Lümmen P, Brandt U (1999) Three classes of inhibitor share a common binding domain in mitochondrial complex I (NADH-ubiquinone oxidoreductase). J Biol Chem 274:2625–2630

Purohit A, Radeke H, Azure M, Hanson K, Benetti R, Su F, Yalamanchili P, Yu M, Hayes M, Guaraldi M, Kagan M, Robinson S, Casebier D (2008) Synthesis and biological evaluation of pyridazinone analogues as potential cardiac positron emmision tomography tracers. J Med Chem 54:2954–2970

Satoh T, Miyoshi H, Sakamoto K, Iwamura H (1996) Comparison of the inhibitory action of synthetic capsaicin analogues with various NADH-ubiquinone oxidoreductases. Biochim Biophys Acta 1273:21–30

Sazanov LA (2007) Respiratory complex I: mechanism and structural insights provided by the crystal structure of the hydrophilic domain. Biochemistry 46:2275–2288

Sazanov LA, Hinchliffe P (2006) Structure of the hydrophilic domain of respiratory complex I from *Thermus thermophilus*. Science 311:1430–1436

Schnermann MJ, Romero FA, Hwang I, Nakamaru-Ogiso E, Yagi T, Goger DL (2006) Total synthesis of piericidin A1 and B1 and key analogues. J Am Chem Soc 128:11799–11807

Schuler F, Yano T, Bernardo SD, Yagi T, Yankovskaya V, Singer TP, Casida JE (1999) NADH-quinone oxidoreductase: PSST subunit couples electron transfer from iron-sulfur cluster N2 to quinone. Proc Natl Acad Sci USA 96:4149–4153

Sekiguchi K, Murai M, Miyoshi H (2009) Exploring the binding site of acetogenin in the ND1 subunit of bovine mitochondrial complex I. Biochim Biophys Acta 1781:1106–1111

Shinkai S, Nakaji T, Nishida Y, Ogawa T, Manabe O (1980) Photoresponsive crown ethers: *cis-trans* isomerism of azobenzene as a tool to enforce conformational changes of crown ethers and polymers. J Am Chem Soc 102:5860–5865

Shiomi K, Ui H, Suzuki H, Hatano H, Nagamitsu T, Takano D, Miyadera H, Yamashita T, Kita K, Miyoshi H, Harder A, Tomoda H, Omura S (2005) A γ-lactone form nafuredin, nafuredin-γ, also inhibits helminth complex I. J Antibiot 58:50–55

Surup F, Shojaei H, von Zezschwitz P, Kunze B, Grond S (2008) Iromycins from *Streptomyces* sp. and from synthesis: new inhibitors of the mitochondrial electron transport chain. Bioorg Med Chem 16:1738–1746

Takada M, Kuwabara K, Nakato H, Tanaka A, Iwamura H, Miyoshi H (2000) Definition of crucial structural factors of acetogenins, potent inhibitors of mitochondrial complex I. Biochim Biophys Acta 1460:302–310

Tocilescu MA, Fendel U, Zwicker K, Kerscher S, Brandt U (2007) Exploring the ubiquinone binding cavity of respiratory complex I. J Biol Chem 282:29514–29520

Tocilescu MA, Fendel U, Zwicker K, Dröse S, Kerscher S, Brandt U (2010) The role of a conserved tyrosine in the 49-kDa subunit of complex I for ubiquinone binding and reduction. Biochim Biophys Acta 1797:625–632

Tormo JR, Carmen Gonzalez M, Cortes D, Estornell E (1999a) Kinetic characterization of mitochondrial complex I inhibitors using annonaceous acetogenins. Arch Biochem Biophys 369: 119–126

Tormo JR, Gallardo T, Aragon R, Cortes D, Estornell E (1999b) Specific interactions of monotetrahydrofuranic annonaceous acetogenins as inhibitors of mitochondrial complex I. Chem Biol Interact 122:171–183

Tormo JR, Estornell E, Gallardo T, Gonzalez MC, Cave A, Granell S, Cortes D, Zafra-Polo MC (2001) γ-Lactone-functionalized antitumal acetogenins are the most potent inhibitors of mitochondrial complex I. Bioorg Med Chem Lett 11:681–684

Tormo JR, Gallardo T, Peris E, Bermejo A, Cabedo N, Estornell E, Zafra-Polo MC, Cortes D (2003) Inhibitory effects on mitochondrial complex I of semisynthetic mono-tetrahydrofuran acetogenin derivatives. Bioorg Med Chem Lett 13:4101–4105

Yabunaka H, Abe M, Kenmochi A, Hamada T, Nishioka T, Miyoshi H (2003) Synthesis and inhibitory activity of ubiquinone-acetogenin hybrid inhibitor with bovine mitochondrial complex I. Bioorg Med Chem Lett 13:2385–2388

Yagi T, Matsuno-Yagi A (2003) The proton-translocating NADH-quinone oxidoreductase in the respiratory chain: the secret unlocked. Biochemistry 42:2266–2274

Yamaguchi M, Belogrudov GI, Hatefi Y (1998) Mitochondrial NADH-ubiquinone oxidoreductase (complex I): effects of substrates on the fragmentation of subunits by trypsin. J Biol Chem 273:8094–8098

Yamashita T, Ino T, Miyoshi H, Sakamoto K, Osanai A, Nakamaru-Ogiso E, Kita K (2004) Rhodoquinone reaction site of mitochondrial complex I in parasitic helminth, Ascaris suum. Biochim Biophys Acta 1608:97–103

Yoshida T, Murai M, Abe M, Ichimaru N, Harada T, Nishioka T, Miyoshi H (2007) Crucial structural factors and mode of action of polyene amides as inhibitors for mitochondrial NADH-ubiquinone oxidoreductase (complex I). Biochemistry 46:10365–10372

Zeng L, Ye Q, Oberlies NH, Shi G, Gu ZM, He K, McLaughlin JL (1996) Recent advances in annonaceous acetogenins. Nat Prod Rep 13:275–306

Chapter 5
My Fifty Years Association with Complex I Study

Tomoko Ohnishi

Abstract I summarized some key steps in my 50 years association with complex I study. First, discovered the absence of energy-coupling at complex I in domestic yeast. Then, found a close correlation among complex I energy-coupling, rotenone sensitivity and g = 1.94 iron-sulfur cluster. To scrutinize this correlation, I performed EPR at 4.2–40K, and detected signals from previously unknown multiple clusters in complex I of wild yeast and bovine heart *in situ*. Named them as N1a, N2, N3, N4, N5 and N1b. Discovered and characterized complex I counterpart in *Escherichia coli*, *Thermus thermophilus* HB-8, *Paracoccus denitrificans* in the membrane system, and made important foundation for the later structural studies by X-ray crystallographers. Characterized two protein-associated semiquinone species.

Keywords NADH: ubiquinone oxidoreductase (complex I) • Iron-sulfur cluster • Flavin • Quinone • Energy-coupling

I would like to introduce some historical events in my complex I studies.

For the sake of simplicity, expression, such as Dr. or Professor, is omitted.

(A) After finishing my Ph.D. work, I selected an isolation and characterization of intact mitochondria from *Saccharomyces carlsbergensis* (used for beer production) as my first research project. It was under Bunji Hagihara of Osaka University, who had just returned from Britton Chance's laboratory. At that time, we did not know correlations between bacterial and mitochondrial systems, but we thought that this is a simplest mitochondrial model system, and

---

T. Ohnishi (✉)
Perelman School of Medicine, University of Pennsylvania, Department of Biochemistry and Biophysics, Johnson Research Foundation, 36th and Hamilton Walk, Philadelphia, PA 19104-6059, USA
e-mail: ohnishi@mail.med.upenn.edu

L. Sazanov (ed.), *A Structural Perspective on Respiratory Complex I: Structure and Function of NADH:ubiquinone oxidoreductase*, DOI 10.1007/978-94-007-4138-6_5,

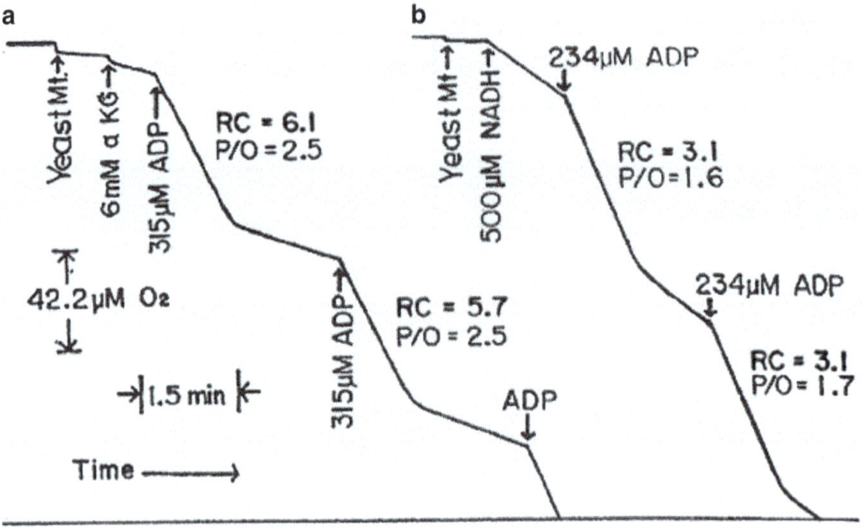

Fig. 5.1 Polarographic tracings of respiratory pattern, with (a) α-ketoglutarate or (b) NADH as substrate. The reaction medium contained 0.5 M mannitol, 10 mM potassium phosphate buffer (pH 6.5), 20 mM Tris-maleate buffer (pH 6.5), 10 mM KCl, and 0.1 mM EDTA. *a*, final protein concentration, 0.28 mg per ml of reaction mixture; *b*, 0.13 mg per ml. Other additions were indicated by *arrows* in the figure. The reaction was carried out at 25°. *Yeast Mt.*, yeast mitochondria (Ohnishi et al. 1966)

therefore, it must be relatively easy to alter mitochondrial properties and functions by changing its growth conditions. After almost 3 years of steady and labor-intensive work (1962–1964), I succeeded for the first time in isolating intact mitochondria which exhibited excellent respiratory control on addition of ADP (Ohnishi et al. 1966) (Fig. 5.1).

We made the following findings: (i) This domestic yeast mitochondria specifically lack the first energy-coupling site in the NADH-ubiquinone oxidoreductase (more known as complex I) segment in the aerobic respiratory chain (Table 5.1). The second and the third energy coupling sites are functional as in the normal mammalian mitochondria.

The “absence of energy-transducing complex I” was important but a negative data. However, the fact that externally added NADH was oxidized with the respiratory control and with the energy transducing capability in both complex III (cytochrome bc_1 complex) and IV (cytochrome *c* oxidase) was a novel positive finding.

This indicated that in domestic yeast mitochondria, there are two non-energy-transducing NADH-ubiquinone oxidoreductases, which are facing to the intermembrane side (positive side) and matrix side (negative side) of the mitochondrial inner membrane, respectively.

(B) After working in Ernster’s lab (1964) and Klingenberg’s lab (1965), I settled at the Johnson Research Foundation in 1966. I discovered that the proton-pumping process is closely correlated with the rotenone sensitivity of the electron

Table 5.1 Q_{O_2}[a] (State 3), RC ratio, and ADP:O ratio in various substrate oxidations

Substrate	Q_{O_2}[a] (State 3)	RC ratio	ADP:O ratio
α-Ketoglutarate	0.26–0.41	5.5–6.0	2.3–2.6
Citrate	0.20–0.45	1.5–1.7	1.7–1.8
Isocitrate	0.14–0.20	1.2–1.5	1.6–1.8
Pyruvate + malate	0.15–0.35	1.6–2.0	1.6–1.8
Ethanol	0.11–0.30	1.4–1.5	1.6–1.8
NADH	0.60–1.10	3.0–3.6	1.5–1.8
Succinate	0.25–0.38	1.4–1.7	1.6–1.8
D-Lactate	0.50–0.75	1.1–1.5	0.9–1.1
L-Lactate	0.15–0.20	1.1–1.3	0.9–1.1
Tetramethyl-*p*-phenylenediamine + ascorbate	0.50–0.70	1.2–1.5	0.8–1.0
α-Glycerophosphate	0.80–0.30		

The reaction medium contained 0.65 M mannitol, 10 mM potassium phosphate buffer (pH 6.5), 10 mM Tris-maleate buffer (pH 6.5), 10 mM KCl, and 0.1 mM EDTA. Substrates were added at a final concentration of 6 mM, except in the cases of NADH and tetramethyl-*p*-phenylenediamine when 560 and 100 μM were used, respectively.

Numbers in column ADP:O ratio (from Citrate to NADH) indicate that there was no oxidative phosphorylation in the complex I of the respiratory chain in *S. carlsbergensis* (Ohnishi et al. 1966).

[a] Microatoms of oxygen uptake per min per mg of protein.

Table 5.2 Strong correlation was indicated among site I energy-coupling, rotenone sensitivity, and g = 1.94 signal of iron-sulfur cluster in the complex I segment of the aerobic respiratory chain in various yeasts and animal tissue (Ohnishi 1973)

	Site 1[a]	Rotenone sensitivity[a]	"g = 1.94" in the NADH dehydrogenase region[a]	*Flavin*	Q, number of side-chain isoprenoid units
S. carlsbergensis	–	–	–	N.T.	6
S. cervisiae	–	–	–	FAD	6
C. utilis	+	+	+	FMN + FAD	7 and 9
E. magnusii, yeast type	+	+	+	N.T.	9 or 10
Animal tissue	+	+	+	FMN	9–10

[a] + present, – absent, *NT* not tested, *Q* ubiquinone

transfer in complex I and the presence of EPR-active iron-sulfur (Fe/S) cluster. We measured EPR signals at 77K (liquid nitrogen temperature) (Table 5.2) in the complex I segment of the obligatory aerobic yeast, such as *Candida utilis,* and bovine heart mitochondrial inner membrane.

It was known, however, that some tetranuclear [4Fe–4S] clusters in the (Ni/Fe)-hydrogenase were detectable only at a much lower temperature (below 15K). Thus, in order to ascertain the correlation between the presence of Fe/S cluster and the energy-coupling site in complex I, I pioneered EPR analysis of mitochondrial Fe/S clusters at extremely low temperature (4.2–40K).

Unexpectedly, this experiment led to the discovery of EPR signals of previously unrecognized 6 Fe/S clusters, namely, N1a, N2, N3, N4, N5 (N6a and N6b signals are not fully resolved in the intact complex I level) in addition to N1b.

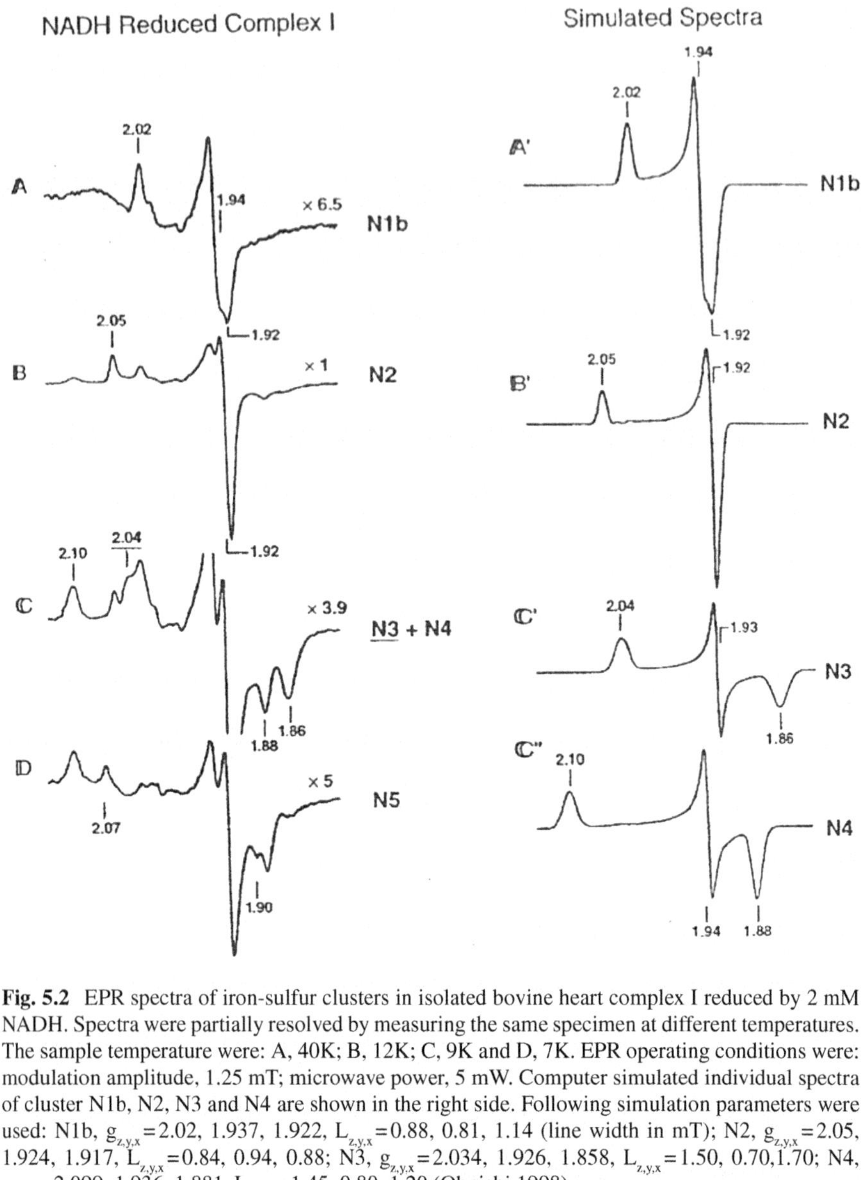

Fig. 5.2 EPR spectra of iron-sulfur clusters in isolated bovine heart complex I reduced by 2 mM NADH. Spectra were partially resolved by measuring the same specimen at different temperatures. The sample temperature were: A, 40K; B, 12K; C, 9K and D, 7K. EPR operating conditions were: modulation amplitude, 1.25 mT; microwave power, 5 mW. Computer simulated individual spectra of cluster N1b, N2, N3 and N4 are shown in the right side. Following simulation parameters were used: N1b, $g_{z,y,x}=2.02$, 1.937, 1.922, $L_{z,y,x}=0.88$, 0.81, 1.14 (line width in mT); N2, $g_{z,y,x}=2.05$, 1.924, 1.917, $L_{z,y,x}=0.84$, 0.94, 0.88; N3, $g_{z,y,x}=2.034$, 1.926, 1.858, $L_{z,y,x}=1.50$, 0.70,1.70; N4, $g_{z,y,x}=2.099$, 1.936, 1.881, $L_{z,y,x}=1.45$, 0.80, 1.20 (Ohnishi 1998)

N1b was the only Fe/S cluster known before in the energy transducing complex I segment of the electron transfer chain. Since the freezing time was slower than the rate limiting step of the substrate reaction with complex I, kinetic resolution of the reduction of the individual clusters was impossible at that time. Therefore, I determined EPR spectral profiles (as shown in Fig. 5.2 as an example) and thermodynamic parameters of individual Fe/S clusters (Ohnishi 1998).

(C) **Discovery of NDH-1 and NDH-2 in *E. coli***

Subsequently, my research group and collaborators studied various bacterial electron/proton transfer systems *in situ* in the cytoplasmic membrane, which are the counterpart of the inner mitochondrial membrane (the location of complex I).

As opposed to the common belief, I revealed that *E. coli* has an energy-transducing complex I counterpart in collaboration with Matsushita and Kaback. We originally designated it NADH dehydrogenase I (abbreviated as NADH dh I) (Matsushita et al. 1987). This became more known as NDH-1 by an abbreviated name given to our enzyme by T. Yagi, in addition to the non-energy-transducing single subunit NADH dh II (NDH-2). Both enzymes are facing to the same (negative) side surface of the cytoplasmic membrane. Only the latter enzyme had been known for a long time since 1981. NDH-1 was specifically reduced by deamino-NADH. We also characterized the *Paracoccus denitrificans* complex I *in situ* and *in vitro* system and demonstrated that its EPR spectral profiles are very similar to the mitochondrial counterpart (Meinhardt et al. 1987).

(D) **T. Oshima isolated the thermophilic bacterium HB-8 from a Japanese spa**

I also requested T. Oshima, a godfather of thermophilic bacteria in Japan, to isolate a rotenone-sensitive thermophilic bacterial strain for the complex I study. He isolated *Thermus thermophilus* HB-8 (this name came from the name of the hot spring from which this bacterial strain was isolated). It showed somewhat different EPR spectral line shape of Fe/S clusters from the other two bacterial enzymes (Yagi et al. 1988).

These early works in my laboratory provided important foundations for recent structure/function studies of complex I on an atomic level, as summarized below.

(E) **Current status of complex I study**

It is known that bovine heart complex I has the largest (980 kDa) and most complicated structure (45 distinct subunits) among energy-transducing complexes in the aerobic electron transfer chain (Carroll et al. 2006). It contains 1 non-covalently bound FMN, 8 distinct Fe/S clusters (9 or 10 in some bacteria), and 2 protein-associated ubiquinone molecules. FMN and ubiquinone are 2 electron ($n=2$) carriers. Flavosemiquinone (here abbreviated as SF in order to clearly distinguish from SQ) and two SQ species, SQ_{Nf} and SQ_{Ns} molecules (because of their **f**ast and **s**low spin relaxation rate, respectively) can be detected in the intermediate redox states as $g=2.00$ free radical signals. These SQ_{Nf} and SQ_{Ns} molecules differ from ubiquinone molecules in the Q-pool, which react only as the oxidized quinone or fully reduced hydroquinone as two electron transfer ($n=2$) system. In the tightly coupled bovine heart submitochondrial particles (SMP) with respiratory control ratio >5, SQ_{Nf} is very sensitive to proton motive force (P), while SQ_{Ns} is insensitive. Both SQ species are equally sensitive to piericidin A, but SQ_{Nf} is much more sensitive to rotenone than $SQ_{Ns.}$ These issues are still highly controversial among different research groups of complex I. However, protein-associated Q molecules are known to play important roles in the energy transducing systems such as bacterial Reaction Center (Q_A and Q_B) (Okamura et al. 2000) and *E. coli* quinol oxidase (Q_L and Q_H) (Sato-Watanabe et al. 1994). In these proteins, semiquinone binds more strongly to its binding site than quinone in the fully oxidized or fully reduced state. Therefore, we can detect semiquinone EPR signals. The roles of semiflavin and semiquinone in a see-saw fashion were presented (Ohnishi 2010).

P. Mitchell approximated stability constant of SQ in the Q-pool as $K_{stability} = 10^{-10}$ (Mitchell 1979). Protein-associated quinone molecules function as a converter between n = 1 and n = 2 electron transfer processes. The converter SQ species are known with a wide range of $K_{stability}$ values of $10^{-3} \sim 10^{1}$. This species of redox intermediate state of n = 2 component (such as flavin and quinone) do not permanently bind to protein. They undergo redox reactions and binding or release occurs in the oxidized or fully reduced state.

Complex I is composed of an extra-and intra-membrane arms showing a unique L-shape (Friedrich 1998). Bacterial complex I has similar and conserved redox co-factors as those of bovine heart complex I, but the molecular weight is ~500kDa (generally 14 subunits). Thus, bacterial complex I can be used as a minimal core model of mitochondrial complex I.

In 2006, L. Sazanov and P. Hinchliffe determined X-ray structure of the hydrophilic extramembrane arm of *T. thermophilus* HB-8 (the strain which I asked Oshima to isolate from a Japanese spa) complex I at 3.3 Å resolution (Sazanov and Hinchliffe 2006). They confirmed all the Fe/S clusters predicted by EPR analysis and further revealed that FMN and 7 Fe/S clusters make approximately 95 Å long electron transfer chain from NADH-binding site to the Fe/S cluster N2 (with the highest redox midpoint potential, E_m) with less than 14 Å edge to edge inter redox center distance (Page et al. 1999). Low E_m cluster N1a is located sideway from FMN and Sazanov's group predicted its possible antioxidant role for the SF site. In May 2010, Sazanov and colleagues (Efremov et al. 2010) also published a long-awaited X-ray structure of the whole complex I at 4.5 Å resolution. I was invited by Nature to write "News & Views" of their exciting new proposal of "Piston drives a proton pump from afar" (Ohnishi 2010). The X-ray analysis at 6.3 Å resolution of mitochondrial complex I was reported more recently by U. Brandt and his colleagues, on obligatory aerobic yeast (*Yarrowia lipolytica*) mitochondrial complex I (Hunte et al. 2010).

5.1 Epilogue

Currently, the complex I field is in its most exciting state. Every month, papers with new structural and functional information are being published. It welcomes participation of new investigators with fresh mind and flexible ideas.

Acknowledgement I would like to thank my husband, S. Tsuyoshi Ohnishi, for his insightful discussions and encouragements during these 50 years.

References

Carroll J, Fearnley IM, Skehel JM, Shannon RJ, Hirst J, Walker JE (2006) Bovine complex I is a complex of 45 different subunits. J Biol Chem 281:32724–32727

Efremov RG, Baradaran R, Sazanov LA (2010) The architecture of respiratory complex I. Nature 465:441–445

Friedrich T (1998) The NADH: ubiquinone oxidoreductase (complex I) from *Escherichia coli*. Biochim Biophys Acta 1364:134–146

Hunte C, Zickermann V, Brandt U (2010) Functional modules and structural basis of conformational coupling in mitochondrial complex I. Science 329:448–451

Matsushita K, Ohnishi T, Kaback HR (1987) NADH-ubiquinone oxidoreductases of the Escherichia coli aerobic respiratory chain. Biochemistry 26:7732–7737

Meinhardt SW, Kula T, Yagi T, Lillich T, Ohnishi T (1987) EPR characterization of the iron-sulfur clusters in the NADH: ubiquinone oxidoreductase segment of the respiratory chain in Paracoccus denitrificans. J Biol Chem 262:9147–9153

Mitchell P (1979) Keilin's respiratory chain concept and its chemiosmotic consequences. Science 206:1148–1159

Ohnishi T (1973) Mechanism of electron transport and energy conservation in the site I region of the respiratory chain. Biochim Biophys Acta 301:105–128

Ohnishi T (1998) Iron-sulfur clusters/semiquinones in complex I. Biochim Biophys Acta 1364:186–206

Ohnishi T (2010) Structural biology: piston drives a proton pump. Nature 465:428–429

Ohnishi T, Kawaguchi K, Hagihara B (1966) Preparation and some properties of yeast mitochondria. J Biol Chem 241:1797–1806

Ohnishi ST, Shinzawa-Itoh K, Ohta K, Yoshikawa S, Ohnishi T (2010) New insights into the superoxide generation sites in bovine heart NADH-ubiquinone oxidoreductase (complex I): the significance of protein-associated ubiquinone and the dynamic shifting of generation sites between semiflavin and semiquinone radicals. Biochim Biophys Acta 1797:1901–1909

Okamura MY, Paddock ML, Graige MS, Feher G (2000) Proton and electron transfer in bacterial reaction centers. Biochim Biophys Acta 1458:148–163

Page CC, Moser CC, Chen X, Dutton PL (1999) Natural engineering principles of electron tunnelling in biological oxidation-reduction. Nature 402:47–52

Sato-Watanabe M, Mogi T, Ogura T, Kitagawa T, Miyoshi H, Iwamura H, Anraku Y (1994) Identification of a novel quinone-binding site in the cytochrome bo complex from Escherichia coli. J Biol Chem 269:28908–28912

Sazanov LA, Hinchliffe P (2006) Structure of the hydrophilic domain of respiratory complex I from *Thermus thermophilus*. Science 311:1430–1436

Yagi T, Hon-nami K, Ohnishi T (1988) Purification and characterization of two types of NADH-quinone reductase from *Thermus thermophilus HB-8*. Biochemistry 27:2008–2013

Part II
Evolution of Complex I

Chapter 6
The Evolutionary Relationship Between Complex I and [NiFe]-Hydrogenase

Anne Volbeda and Juan C. Fontecilla-Camps

Abstract Complex I (NADH:ubiquinone oxidoreductase), the first component of the respiratory chain in aerobic microorganisms and mitochondria, has been shown to be structurally related to [NiFe]-hydrogenases. A plausible theory postulates that the ancestor of complex I was formed by the association of a soluble hydrogenase with membrane-bound antiporters which had already acquired proton pumping capabilities. Contemporary examples of this type of hydrogenases are the energy-conserving enzymes found in several microorganisms. In this chapter we discuss this process and speculate on the origin of aerobic respiration and its relation to oxygen levels in the early atmosphere. We also compare the known crystal structures of the hydrophilic domain of complex I and [NiFe]-hydrogenase and postulate that a putative redox-induced rearrangement of a common long four-helical bundle between a kinked and a straight conformation may be involved in proton pumping.

Keywords NADH:ubiquinone oxidoreductase • Energy-Conserving Hydrogenases • Evolution • Ferredoxin • Flavodoxin • Proton transfer

6.1 Introduction

In the 1990s, unexpected relationships were noted between [NiFe]-hydrogenase and complex I. This is especially true in the case of the energy-converting hydrogenases (Ech), which couple the oxidation of H_2 with the translocation of protons through a membrane (Hedderich 2004). The first sequence similarities between [NiFe]-hydrogenase and complex I were reported for three open reading frames of

A. Volbeda • J.C. Fontecilla-Camps (✉)
Laboratoire de Cristallographie et de Cristallogenèse des Protéines,
CEA, CNRS, Université Joseph Fourier, 41, rue Jules Horowitz,
38027 Grenoble Cedex 1, France
e-mail: anne.volbeda@ibs.fr; juan.fontecilla@ibs.fr

L. Sazanov (ed.), *A Structural Perspective on Respiratory Complex I: Structure and Function of NADH:ubiquinone oxidoreductase*, DOI 10.1007/978-94-007-4138-6_6,

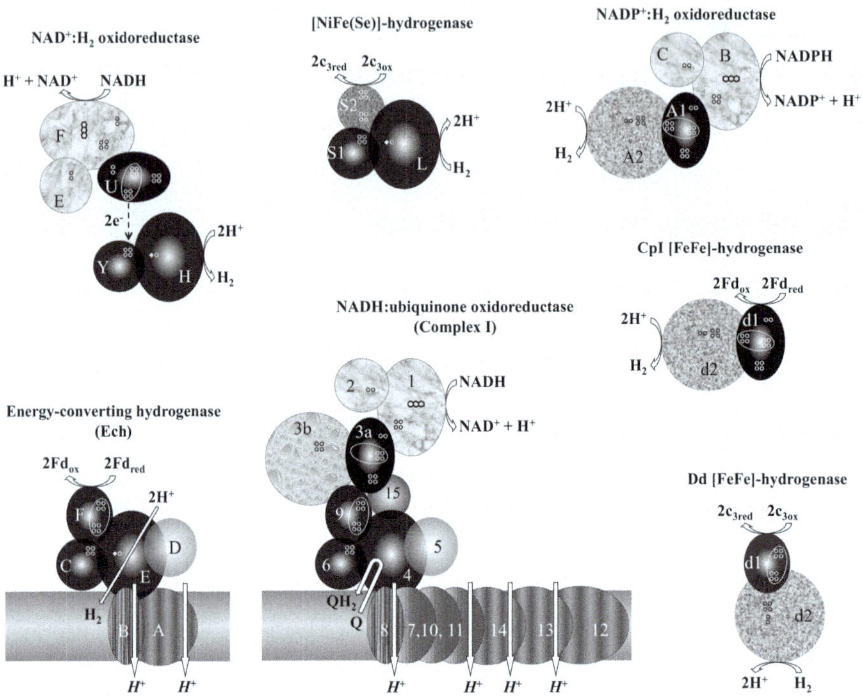

Fig. 6.1 Schematic depiction of complex I – hydrogenase relationships. The complete *T. thermophilus* complex I model, consisting of 15 Nqo subunits (Efremov et al. 2010), is surrounded, clockwise from *left to right*, by the energy converting (Ech) hydrogenase from *M. barkeri* (Hedderich 2004); the five Hox subunits of NAD(P)$^+$-reducing [NiFe]-hydrogenase from *Synechocystis* sp. strain PCC 6803 (Cournac et al. 2004); heterodimeric [NiFe(Se)] hydrogenase (Volbeda et al. 1995; Garcin et al. 1999); three subunits of NADP$^+$-reducing [FeFe]-hydrogenase from *T. maritima* (Verhagen et al. 1999); and two types of soluble [FeFe]-hydrogenases, from *Clostridium pasteurianum* (CpI) (Peters et al. 1998) and *Desulfovibrio desulfuricans* (Dd) (Nicolet et al. 1999). Each subunit/domain is represented by a specifically marked *ellipsoid*. The *gray rectangle* represents the membrane. FeS clusters are specified by *small circles*, Ni by a white diamond and FMN by three connected hexagons. *Small white ellipsoids* indicate ferredoxin-like domains containing up to two Fe_4S_4-clusters

the *Escherichia coli* formate hydrogenlyase operon (Böhm et al. 1990; Fearnley and Walker 1992), one of which corresponds to the enzyme's large subunit. In addition, the 75- and 51-kDa subunits of bovine mitochondrial complex I were shown to be related to the two diaphorase subunits of the soluble NAD$^+$-reducing hydrogenase from *Ralstonia eutropha* (Pilkington et al. 1991). Other significant similarities are found between complex I and the N-terminal flavodoxin-like domain of the [NiFe]-hydrogenase small subunit (Albracht 1993), two diaphorase subunits of the heterotetrameric NADP-reducing *Desulfovibrio fructosovorans* [FeFe] hydrogenase, and the N-terminal region of its H-cluster-containing subunit (Malki et al. 1995; Albracht and De Jong 1997).

As could be expected, similarities of complex I to many other hydrogenases have subsequently been noted (Fig. 6.1). Examples are: the heterotrimeric [FeFe]-hydrogenase from *Thermotoga maritima* (Verhagen et al. 1999), the heteropentameric

Table 6.1 Nomenclature of related complex I and hydrogenase subunits (Adapted from Fontecilla et al. 2007)

NAD^+:Q oxidoreductase[a]	Ech hydrogenase[b]	H_2-NAD(P)$^+$ oxidoreductase[c]	H_2-NADP$^+$ oxidoreductase[d]	Cofactors/reactants, special features
Nqo1-NuoF-51K		HoxF	HydB	NADH, FMN, $[Fe_4S_4]$
Nqo2-NuoE-24K		HoxE	HydC	$[Fe_2S_2]$
Nqo3-NuoG-75K		HoxU	HydA	$[Fe_2S_2]$, $2[Fe_4S_4]$ [e]
Nqo4-NuoD-49K	E	HoxH		Quinone or [NiFe]
Nqo5-NuoC-30K	D			
Nqo6-NuoB-PSST	C	HoxY		$[Fe_4S_4]$
Nqo9-NuoI-TYKY	F			$2[Fe_4S_4]$
Nqo15[f]				
Nqo7-NuoA-ND3				Membrane-bound
Nqo8-NuoH-ND1	B			Membrane-bound
Nqo10-NuoJ-ND6				Membrane-bound
Nqo11-NuoK-ND4L				Membrane-bound
Nqo12-NuoL-ND5	A			Membrane-bound
Nqo13-NuoM-ND4	A			Membrane-bound
Nqo14-NuoN-ND2	A			Membrane-bound

[a] Complex I subunits in *T. thermophilus*, *E. coli* and bovine mitochondrion (the latter has at least 32 other subunits) (Yagi and Matsuno-Yagin 2003)
[b] *M. barkeri* [NiFe]-hydrogenase (Hedderich 2004)
[c] *Synechocystis* [NiFe]-hydrogenase (Cournac et al. 2004)
[d] *T. maritima* [FeFe] hydrogenase (Verhagen et al. 1999)
[e] In HydA these clusters are located in the N-terminal domain, the C-terminal domain containing the active site H-cluster (Malki et al. 1995)
[f] This subunit is restricted to thermophiles (Sazanov and Hinchliffe 2006)

NAD(P)$^+$-reducing hydrogenase from *Synechocystis* sp. strain PCC 6803 (Cournac et al. 2004), and the membrane-bound multisubunit [NiFe]-hydrogenases. The latter have six conserved core subunits, which are related to complex I (Friedrich and Scheide 2000; Hedderich 2004). Representatives of this class include hydrogenases 3 and 4 of *E. coli*, a CO-induced hydrogenase of *Rhodospirillum rubrum* that forms a complex with carbon monoxide dehydrogenase, and Ech hydrogenases from archaea. Like complex I, these enzymes function as ion pumps. Proton translocation is also used by the archaeal $F_{420}H_2$ dehydrogenase complex, where 11 of the 13 subunits are homologous to complex I subunits (Deppenmeier 2004). Instead of NAD(P)H as hydride donor the archaeal enzyme uses $F_{420}H_2$ and the electron acceptor is methanophenazine instead of ubiquinone. Related complex I and [NiFe]-hydrogenase subunits are shown in Table 6.1.

6.2 Origin of the Homologies Between Complex I and FeNi-Hydrogenases

The evolutionary relationship between complex I and [NiFe]-hydrogenases is intimately connected to energy metabolism and to atmospheric oxygen concentration levels. The initial step in the process leading to complex I is postulated to have been the association of a soluble heterodimeric reversible [NiFe]-hydrogenase with membrane-bound K^+/H^+ or Na^+ antiporters (Friedrich and Weiss 1997; Friedrich and Scheide 2000; Mathiesen and Hägerhäll 2003). Subsequent acquisition of subunits with additional redox properties would have transformed this ancestral enzyme in a proton-pumping hydrogen:quinone oxidoreductase. This protein would have been the common ancestor of bacterial complex I, formate hydrogenlyase, archeal $F_{420}H_2$:quinone reductase and contemporary membrane-bound Ech hydrogenases. The fact that the small and large subunits of the latter show a higher degree of sequence identity to their complex I counterparts than to the soluble [NiFe]-hydrogenases (Friedrich and Weiss 1997) supports this evolutionary model. Subsequent multiplication and/or recruitment of antiporter genes and addition of other subunits, along with loss of hydrogenase activity and gain of diaphorase activity, would have given rise to complex I. The quinone-binding site is located in the *Thermus thermophilus* complex I Nqo4 subunit and is topologically equivalent to the ancestral hydrogen-metabolizing NiFe active site (Fig. 6.2). Also, highly conserved Glu and Asp residues in both the small membrane-bound subunit NuoK from *E. coli*'s complex I and the [NiFe]-hydrogenase large subunit HycE (Magalon and Böck 2000) from membrane-bound Ech hydrogenases (see Table 6.1 for nomenclature) may be of either functional or structural relevance (Mathiesen and Hägerhäll 2003). The association of a soluble [NiFe]-hydrogenase with membrane-bound antiporter subunits may have been caused by the need of highly reducing electrons in several key reactions. It has been postulated that hydrogen may have been very abundant in the primordial atmosphere (Tian et al. 2005) explaining the early appearance of soluble [NiFe]-hydrogenases. However, this gas can easily escape to space and its atmospheric concentration may have dropped rather rapidly at the geological scale. At present, and because of the typically very low hydrogen concentration in the atmosphere, the coupling of its oxidation to CO_2 reduction is endergonic, requiring an additional energy input to proceed. There is evidence that in methanogens this process is driven by an electrochemical proton or sodium ion gradient (Kaesler and Schönheit 1989; Winner and Gottschalk 1989; Hedderich 2004). The coupling of thermodynamically unfavorable redox reactions to the consumption of a membrane-associated ion gradient is often referred to as reverse electron transport. In this case, the Ech hydrogenase functions as the ion pump (Meuer et al. 2002). Unlike complex I, these enzymes do not use quinones in their electron transfer reactions (Hedderich 2004). Reverse electron transport coupled to CO_2 reduction may have been the selective advantage that favored the formation of the hydrogenase:antiporter complex.

When and for what use did complex I and the rest of the respiratory chain evolve? This question is intimately related to the presence of molecular oxygen in Earth's early atmosphere. A commonly accepted view has been that aerobic respiration evolved after the release of large amounts of oxygen by photosynthetic cyanobacteria (Castresana and Moreira 1999). This great oxidation event (GOE) took place some 2.3 billion years ago (Konhauser et al. 2009). However, cytochrome oxidase, the key enzyme of aerobic metabolism that catalyzes the reduction of oxygen to water, is present in both bacteria and archaea. Assuming the absence of horizontal gene transfer, this suggests that the last universal common ancestor (LUCA) may

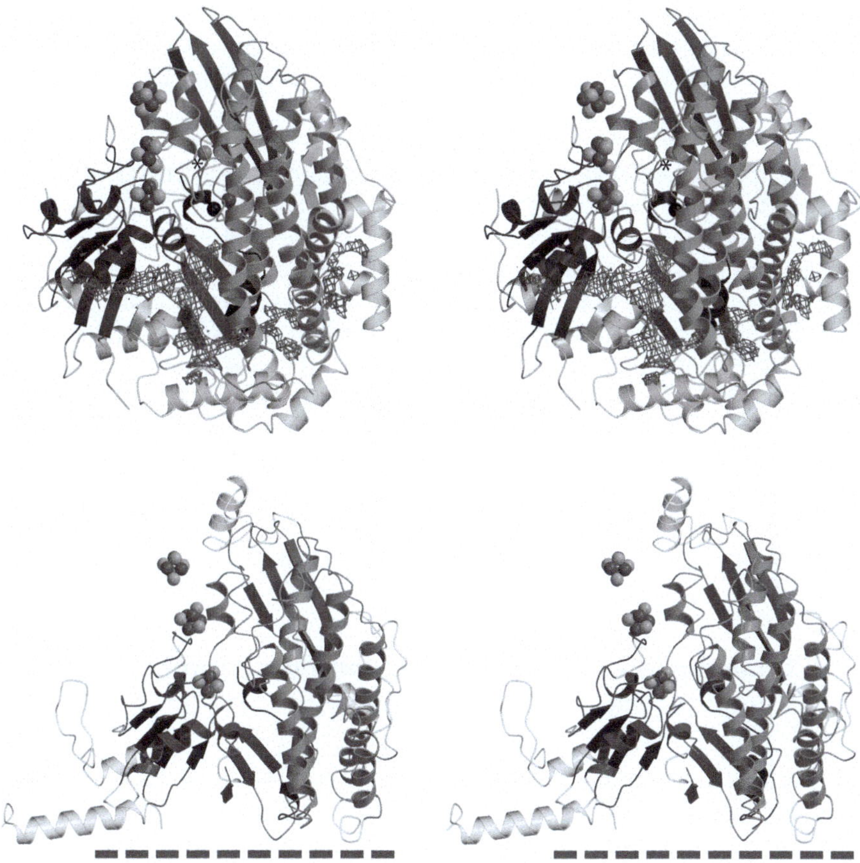

Fig. 6.2 Stereo pair of *Df* [NiFe]-hydrogenase (*top*) and relevant regions of the complex I Nqo4 and Nqo6 subunits (*bottom*). Arrows depict β-strands, ribbons $\alpha/3_{10}$-helices, whereas spheres denote Ni (*black*), Fe (*dark gray*) and S (*light gray*) atoms. An *asterisk (*) indicates* a metal (assigned to Mg or Fe) bound to the C-terminal histidine residue of the hydrogenase large subunit. The two upper FeS clusters in both structures are shown without their protein environment, i.e. the C-terminal domain of the hydrogenase small subunit and the non-homologous Nqo9 subunit of complex I. The hydrophobic tunnel in the hydrogenase is shown as chicken wire. The *dashed line* represents the membrane surface

have used an aerobic respiratory chain under certain conditions. Because LUCA is supposed to be older than two billion years, aerobic respiration must have predated the GOE. Traces of atmospheric oxygen, generated by the photolysis of water, may have existed before the emergence of oxygenic photosynthesis, allowing for the existence of localized, relatively oxygen-rich environments that would have favored the evolution of aerobic respiration (Kasting 1993). Even in the contemporary atmosphere, oxygen is 2.3% heavier by isotope composition than photosynthetic oxygen, due to the fraction of abiogenically-produced gas (Samuilov 2000). In spite of the aggressive nature of oxygen species such as singlet oxygen, superoxide, peroxide and the hydroxyl radical, there was probably significant evolutionary pressure to use oxygen as the final electron acceptor given the very positive potential of the O_2/H_2O couple. This, in turn, allows for a more efficient energy extraction from low-potential electrons.

Martin and Muller have proposed that the emergence of the eukaryotic cell resulted from the association of an archaeon and an α-proteobacterium that provided the former with hydrogen from the latter (Martin and Mueller 1998). However, this view has been contested because there is no genetic evidence for a hydrogen-metabolizing α-proteobacterium (Kurland and Andersson 2000). Instead, it has been proposed that the eukaryotic cell arose from the association of an archaeon and an α-proteobacterium endowed with hydrogen-based metabolism and aerobic respiration, respectively (Kurland and Andersson 2000). There is ample evidence that the α-proteobacterium became the mitochondrion (Kurland and Andersson 2000). It has been postulated that, thanks to its respiratory chain, the α-proteobacterial symbiont afforded protection to the anaerobic archaeal host by reducing oxygen to water. Thus in this model, aerobic respiration evolved very early for energy reasons and was later instrumental in the emergence of the eukaryotic cell. As the first component of the respiratory chain, complex I played a central role in this process.

6.3 Contemporary Structural Relationships Between Complex I and FeNi-Hydrogenases

The 3.3 Å resolution crystal structure of the hydrophilic domain of *T. thermophilus* complex I (Sazanov and Hinchliffe 2006) has shown that its Nqo3, 4, 6 and 9 subunits have very similar three-dimensional structures as [NiFe]- and [FeFe]-hydrogenase subunits (Table 6.2, Figs. 6.1 and 6.2), as it was expected from amino acid sequence homologies. As depicted in Fig. 6.1, seven evenly spaced FeS clusters form an electron transfer pathway between the FMN of the Nqo1 diaphorase active site and the ubiquinone-reducing site in Nqo4. The distal [Fe_2S_2] cluster in Nqo2, close to the FMN, may function as an antioxidant (Hinchliffe and Sazanov 2005). An additional FeS cluster in the Nqo3 C-terminal domain is too far away to participate in fast electron transfer. However, in some other species this cluster may be involved in electron transfer with an external redox partner if, as predicted by sequence similarities with the medial [Fe_4S_4] cluster ligands in [FeFe]-hydrogenase, there is an additional intervening [Fe_4S_4] cluster (Nicolet et al. 2000).

The complete structure of complex I has now been reported at 4.5 Å resolution (Efremov et al. 2010). Nqo12 and Nqo8, which are two of the seven hydrophobic membrane-bound subunits of complex I, are homologous to *Methanosarcina barkeri* Ech hydrogenase subunits A and B, respectively (Table 6.1, Fig. 6.1). Nqo12 and Nqo8 are also homologous to bacterial K^+ or Na^+/H^+ antiporters, in agreement with their proton pumping function (Sazanov and Hinchliffe 2006). In addition, Nqo12 is homologous to Nqo13 and Nqo14 (Efremov et al. 2010). The similarities between complex I and NAD^+-reducing [NiFe]-hydrogenases have been studied using

Table 6.2 Structural superposition of complex I subunits to various hydrogenases

Complex I subunit	pdb code	Residue range	Hydrogenase	pdb code	Residue range	Nseg[a]	NCα[b]	Δrms[c] (Å)	Identity (%)
Nqo3	3iam	1–239	FeFe-CpI	3c8y	1–209	8	179	1.94	24.0
Nqo3	3iam	1–240	FeFe-Dd	1hfe	L26-86	4	60	2.24	20.0
Nqo46 (2 subunits)	3iam	D26-409	NiFe-Df (2 subunits)	1yqw	L14-549	23	389	2.31	19.0
		F36-152			S7-159				
Nqo46 (2 subunits)	3iam	D26-409	NiFeSe-Db (2 subunits)	1cc1	L12-498	23	380	2.12	18.9
		F36-152			S8-176				
Nqo9	3iam	39–116	FeFe-CpI	3c8y	138–209	5	54	1.63	29.6
Nqo9	3iam	39–117	FeFe-Dd	1hfe	L26-85	3	57	1.92	36.8

[a] Superimposed peptide fragments
[b] Superimposed Cα positions
[c] Root mean square deviation

the *A. vinosum* HoxEFUYH-type enzyme (Long et al. 2007). According to its amino acid sequence, this enzyme should contain two additional FeS clusters. The one in HoxU corresponds to the second [4Fe-4S] cluster of the ferredoxin-like domain of Nqo3, which is also found in Nqo9. This ferredoxin-like domain could link the diaphorase with the hydrogenase moiety. In the case of $NADP^+$-reducing [FeFe]-hydrogenases, the same domain links the diaphorase to the H-cluster domain (Malki et al. 1995), like in the simpler soluble [FeFe]-hydrogenases of known structure (Peters et al. 1998; Nicolet et al. 1999).

The ubiquinone-binding Nqo4 subunit and the [4Fe-4S]-cluster-carrying Nqo6 subunit of complex I are to a great extent structurally homologous to the [NiFe(Se)]-hydrogenase large subunit and the N-terminal domain of its small subunit, respectively (Table 6.2, Figs. 6.2 and 6.4). These homologies, which include most of the secondary structure elements of the Nqo4 and Nqo6 subunits, are outstanding because the enzymes catalyze widely different reactions. [NiFe]-hydrogenase has a hydrophobic gas tunnel that runs between the small and the large subunit and connects the active site with the molecular surface. This tunnel coincides topologically with a wide funnel that faces the membrane surface in complex I (Fig. 6.2), and is probably the binding site for the ubiquinone substrate (Sazanov and Hinchliffe 2006). Albracht and coworkers have reported evidence for the presence of a second, labile FMN in bovine complex I (Albracht and De Jong 1997; Albracht and Hedderich 2000). This second FMN is proposed to bind to the complex I Nqo4 subunit, based on its similarity to the flavodoxin-like N-terminal domain of [NiFe]-hydrogenase. A labile FMN was also reported to be present, next to the diaphorase-linked FMN, in the soluble [NiFe]-hydrogenase of *R. eutropha* (Van der Linden et al. 2006). The presence or absence of a second FMN has important consequences for the functional understanding of complex I (Albracht 2010a). Its detection appears to be strongly dependent on the used purification protocol and is favored when conditions remain as close as possible to the natural environment of the bifunctional enzyme (Albracht 2010b). The wide funnel facing the membrane surface between subunits Nqo4 and Nqo6 might be large enough to allow the binding of both the ubiquinone substrate and FMN. However, the exact shape and size of this cavity cannot be ascertained because some of the protein regions in its vicinity are not resolved (Fig. 6.5). In any case, according to a structural superposition (not shown), the putative second FMN cannot bind as in flavodoxin, because this would cause collisions with other parts of the structure, including the bound [4Fe-4S] cluster.

A sequence alignment based on the structural superposition of the complex I Nqo4 and Nqo6 subunits to heterodimeric [NiFe]- and [NiFeSe]-hydrogenases reveals a significant number of conserved residues (Figs. 6.3 and 6.4). These are more or less evenly distributed over the entire polypeptide chain, with a slightly higher conservation located in the immediate environment of the Ni-Fe active site and the proximal [4Fe-4S] cluster. Most of the 13 invariant glycines are positioned just before or after β-strands or $\alpha/3_{10}$-helices, probably allowing main chain conformations that would not be possible with other amino acids. As expected, most of the conserved hydrophobic residues are located in hydrophobic internal regions. Three of the four cysteine ligands of the hydrogenase proximal [4Fe-4S] cluster are conserved in the Nqo6 subunit, whereas the closest one to the N-terminus is at a

```
prop .......... .......... ...735.... .......... .......... .6.9...... .
HdfL -----ptpqs-tftgpivvdp-itrieghlri-mvevengkvk-dawsssql-frgLEIILkgrd  60
Nqo4 -------------vmtlnvgpqhpsthgvlrl-mvtlsgeevl-evvphigylhtgFEKTMEHrt  74
HdbL --------------vkisidp-ltrveghlki-evevkdgkvv-dakcsggm-frgFEQIlrgrd  58

prop .......988 5033068383. 97........ .......... .......7..6 7..47..9..
HdfL PRDAQHFTQr-acgvctyVHA-LASSRCVDDA-VkvsipANAR-MMRNLVMASQ-YLHDHLVHFY 120
Nqo4 ylQNITyTPR-mdylhSFAHD-LAYALAVEKL-lgavvpPRAE-TIRVILNELS-RLASHLVFLG 134
HdbL PRDSSQIVQR-IcgvcPTAHC-TASVMAQDDA-FgvkvtTNGR-ITRNLIFGAN-YLQSHILHFY 118

prop .8........ .......... .......... .......... .......... ..........
HdfL HlhALDwvdv-TAALKadPNK-AAKLAAsiap-arPGNsAKAL-KAVQDKLKAF-VEsgqlGIFt 180
Nqo4 -TGLldlg---------------------------------------------------------- 141
HdbL HlaALDyvkgp-------------------------------------------------dvspfv 135

prop .......... .......... .......... .......9. .....97.. ..........
HdfL nayflgghka-yylpPEVNLI-ATAHYLEALH-MQVKAASAMA-ILGgknphtq-ftvvggcsnY 240
Nqo4 ----------------------alTPFFYAFR-ERETILDLFE-WVTgqrfhhn-yiriggvked 181
HdbL prya-*-ikd-GAKADATNTY-GLNQYLKALE-IRRICHEMVA-MFGgrmphvq-gmvvggatei 180

     .......... .......... .......... .......... .........9.9......
HdfL QGLtKDPLAN-YLALSKEVCQ-FVNEcYIPDL-LAVAGFyKDW-GGiggt-snylafgefatdds 300
Nqo4 l-pe-efVPE-LKKllevLPH-RIDE-YEALF-AespiFYERA-rgvgvipPE------------ 226
HdbL p-tA-DKVAE-YAARFKEVQK-FVIEeYLPLI-YTLGSVyTDL-Fetgigwknviafgvfpeddd 259

     .......... .......... .......... .......... .......... ..........
HdfL sPSKHLATsq-fpsgvitgrd-lgkvdnvDLG-aiyedvkyswya-pggdgkhP-YDgvtdpkyt 360
Nqo4 ------------------------------------------------------------------
HdbL ----y-ktfl-lkpgvyidg----kdeefdSK-LvkeyvghsffdhsapgglhY-SVgetnpnpd 312

     .......... 7......... ..677..7.. .......... .......... ..........
HdfL klddkdhysw-mkaprykgka-mevgPLARTF-IAYAKgqPDF-KKVVDMVLGK-LsvpATALhs 420
Nqo4 ----------------VAIDL-gltggslRAS--gvnydVRKA-ypysgYETyt-fdvplgergD 271
HdbL --kp-gaysf-vkaprykdkp-cevgPLARMW-VQnpelsPVG-QKLLKELY**-FRDLGDKAfs 373

     ........8.. 8......... .......... .......... .........9 .78333483 9
HdfL TLGRTAARGI-ETAIVCANME-KWIKEMADSG-AKdntlcakw-empeeskgvg-ladaprgals 480
Nqo4 VFDRMLVRIR-EMRESVKIIK-QALERl------epgpvrdpn**hppkgevyv-ptesargelg 354
HdbL IMGRHVLRAE-ETWLTAVAVE-KWLKQv-----qpgaetyvks-eipdaaegtg-fteaprgall 429

     6....... ......793336 8669...... .......... .......... .........8
HdfL hwirikgk-kidnfqlvvPAT-WNLgprgaqg-dksPVEEALi-gtpiadpkrp-VEILRTVHAf 540
Nqo4 yyivsdggsmpyrvkvraPSF-Vnl-----------qsLPYAck-geq------v-pDMVAIIAsl 400
HdbL hylkik-dkkienyqivsatl-wnanprddmg-qrgPIEEALi-gvpvpdiknp-VNVGRLVRSy 489

     670640497
HdfL dpcIACGVH 549
Nqo4 dpvMGDVDR 409
HdbL dpuLGCAVH 498
```

Fig. 6.3 Structure-based sequence alignment of the complex I Nqo4 subunit with the large subunits of *Df* [NiFe]-hydrogenase (HdfL) and of *Db* [NiFeSe]-hydrogenase (HdbL). *Skew underlined and capital letters* indicate a location in β-strands and α– (or 3_{10}-) helices, respectively. The *property (prop) line* shows nearest distances <10 Å to the NiFe-site (including the metal ligands). *Underlinings* indicate a distance <5 Å from any residue of HdfS. The four-helix bundle is highlighted with *gray bars*

different position. The most C-terminal Cys-ligand is located in an invariant PGCPPxP motif, with the second proline being in a *cis* conformation. Another *cis*-proline in the hydrogenase large subunit, P20L in HdfL (Fig. 6.3), corresponds to the first residue of a short disordered segment in Nqo4, located in the putative

```
prop ......... 7.88650530 7757...... ............ ........... ..........
HdfS --akhrpsvv-wlhnaectGC-TEAAirtikp-yIDALILDti-sldyqetima-aaGEAAEAAL  60
Nqo6 *GRSNslwpa-tfgl---acC-AIEMmas-----tda---------------------------  56
HdbS ----kkapvi-wvqgqgctGC-SVSLLnavhp-rIKEILLDvi-slefhptvma-seGEMALAHM  61

prop .......... ...89444... ............ ......... ...    .....9843506884
HdfS HQALegk-dgyylvveggl pt-IDGGqwgmva-ghpMIETTKK-AAA---kakgiicigTCSAYg 118
Nqo6 ---#74--qadvmivagrlsk--------------kmAPVMRR-VWEqmpdpkwvismGACASsg 115
HdbS YEIAEKfngnfflIveg aipt-AKEGrycivg*evtMMELIRD-LAp---kslatvavgTCSAYg 130

prop 34379.......8 .89.....  ... ..784033. 9..7...... ...
HdfS gvqkakpnpsqa-kgVSEALg--vkt-inipgcppn-PINFVGAVVH-VLT 163
Nqo6 gmfn---nyaiv-qnvdsvv---pvd-vyvpgcppr-PEALIYAVMQ-LQK 156
HdbS gipaaegnvtgs-ksVRDFFA*ekll-vnvpgcpph-PDWMVGTLVA-AWS
```

Fig. 6.4 Structure-based sequence alignment of the complex I Nqo6 subunit with the flavodoxin-like domains of the small subunit of *Desulfovibrio fructosovorans* [NiFe]-hydrogenase (HdfS) and of *Desulfomicrobium baculatum* [NiFeSe]-hydrogenase (HdbS). The *property (prop) line* above HdfS shows nearest distances <10 Å to the proximal FeS-cluster and its S-ligands. *Underlinings* indicate a distance <5 Å from any residue of the large subunit (HdfL)

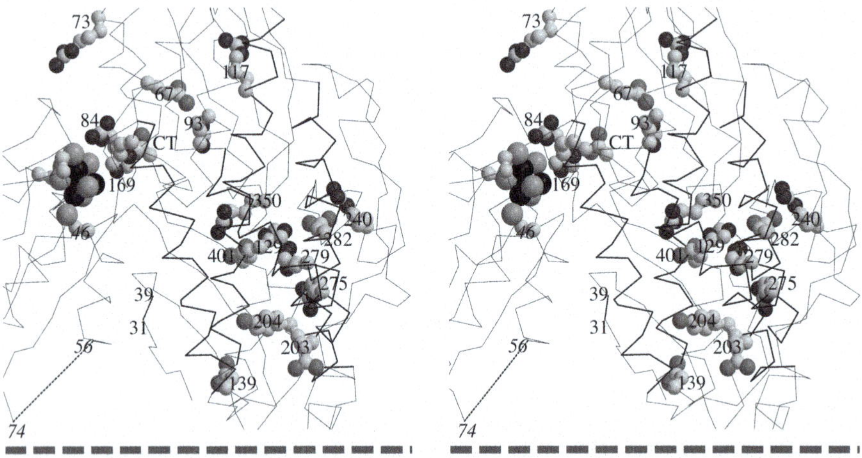

Fig. 6.5 Stereoscopic zoom of the NiFe-hydrogenase-like subunits of complex I (pdb code 3IAM), highlighting conserved Arg (73, 84, 117, 240, 275, 279 and 350), Asp (139 and 401), His (93, 129 and 169), Cys (*46*), Glu (67 and 282) and Tyr (204) residues. CT indicates the conserved C-terminal carboxylate moiety. The four-helix bundle is shown with *thicker lines*. Atoms are shown as large spheres for Fe (*black*) and S (*light-gray*) and smaller ones for C (*white*), N (*dark gray*) and O (*intermediate gray*)

quinone-binding funnel region. Furthermore, a remarkably large number of hydrophilic amino acids is conserved and most of these are buried in the protein (Fig. 6.5). It is tempting to speculate that these are somehow involved in proton transfer.

The two-electron reduction of the ubiquinone substrate of complex I is accompanied by its double protonation that is expected to involve two spatially independent proton transfer pathways. At least one additional pathway is required for proton pumping. Complex I and [NiFe]-hydrogenase might have similar pathways for proton transfer (Albracht and Hedderich 2000). Although the improved 3.1 Å resolution of the hydrophilic domain structure of complex I (Berrisford and Sazanov 2009) does not allow the modeling of water molecules, there may be a hydrophilic cavity between the conserved *Asp401* and *Asp139* (Fig. 6.5). Hydrogenase does not have such a cavity because the corresponding space is occupied by three partially conserved histidine residues. The Asp residue corresponding to *Asp401* has been proposed to be involved in proton transfer in [NiFe]-hydrogenase (Matias et al. 2001). *Asp139* faces the membrane surface, where it may interact with proton-translocating subunits. Domain movements in this region could help proton transfer.

In complex I and Ech hydrogenases protons are supposed to transit through a pathway connected to the cytoplasm and depart through a pathway connected to the membrane. To avoid proton leakage these pathways should not be active simultaneously. Consequently, their function needs to be regulated, probably through a redox change, which in turn would trigger a conformational change in the protein structure (Hosler et al. 2006). [NiFe]-hydrogenase has three kinked α-helices in the long four-helix bundle of the large subunit (Fig. 6.5). A change of these kinked helices into a straighter conformation would separate two structural domains, generating a hydrophilic cavity between *Asp528* and *Asp129*, as seen between the corresponding *Asp401* and *Asp139* in complex I. This might facilitate proton transit towards the membrane.

Other pathways may also require domain movements. For instance, the invariant large subunit *His219* and *Arg63* of [NiFe]-hydrogenase (corresponding to *His169* and *Arg84* in complex I, see Fig. 6.3) connect the proximal cluster to the solvent medium through a water network found at the interface between the small subunit C-terminal and flavodoxin-like domains. The former is replaced by a ferredoxin-like subunit in complex I. It has been shown that the pH dependence of the $[Fe_4S_4]$ cluster midpoint potential in complex I is modulated by *His169* (Zwicker et al. 2006). Its mutation to methionine removed this pH dependence but had no effect on the proton pumping efficiency. Therefore, proton pumping does not seem to depend on the redox state of this cluster and the various redox and protonation states of the ubiquinone substrate in complex I probably control proton flow. A similar mechanism is likely to operate in the Ni-Fe active site of hydrogenase.

6.4 Conclusions

There is now strong evidence that complex I and of [NiFe]-hydrogenases are phylogenetically related. Furthermore, extant energy-converting hydrogenases provide the basis for a sequential process where the initial step was the formation of a complex between a soluble hydrogenase and proton-pumping, membrane-bound antiporters. Several lines of evidence suggest that aerobic respiration evolved long before oxygen became a major component of Earth's atmosphere. In this context it is worth noting that most likely both complex I and [NiFe]-hydrogenase were actors in the emergence of the eukaryotic cell.

References

Albracht SPJ (1993) Intimate relationships of the large and the small subunits of all nickel hydrogenases with two nuclear-encoded subunits of mitochondrial NADH:ubiquinone oxidoreductase. Biochim Biophys Acta 1144:221–224

Albracht SPJ (2010a) The reaction of NADPH with bovine mitochondrial NADH:ubiquinone oxidoreductase revisited. I. Proposed consequences for electron transfer in the enzyme. J Bioenerg Biomembr 42:261–278

Albracht SPJ (2010b) The reaction of NADPH with bovine mitochondrial NADH:ubiquinone oxidoreductase revisited. II. Comparison of the proposed working hypothesis with literature data. J Bioenerg Biomembr 42:279–292

Albracht SPJ, De Jong AMP (1997) Hypothesis: bovine-heart NADH:ubiquinone oxidoreductase is a monomer with 8 Fe-S clusters and 2 FMN groups. Biochim Biophys Acta 1318:92–106

Albracht SP, Hedderich R (2000) Learning from hydrogenases: location of a proton pump and of a second FMN in bovine NAD-ubiquinone oxidoreductase (complex I). FEBS Lett 485:1–6

Berrisford JM, Sazanov LA (2009) Structural basis for the mechanism of respiratory complex I. J Biol Chem 284:29773–29783

Böhm R, Sauter M, Böck A (1990) Nucleotide sequence and expression of an operon in *Escherichia coli* coding for formate hydrogenlyase components. Mol Microbiol 4:231–243

Castresana J, Moreira D (1999) Respiratory chains in the last common ancestor of living organisms. J Mol Evol 49:453–460

Cournac L, Guedeney G, Peltier G, Vignais PM (2004) Sustained photoevolution of molecular hydrogen in a mutant of *Synechocystis* sp. strain PCC6803 deficient in the type I NAPDH-dehydrogenase complex. J Bacteriol 186:1737–1746

Deppenmeier U (2004) The membrane-bound electron transport system of *Methanosarcina* species. J Bioenerg Biomembr 36:55–64

Efremov RG, Baradaran R, Sazanov LA (2010) The architecture of respiratory complex I. Nature 465:441–447

Fearnley IM, Walker JE (1992) Conservation of sequences of subunits of mitochondrial complex I and their relationships with other proteins. Biochim Biophys Acta 1140:105–134

Fontecilla-Camps JC, Volbeda A, Cavazza C, Nicolet Y (2007) Structure/function relationships of [NiFe]- and [FeFe]-hydrogenases. Chem Rev 107:4273–4303

Friedrich T, Scheide D (2000) The respiratory complex I of bacteria, archaea and eukarya and its module common with membrane-bound multisubunit hydrogenases. FEBS Lett 479:1–5

Friedrich T, Weiss H (1997) Modular evolution of the respiratory NADH: ubiquinone oxidoreductase and the origin of its modules. J Theor Biol 187:418–439

Garcin E, Vernede X, Hatchikian EC, Volbeda A, Frey M, Fontecilla-Camps JC (1999) The crystal structure of a reduced [NiFeSe] hydrogenase provides an image of the activated catalytic center. Structure 7:557–566

Hedderich R (2004) Energy-converting [NiFe] hydrogenases from archaea and extremophiles: ancestors of complex I. J Bioenerg Biomembr 36:65–75

Hinchliffe P, Sazanov LA (2005) Organization of iron-sulfur clusters in respiratory complex I. Science 309:771–774

Hosler JP, Ferguson-Miller S, Mills DA (2006) Energy transduction: proton transfer through the respiratory complexes. Annu Rev Biochem 75:165–187

Kaesler B, Schönheit P (1989) The sodium cycle in methanogenesis. CO_2 reduction to the formaldehyde level in methanogenic bacteria is driven by a primary electrochemical potential of Na^+ generated by formaldehyde reduction to CH_4. Eur J Biochem 186:309–316

Kasting JF (1993) Earth's early atmosphere. Science 259:920–926

Konhauser KO, Pecoits E, Lalonde SV, Papineau D, Barley ME, Arndt NT, Zahnle K, Kamber BS (2009) Oceanic nickel depletion and a methanogen famine before the Great Oxidation Event. Nature 458:750–751

Kurland CG, Andersson SGE (2000) Origin and evolution of the mitochondrial proteome. Microbiol Mol Biol Rev 64(4):786–820

Long M, Liu J, Chen Z, Bleijlevens B, Roseboom W, Albracht SP (2007) Characterization of a HoxEFUYH type of [NiFe] hydrogenase from *Allochromatium vinosum* and some EPR and IR properties of the hydrogenase module. J Biol Inorg Chem 12:62–78
Magalon A, Böck A (2000) Analysis of the HypC-HycE complex, a key intermediate in the assembly of the metal center of the *Escherichia coli* hydrogenase 3. J Biol Chem 275:21114–21120
Malki S, Saimmaime I, De Luca G, Rousset M, Dermoun Z, Belaich J-P (1995) Characterization of an operon encoding an NADP-reducing hydrogenase in *Desulfovibrio fructosovorans*. J Bacteriol 177:2628–2636
Martin W, Mueller M (1998) The hydrogen hypothesis for the first eukaryote. Nature 392:37–41
Mathiesen C, Hägerhäll C (2003) The 'antiporter module' of respiratory chain complex I includes the MrpC/NuoK subunit – a revision of the modular evolution scheme. FEBS Lett 549:7–13
Matias PM, Soares CM, Saraiva LM, Coelho R, Morais J, Le Gall J, Carrondo MA (2001) [NiFe] hydrogenase from *Desulfovibrio desulfuricans* ATCC 27774: gene sequencing, three-dimensional structure determination and refinement at 1.8 Å and modelling studies of its interaction with the tetrahaem cytochrome c_3. J Biol Inorg Chem 6:63–81
Meuer J, Kuettner HC, Zhang JK, Hedderich R, Metcalf WW (2002) Genetic analysis of the archaeon *Methanosarcina barkeri* Fusaro reveals a central role for Ech hydrogenase and ferredoxin in methanogenesis and carbon fixation. Proc Natl Acad Sci USA 99:5632–5637
Nicolet Y, Piras C, Legrand P, Hatchikian CE, Fontecilla-Camps JC (1999) *Desulfovibrio desulfuricans* iron hydrogenase: the structure shows unusual coordination to an active site Fe binuclear center. Structure 7:13–23
Nicolet Y, Lemon BJ, Fontecilla-Camps JC, Peters JW (2000) A novel FeS cluster in Fe-only hydrogenases. Trends Biochem Sci 25:138–143
Peters JW, Lanzilotta WN, Lemon BJ, Seefeldt LC (1998) X-ray crystal structure of the Fe-only hydrogenase (CpI) from *Clostridium pasteurianum*. Science 282:1853–1858
Pilkington SJ, Skehel JM, Gennis RB, Walker JE (1991) Relationship between mitochondrial NADH-ubiquinone reductase and a bacterial NAD-reducing hydrogenase. Biochemistry 30:2166–2175
Samuilov VD (2005) Energy problems in life evolution. Biochemistry (Moscow) 70:246–250
Sazanov LA, Hinchliffe P (2006) Structure of the hydrophilic domain of respiratory complex I from *Thermus thermophilus*. Science 311:1430–1436
Tian F, Too OB, Pavlov AA, De Sterck H (2005) A hydrogen-rich early earth atmosphere. Science 308:1014–1017
Van der Linden E, Burgdorf T, de Lacey AL, Buhrke T, Scholte M, Fernandez VM, Friedrich B, Albracht SPJ (2006) An improved purification procedure for the soluble [NiFe]-hydrogenase of *Ralstonia eutropha*: new insights into its (in) stability and spectroscopic properties. J Biol Inorg Chem 11:247–260
Verhagen MF, O'Rourke T, Adams MW (1999) The hyperthermophilic bacterium, *Thermotoga maritima*, contains an unusually complex iron-hydrogenase: amino acid sequence analyses versus biochemical characterization. Biochim Biophys Acta 1412:212–229
Volbeda A, Charon MH, Piras C, Hatchikian EC, Frey M, Fontecilla-Camps JC (1995) Crystal structure of the nickel-iron hydrogenase from *Desulfovibrio gigas*. Nature 373:580–587
Winner C, Gottschalk G (1989) H_2 and CO_2 production from methanol or formaldehyde by the methanogenic bacterium strain Gö1 treated with 2-bromoethanesulfonic acid. FEMS Microbiol Lett 65:259–264
Yagi T, Matsuno-Yagin A (2003) The proton-translocating NADH-quinone oxidoreductase in the respiratory chain: the secret unlocked. Biochemistry 47:2266–2274
Zwicker K, Galkin A, Dröse S, Grgic L, Kerscher S, Brandt U (2006) The redox-Bohr group associated with iron-sulfur cluster N2 of complex I. J Biol Chem 281:23013–23017

Chapter 7
Recruitment of the Antiporter Module – A Key Event in Complex I Evolution

Vamsi Krishna Moparthi and Cecilia Hägerhäll

Abstract It has long been known that NADH:quinone oxidoreductase or complex I is composed of proteins homologous to other smaller enzymes, electron carriers and transporters. By understanding the recruitment process of each of the smaller functional units, their individual functions and at what stage of complex I building the different functions were needed, we can also better understand the current functional mechanism of this large and intricate molecular machine. This work is focused on the role of the ion translocation proteins, important components of the energy coupling mechanism. We postulate that only one ion transporter protein is actually used for pumping protons in present day complex I, whereas the other two ion transporter proteins form a unit that is used to modulate the output from the proton pumping engine. When recruited to primordial complex I, the modulator unit allowed the proton pumping engine to adapt the stoichiometric output depending on the amount of energy available in different redox couple substrates. This critical step made the emerging complex I a vastly more useful and versatile enzyme, ensuring its evolutionary success. Today, when many members of the complex I family reside in more constant environments and need to use only a limited number of substrates, the most important function of the modulator unit is to function as an antiporter, regulating which type of ion gradient is maintained across the membrane.

Keywords Mrp antiporter • Ion transport • Proton pumping • Protein evolution • NuoL • NuoM • NuoN

V.K. Moparthi • C. Hägerhäll (✉)
Department of Biochemistry and Structural Biology,
Center for Molecular Protein Science, Lund University,
Box 124, 22100 Lund, Sweden
e-mail: Cecilia.Hagerhall@biochemistry.lu.se

L. Sazanov (ed.), *A Structural Perspective on Respiratory Complex I: Structure and Function of NADH:ubiquinone oxidoreductase*, DOI 10.1007/978-94-007-4138-6_7,

7.1 An Introduction to Complex I and Complex I Evolution

The classical bacterial NADH:quinone oxidoreductase or complex I is composed of 14 protein subunits, making up an intricate, energy conserving bioenergetic device, capable of both NADH driven proton pumping and $\Delta\mu H^+$ supported NAD^+ reduction (Friedrich 1998; Yagi et al. 1998). In addition to these core subunits, that can be regarded as the actual machine parts, the mammalian complex I contains 31 additional, accessory subunits (Brandt 2006). Most enzyme complexes in the mitochondrial respiratory chain have such accessory subunits, albeit not as many as complex I (Saraste 1999; Carroll et al. 2006). Seven of the core subunits are soluble proteins, harbor FMN and iron-sulfur clusters, and protrude into the bacterial cytoplasm or mitochondrial matrix. The remaining seven subunits are membrane spanning, and lack known prosthetic groups (Fig. 7.1). In mammals, these latter seven subunits are encoded by mitochondrial DNA, whereas the remaining proteins in mammalian complex I are nuclear encoded. In bacteria, complex I is commonly expressed from genes organized into one operon, but some exceptions exist (Brandt 2006; Yagi et al. 1998; Friedrich 1998). The first structure of the seven water-soluble proteins of complex I became available in 2006 (Sazanov and Hinchliffe 2006), followed by structures of the whole enzyme from bacteria (Efremov et al. 2010) and the yeast *Yarrowia lipolytica* (Hunte et al. 2010). Recently, the structure of most part of the membrane domain of complex I from *Escherichia coli*, including the subunits most relevant to this paper, was solved at a higher resolution than previously (Efremov and Sazanov 2011).

The energy coupling mechanism of complex I remains unknown, although many creative models and schemes have been devised over the years (Brandt 1997; Hirst 2003; Brandt et al. 2003; Stolpe and Friedrich 2004; Ohnishi et al. 2010b, a; Steuber 2001; Dutton et al. 1998; Berrisford and Sazanov 2009; Friedrich 2001). The progress in understanding the complex I functional mechanism has, however, more than anything else been hampered by the actual lack of data. The lack of high resolution structural information was until recently a major obstacle, just as the scarcity of spectroscopically detectable features in the membrane spanning part of the enzymes. The proposed number of quinone binding sites range from one large binding pocket up to five specific sites (Okun et al. 1999; Magnitsky et al. 2002; Kurki et al. 2000; Gong et al. 2003; Nakamaru-Ogiso et al. 2003, 2010a; Tocilescu et al. 2010), but the actual number and location remain somewhat elusive. Another unresolved issue is whether complex I pumps only protons or if it can translocate sodium as well. The proton/electron stoichiometry of complex I has been scrutinized in many labs over the years (summarized in Galkin 1999), reaching a consensus of $4H^+/2e^-$ in complex I in tightly coupled submitochondrial particles from bovine heart. But it was also demonstrated that complex I from *Klebsiella pneumonie* translocated Na^+ (Gemperli et al. 2002), and other experiments suggested that complex I from *E. coli* (Steuber et al. 2000) was capable of working as a primary sodium pump. These results were under much debate (Hirst 2003) and it was suggested that the complex I preparations had been contaminated with a Na^+ pumping NQR enzyme (Bertsova and

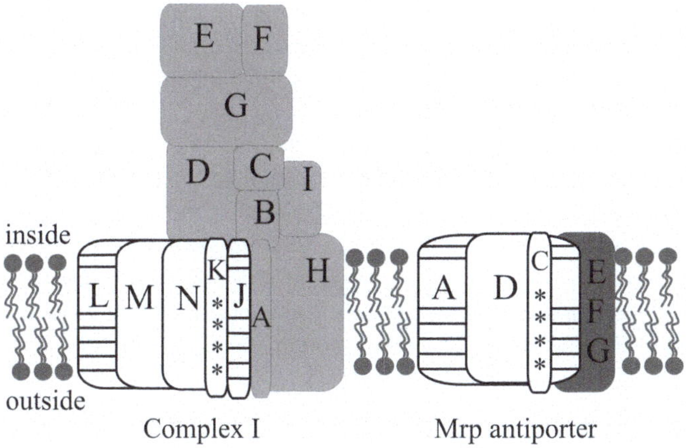

Fig. 7.1 A schematic representation of respiratory chain complex I and the Mrp antiporter complex. The ion transporter proteins MrpA and NuoL harboring a Na^+ channel are indicated with stripes, whereas the ion transporter proteins MrpD and NuoM/NuoN containing a H^+ channel are shown in plain white. The small homologous proteins MrpC and NuoK are marked with asterisks. In Mrp complexes from some organisms MrpB is a separate subunit, but often MrpA and MrpB comprises only one fused protein, denoted MrpA (Swartz et al. 2005). Other complex I proteins are shown in *light gray* and other Mrp proteins in *dark grey*

Bogachev 2004) or alternatively, that *E. coli* complex I is capable of secondary Na^+/H^+ antiport activity (Stolpe and Friedrich 2004). Recently, sodium dependence of energy transduction was observed in *Rhodothermus marinus* and *E. coli*, indicating that complex I in these organisms contain one Na^+ dependent and one Na^+ independent coupling site (Batista et al. 2010; Batista and Pereira 2011).

The evolution of the large complex I enzyme from smaller functional building blocks has long been recognized (Friedrich et al. 1993; Friedrich and Weiss 1997). The NADH dehydrogenase module (N-module) of complex I consists of the NuoE, NuoF and NuoG. The two former proteins build up the so called FP fragment, that harbor the NADH binding site and have FMN and iron-sulfur clusters as prosthetic groups (Yano et al. 1996). The NuoG subunit resembles Fe-only hydrogenase and molybdenopterin containing enzymes such as nitrate reductase and formate dehydrogenase, and contribute additional iron-sulfur clusters to the N-module (Yano et al. 1995; Rothery et al. 2008; Sazanov and Hinchliffe 2006). The quinone module (Q-module) is composed of NuoB, NuoC, NuoD and NuoI. The NuoB and NuoD are homologous to the small and large subunit of soluble NiFe hydrogenase, but only the proximal iron-sulfur cluster remains in NuoB, and the NiFe active site has been lost in NuoD. NuoI looks like a bacterial ferredoxin with two FeS clusters. NuoC has no prosthetic group and seems to be unique for complex I (Brandt 2006; Kerscher et al. 2001). Finally, the proton translocation module (P-module) is composed of the seven membrane-spanning subunits NuoAHJKLMN (Brandt 2006). All the three complex I subunits NuoL, NuoM and NuoN are homologous to protein subunits in one particular type of Na^+/H^+ or K^+/H^+ antiporter (Fearnley and Walker

1992; Hamamoto et al. 1994, Fig. 7.1, see also Sect. 7.2). This type of antiporter is built up by seven proteins, MrpABCDEFG (Swartz et al. 2005), of which MrpA was shown to have higher sequence similarity to NuoL whereas MrpD showed more similarity to NuoM and NuoN (Mathiesen and Hägerhäll 2002). The small NuoK is homologous to MrpC (Mathiesen and Hagerhall 2003). In an alternative module terminology (Friedrich and Scheide 2000; Friedrich 2001), the hydrogenase module is formed from the Q-module and two additional membrane-spanning proteins, NuoH and one of the homologous Mrp antiporter-derived proteins. The hydrogenase module composition is thereby equivalent to the protein subunit composition of present day Ech and Hyc hydrogenases (Hedderich 2004; Vignais et al. 2001). The remaining five membrane-spanning proteins comprise the transporter-module (Friedrich and Scheide 2000; Friedrich 2001).

7.2 The Mrp Antiporter

The Mrp antiporter complex (also denoted Mnh, Sha or Pha in different organisms, Fig. 7.1) is structurally and functionally different from better understood antiporters such as NhaA (Hunte et al. 2005). The Mrp antiporters constitute their own family, cation proton antiporter-3, (CPA-3) that hold the transporter classification TC.2.A.63. This particular type of antiporter was first discovered as a mutant in the alkalophile *Bacillus halodurans*, that was unable to grow at alkaline pH but remained viable at neutral pH (Hamamoto et al. 1994). The mutation was single point mutation (G393 to R) in MrpA. The similarity to the NuoL complex I protein was noticed (Hamamoto et al. 1994) and the three NuoL, NuoM and NuoN were already known to be similar to each other (Fearnley and Walker 1992). This discovery thereafter rendered the subunits prime candidates for harboring important parts of the proton translocation machinery in complex I, but for long time there was essentially no biochemical evidence. The progress in understanding the Mrp antiporters was also modest. The polypeptides were large, very hydrophobic and difficult to work with. Good antibodies were scarce and the Mrp complex was usually present in low amounts in cells. The complex I proteins offered similar difficulties. The truncated NuoL was successfully produced in *E. coli* (Steuber 2003), and later the full length protein was made in yeast cells (Gemperli et al. 2007). Only recently, a large amounts of all three subunits, NuoL, NuoM and NuoN, have been produced as cytochrome c fusion proteins (Gustavsson et al. 2010). Fusion protein strategies have also proved fruitful for studies of the Mrp complex (Morino et al. 2008; Kajiyama et al. 2007).

It has been known since long that both the MrpA and the MrpD proteins had a role in Na^+ resistance and Na^+ dependent pH homeostasis (Ito et al. 1999, 2000; Kosono et al. 1999). Both experiments with whole cells and experiments using membrane vesicles showed that Mrp worked as a secondary antiporter that was energized by the membrane potential (Kitada et al. 2000; Padan et al. 2005; Swartz et al. 2005, 2007). The Mrp proteins are typically encoded by one gene cluster or operon, comprising seven genes. Often, *mrpA* and *mrpB* are fused into one gene

(Swartz et al. 2005). The fact that these operons contain two genes encoding homologous proteins, *mrpA* and *mrpD*, together with five genes encoding dissimilar proteins indicated that the homologous MrpA and MrpD were likely to have somewhat specialized functions. However, MrpA and MrpD proteins from the *Bacillus pseudofirmus* Mrp antiporter that were expressed in the *E. coli* KNabc, a strain that has been deleted for endogenous antiporters, showed no activity (Morino et al. 2008; Krulwich et al. 2009). The antiporter activity was demonstrated to depend on the presence of all the protein subunits, indicating that Mrp works in the form of a protein complex (Hiramatsu et al. 1998; Ito et al. 2000; Kajiyama et al. 2007). It has been suggested that the Mrp complex consist of two subcomplexes, one comprising the complex I homologues MrpABCD, and one made up of the MrpEFG, that have no resemblance to complex I proteins. These two Mrp modules can assemble independently, but changes in either module can still influence the other (Morino et al. 2010). In addition to Na^{+}/H^{+} antiporter, the Mrp complex is capable of cholate efflux, an activity seemingly conducted by the MrpF subunit (Ito et al. 2000). The MrpE subunit was shown to be particularly important for Mrp complex assembly and stability (Morino et al. 2010). All the protein subunits in the Mrp complex have been tagged with unique c-terminal fusion domains that allowed individual detection of each protein (Morino et al. 2008). This opened up possibilities to perform detailed studies of point mutations, where stability effects and specific effects on activity can now be readily discriminated (Morino et al. 2010; Kajiyama et al. 2009). There is no crystal structure or electron microscopy image available for the Mrp antiporter complex, and no detailed functional mechanism has been proposed for the antiporter.

7.3 Mrp Ion Transporter Protein Occurrence

The Mrp antiporter complex is widely distributed in the phylogenetic tree of life. It is found both in archea and in eubacteria, and both in extreme alkalophiles and mesophiles with a more normal lifestyle. It seemingly always contains both an MrpA and an MrpD subunit. An excellent summary of the occurrence of Mrp antiporters and their genomic organization in different organisms has been provided by Krulwich and coworkers (Swartz et al. 2005).

In the large eukaryote complex I the ND5, ND4 and ND2 subunits (corresponding to NuoL, NuoM and NuoN) are always encoded by mitochondrial DNA (Brandt 2006). As mentioned in Sect. 7.1, it was demonstrated by phylogenetic analysis that although the antiporter subunits MrpA and MrpD and the complex I proteins NuoL, NuoM and NuoN are all homologous proteins, MrpA is more similar to NuoL and MrpD more similar to NuoM and NuoN (Mathiesen and Hägerhäll 2002). In prokaryotes, complex I also exist in smaller, compact variants comprising 11 or 12 subunits, in addition to the standard bacterial complex I with 14 protein subunits (Moparthi and Hagerhall 2011; Friedrich and Scheide 2000). The former enzymes are lacking all or part of the N-module, and instead are using or are presumed to be using alternative electron donor/acceptor proteins. Nevertheless, the NuoL, NuoM and NuoN subunits are present with such a degree of sequence conservation that they can always be assigned correctly (Mathiesen and Hägerhäll 2002). Interestingly, cyanobacteria, that have a compact, 11-subunit complex I, use multiple copies of the NuoL, NuoM and NuoN proteins, with slightly different primary sequences. These proteins are assembled into different varieties of the compact complex I that are used under different conditions, such as during CO_2 fixation, chlororespiration and cyclic electron flow around PSI (Battchikova and Aro 2007; Battchikova et al. 2011; Peng et al. 2011).

Membrane bound hydrogenases come in two major varieties, one smaller enzyme denoted hydrogenase-3 that contain one Mrp homologous transporter protein and one larger enzyme that harbors three such transporter proteins and an equivalent of MrpC/NuoK. However, in both types of hydrogenases, the primary sequence similarity of the transporter proteins is reduced, in such a way that it is no longer possible to distinguish between the MrpA-type or MrpD-type polypeptides, or to assign them as NuoL, NuoM or NuoN with any high degree of certainty (Mathiesen and Hägerhäll 2002). Therefore, the whole group of homologous proteins, whether MrpA and MrpD in the Mrp antiporter complex or NuoL, NuoM and NuoN in complex I or EchA/HycC in hydrogenase-3 or HyfB, HyfD and HyfF in hydrogenase-4, will in the following be referred to simply as the transporter proteins. The occurrence of the various types of hydrogenases in the kingdoms of life was extensively documented by Vignais and Billoud in a recent review (Vignais and Billoud 2007).

In addition, genes encoding Mrp homologous transporter proteins are present outside an operon or gene cluster arrangement, in such a way that their function remains uncertain. They may also comprise subunits of yet unknown protein complexes. In *B. subtilis*, *Vibrio cholerae* and *Aquifex aeolicus* two conserved, seemingly co-expressed adjacent genes were found, presumably encoding an Mrp homologous transporter protein and a large hydrophilic protein of unknown function (Mathiesen and Hägerhäll 2002).

7.4 Transporter Protein Topology and Structure

In the recently published structures of the membrane-domain of complex I, NuoL, NuoM and NuoN can be resolved as three structurally nearly identical entities (Efremov et al. 2010; Efremov and Sazanov 2011). The conserved part reveal 14 transmembrane helices in each protein. Hydropathy plots (Kyte and Doolittle 1982) also show 14 hydrophobic segments conserved in all the homologous transporter proteins, whereas both MrpA and NuoL have a c-terminal extension that contain additional hydrophobic stretches (Mathiesen and Hägerhäll 2002). In the solved three-dimensional structures, a long, rod-like helix stretches across from the distal end of the P-domain towards the Q-domain, parallel to the inner membrane surface (Efremov et al. 2010; Hunte et al. 2010). This structure originates from the c-terminal domain of NuoL (Efremov and Sazanov 2011). In MrpA, the c-terminal extension is much longer than in NuoL, and contain additional hydrophobic stretches that probably make up transmembrane helices. In some organisms, the MrpB subunit is fused to the MrpA c-terminus, which can make this domain appear even larger than actual (Swartz et al. 2005). The complex I structure also revealed that two of the transmembrane segments in each of the transporter proteins are helices interrupted at the center (Efremov et al. 2010), features that probably are important in the ion translocation mechanism. The interrupted transmembrane helices seen in the structure correspond to TM helix VII and TM helix XII (Fig. 7.2, Efremov and Sazanov 2011).

In previous work, where the transmembrane topology of the transporter proteins was analyzed with alkaline phosphatase fusion proteins, one problematic region between helix IX and XII was revealed. Proteins with a fusion point at the beginning of helix IX and at the end of helix XII were clearly PhoA positive and thus located on the periplasmic side of the membrane. Fusion proteins at the end of helix IX or beginning of helix XII were inactive, suggesting a cytoplasmic location. But fusion points between helix X and XI were ambiguous, resulting in both active and inactive fusion proteins, even when the fusion point differed by only a few residues (Mathiesen and Hägerhäll 2002). Other researchers, using insertion of histidine stretches that can be detected using penta-his antibodies, detected the loop between helix X and XI on the cytoplasmic side of the membrane (Torres-Bacete et al. 2009). The latter strategy has the advantage of not truncating the polypeptide to be analyzed, thereby minimizing the risk of misfolding. In any case, placing helix X and XI outside of the membrane as in (Mathiesen and Hägerhäll 2002) and (Torres-Bacete et al. 2009) results in 12 conserved transmembrane segments in the transporter proteins, which is not compatible with the recent structural data (Efremov and Sazanov 2011). Amarneh and Vik scrutinized the topology in the troublesome region using chemical labeling of single cysteine mutants. In that work, the loop between helix X and helix XI could be labeled with N-maleimidyl-propionyl biocytin in inverted vesicles but not in whole cells indicating a periplasmic location, thus bringing the number of transmembrane helices to 14 (Amarneh and Vik 2010). In all three cases described above, the polypeptides investigated has been overexpressed, and not assembled into complex I when analyzed, and in (Mathiesen and Hägerhäll 2002) the polypeptide was even truncated at the respective fusion points. It remains to be seen if some flexibility or conformational changes in this area in the working enzyme could perhaps explain the earlier results.

7.5 Transporter Protein Subunit Organization

As mentioned in Sect. 7.2, the Mrp antiporter complex was shown to consist of two subcomplexes, one comprising MrpABCD and one MrpEFG, that could assemble independently (Morino et al. 2010). The function of the Mrp complex depended on the presence of all subunits (Hiramatsu et al. 1998; Ito et al. 2000; Kajiyama et al. 2007), but Mrp E was shown to be essential for assembly and/or stability (Morino et al. 2010).

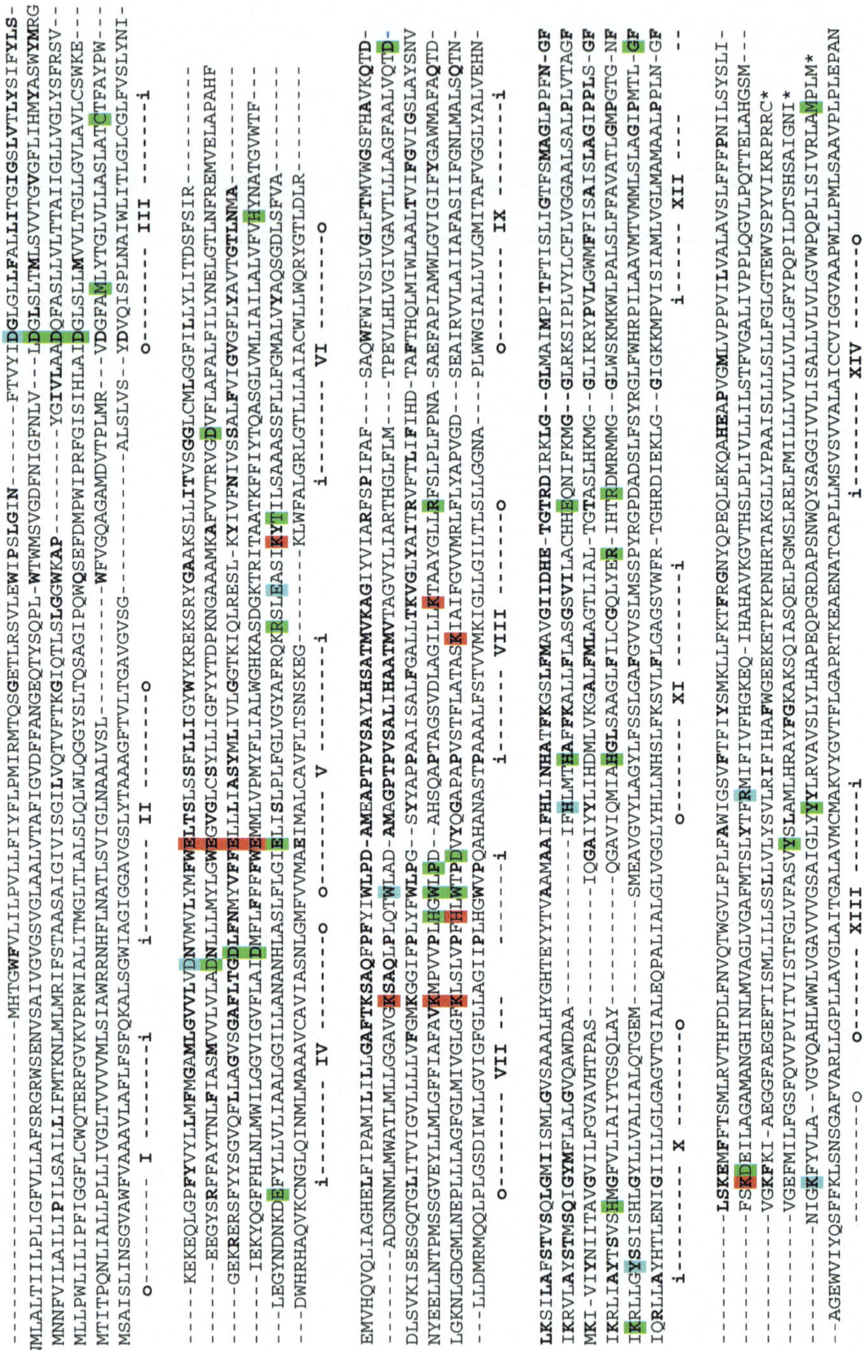

In the recent three-dimensional structure of complex I, the three structurally virtually identical transporter proteins are lined up in the P-domain, one after the other (Efremov et al. 2010). Fractionation studies have indicated that NuoL and NuoM were located at the distal end of the P-domain (Holt et al. 2003), and this was later confirmed with single particle analyses, demonstrating that NuoL and NuoM could be split off from the other membrane spanning subunits, leaving the proximal part of the P- domain intact (Baranova et al. 2007). Two distinct modules in the P-domain were also noticed in the *Y. lipolytica* complex I structure, where the two entities were visible and separated by a narrow centrally located interface (Hunte et al. 2010). The recent, high resolution structure of the membrane domain of *E. coli* complex I have confirmed that the proximal transporter protein subunit correspond to NuoN. In the distally located module, NuoL is positioned at the furthest end of the P-domain (Efremov and Sazanov 2011).

Another prominent feature in both the prokaryote and the eukaryote complex I structure is the long rigid helix structure running parallel to the membrane along the inner membrane surface. This so called "coupling rod" or "piston" is emerging from the most distal NuoL subunit, and is reaching back across the P-domain towards the Q-domain in the protruding part of the enzyme. The structure was immediately recognized as important for conveying the long range conformational changes that are anticipated in the complex I coupling mechanism (Efremov et al. 2010; Hunte et al. 2010). Nevertheless, it remains to be determined experimentally what exact role(s) the long helix structure might have. In a first attempt to clarify this, a fully assembled complex I with a truncated NuoL, lacking the c-terminal part was produced and reconstituted into liposomes. ACMA fluorescence quenching measurements indicated that the proton:electron stoichiometry achieved was about half for enzymes having the truncated version of NuoL compared to for wild type complex I (Steimle et al. 2011). It is also interesting to note that both NuoL and MrpA contain a c-terminal extension that is not present in the other homologous transporter proteins. In NuoL, it is forming the long, membrane parallel helix structure (Efremov et al. 2010; Efremov and Sazanov 2011), but the role in MrpA is unknown. Even if the c-terminal domain is much longer in MrpA compared to in NuoL and there is no obvious sequence similarity, one may wonder if there is some original, evolutionary structural or functional similarity between the two, or if the NuoL-domain has acquired an entirely new function (see also Sects. 7.4 and 7.7).

Fig. 7.2 Primary sequence alignment of the homologous transporter proteins. Conserved residues in the respective protein types are shown in *bold letters*. The transmembrane stretches (Adapted from Mathiesen and Hägerhäll 2002) are indicated below the sequence. The mutated positions are marked in each sequence, and color-coded such that *red* indicates that the mutation had a severe effect, *blue* substitutions resulted in some effect whereas mutations in positions marked in *green* had essentially no effect. The outcomes of mutagenesis of the respective subunits were compiled from Morino et al. (2010) and Kajiyama et al. (2009) for MrpA and MrpD, from Nakamaru-Ogiso et al. (2010b) for NuoL, from Lunardi (1998), Euro et al. (2008) and Torres-Bacete et al. (2007, 2009) for NuoM and from Amarneh and Vik (2003) for NuoN

7.6 Transporter Protein Function

7.6.1 *Quinone Binding Sites*

The presence of quinone binding sites on the transporter proteins was first suggested based on primary sequence signature motifs in NuoL and NuoM (Fisher and Rich 2000). Thereafter, direct labeling with quinone– or specific inhibitor analogues have in fact pinpointed sites on all three complex I proteins. The *E. coli* NuoM subunit was labeled with an azidoquinone (Gong et al. 2003), and the mitochondrial NuoL (i.e. ND5) was labeled with a fenpyroximate analogue (Nakamaru-Ogiso et al. 2003). Recently, both NuoL and NuoN (i.e. ND5 and ND4) from bovine mitochondria were labeled with asimicin, a potent complex I inhibitor (Nakamaru-Ogiso et al. 2010a). Whether these quinone binding sites are catalytic sites or merely serve some regulatory purpose remains to be established. In addition, it is not known whether the Mrp antiporter proteins also contain quinone binding sites, or if these have been added after the subunits were recruited to complex I.

7.6.2 *Mutant Properties*

As already discussed in Sect. 7.2, recent fusion protein strategies applied to the Mrp antiporter complex have made detailed mutagenesis studies possible. Mrp antiporter fusion proteins from *B. subtilis* (Kajiyama et al. 2009) or *B. pseudofirmus* (Morino et al. 2010) are expressed in readily detectable amounts in the *E. coli* strain KNabc where antiporter activity can also be assayed, since this strain has been deleted of its three major endogenous antiporters (Nozaki et al. 1996). In this experimental system it is possible to differentiate between mutations affecting protein complex stability and mutations affecting antiporter activity (Morino et al. 2010; Kajiyama et al. 2009). Site-directed mutagenesis of conserved residues in the NuoL, NuoM and NuoN subunits have been done by several groups (Amarneh and Vik 2003; Torres-Bacete et al. 2007, 2009; Euro et al. 2008; Nakamaru-Ogiso et al. 2010b). The more recent studies have been using the plasmid pKO3, that has a temperature sensitive origin of replication, to make convenient and precise changes in the *E. coli* chromosome. The mutated complex I is therefore expressed in its normal context, with the only change being the introduced single amino acid change. Nevertheless, as often with site-directed mutants, many mutations cause pleiotropic effects and/or affect the stability resulting in lower amounts of enzyme present compared to in wild type cells. That in turn can result in expression of a different overall protein assemblage compared to in wild type cells. The quantification of complex I is often limited to relatively crude methods such as bands detected with western blot or NADH:ferricyanide reductase activity. Therefore, it is difficult to draw conclusions that require a high degree of exactness, such as counting proton electron stoichiometries, from such data, no matter how careful and accurate the researchers have been.

As an overview of currently existing site-directed mutants, a list of mutated positions have been compiled and the results have been plotted in the primary sequence alignment of the transporter proteins shown in Fig. 7.2. The mutants have been roughly categorized into those that have a severe effect on enzyme activity, shown in red, those with moderate effects, shown in blue and those causing essentially no effect, illustrated in green.

Naturally, the initial focus of mutagenesis has been on conserved, charged residues that could be suspected to be involved in the ion translocation mechanism. Some of the mutated positions warrant special mentioning. The first is a fully conserved glutamate located in helix V (Fig. 7.2). In both MrpA and MrpD drastic effects on antiporter activity was seen when this glutamate residue was changed to alanine or glutamine. The E137D mutation in MrpD was active or somewhat active, indicating that the carboxyl group is crucial for function (Kajiyama et al. 2009; Morino et al. 2010). In NuoL and NuoM the mutation resulted in lower enzyme activities and loss of energy coupling, but also to decreased complex I stability (Euro et al. 2008; Nakamaru-Ogiso et al. 2010b; Torres-Bacete et al. 2007, 2009). In some cases, the glutamate could be replaced with aspartate with retained function, similar to what was found in MrpD. Somewhat surprisingly, there was very little effect of mutation at this position in NuoN (Amarneh and Vik 2003), although the glutamate is conserved also in this subunit (Fig. 7.2). Another mutation hot spot is the lysine in helix VII (Fig. 7.2). Mutations at this position caused severe functional impairment in all three complex I transporters (Euro et al. 2008; Nakamaru-Ogiso et al. 2010b; Torres-Bacete et al. 2007, 2009; Amarneh and Vik 2003), in some cases even when changed to arginine. Notably, this lysine is only present in some of the corresponding hydrogenase transporter proteins, but not in all. It should also be mentioned that a number of disease-related mutations have been found in the mitochondrially encoded ND5, ND4 and ND2, (equivalent of NuoL, NuoM and NuoN) as recently summarized by Tuppen et al. (Tuppen et al. 2010). However, the encountered disease mutations often cause relatively mild immediate effects, since they would otherwise have caused demise already at the embryological stage. Therefore, they have not been included in the compilation. An exception is the well studied Leber's disease mutation ND4-11778, that has been reproduced (R368H) and studied in *Rhodobacter capsulatus* complex I (Lunardi 1998).

7.6.3 Ion Transport

As mentioned in Sect. 7.1, the antiporter-like complex I subunits early on became prime candidates for harboring important parts of the proton translocation machinery. If the antiporter-like complex I proteins still retained an antiporter function, or if they for instance had evolved into mere proton channels, remained uncertain. Proton pumping by complex I is a well established activity, with the proton electron stoichiometry being scrutinized in many laboratories over the years, c.f. (Ragan and Hinkle 1975; Wikstrom 1984), reaching a consensus of $4H^+/2e^-$ for the mammalian enzyme (Galkin 1999). Concerning sodium, both a truncated NuoL produced in *E. coli* (Steuber 2003) and the full length NuoL produced in *Saccharomyces cerevisiae* (Gemperli et al. 2007) were shown to translocate sodium. There were also reports showing sodium translocation by intact complex I (Steuber et al. 2000; Gemperli et al. 2002), but these results were disputed by others. Pereira and coworkers recently observed that sodium stimulated the NADH:quinone reductase activity of the *Rhodotermus marinus* (Batista et al. 2010) and *E. coli* complex I, (Batista and Pereira 2011) suggesting that one Na^+ dependent and one Na^+ independent site was present in complex I. In the *bona fide* antiporter, it was anticipated that MrpA and MrpD must have unique functions, as they always occur in pairs in a conserved operon structure. As described in Sect. 7.2, deletion of *mrpA* or *mrpD* from the *B. subtilis* chromosome resulted in salt and pH sensitive growth phenotypes, but to a somewhat different degree, further supporting the notion that the two proteins have somewhat different functions. The *B. subtilis* growth phenotype was recently explored to study the properties of the NuoL, NuoM and NuoN protein function individually. In these deletion strains, MrpA expressed *in trans* could rescue the Δ*mrpA* strain but not the Δ*mrpD* strain and *vice versa* at pH 7.4. Likewise, NuoL from *E. coli* expressed in the *B. subtilis* Δ*mrpA* brought back the wild type growth properties whereas expression of *E. coli* NuoN in the same deletion strain did not improve the growth at all. The opposite was true for the *B. subtilis* Δ*mrpD* strain, that was rescued by NuoN but not by NuoL, that improved the growth only to a much lesser extent (Moparthi et al. 2011). The fact that the complex I proteins could complement the absence of an Mrp protein subunit strongly suggest that Mrp complex formation is not an absolute requirement for Mrp antiporter function. In the *E. coli* KNabc model system, it was demonstrated that the presence of all Mrp subunits was necessary for antiporter activity (Morino et al. 2008), and the same has been concluded in native systems (Hiramatsu et al. 1998; Ito et al. 2000). The simplest explanation, that accounts for all these observations, is that MrpA and MrpD are single ion transporters that together form an antiporter (Moparthi et al. 2011). In the *B. subtilis* deletion strains, all proteins are present, albeit one is expressed *in trans* and not always assembled into a complex, whereas in the other experimental systems, one protein subunit was absent when the antiporter did not work.

So which single ion transporter protein is responsible for transport of which ion species in the Mrp Na^+/H^+ antiporter complex? Sodium transport has previously been demonstrated for the NuoL subunit (Steuber 2003; Gemperli et al. 2007). Unfortunately, the NuoM and NuoN could not be tested in parallel in those experiments. Nevertheless, awaiting more conclusive experimental evidence, the sequence similarity between NuoL and MrpA makes MrpA the more likely candidate for Na^+ transport. From the *B. subtilis* deletion strain experiments, it seems clear that whereas the NuoN subunit is very efficient in rescuing the MrpD deletion, it does not work at all in the MrpA deleted strain. NuoL on the other hand, improves the growth to some extent also in the MrpD deleted strain. This corroborates the notion that the transporter protein accommodating the larger ion could be somewhat more promiscuous than *vice versa*. Taken together, it can be concluded that complex I contain three single ion transporter proteins, where NuoL most likely comprises a

Na^+ transporter and NuoN a H^+ transporter. The *E. coli* NuoM subunit could not complement the *B. subtilis* MrpD deletion strain under the conditions when NuoN restored the wild type growth properties (Moparthi et al. 2011). However, based on the primary sequence similarity of MrpD, NuoM and NuoN, the most likely function of NuoM is H^+ transport.

It is not known if the individual proteins have a preferred transport directionality. In the alkalophiles, the main role of the Mrp antiporter complex is to maintain a neutral cytoplasmic pH, and in the mesophile *B. subtilis* salt tolerance was used as an easily monitored phenotype (see Sect. 7.2). In complex I, we know that the enzyme is capable of both energy conserving NADH oxidation and $\Delta\mu H^+$ supported NAD^+ reduction (Sect. 7.1). Awaiting more experimental data on individual proteins reconstituted into liposomes, we can probably assume that the proteins are capable of ion transport in either direction, depending on energy availability, presence of ion gradients and other such conditions.

7.7 Role of the Antiporter-Module in Complex I Evolution

7.7.1 Recruitment of the Transporter Proteins – A Two Step Process

The presence of three antiporter-like proteins in complex I was first imagined to result from the triplication of one primordial gene (Friedrich and Scheide 2000; Friedrich and Weiss 1997). The later realization that features of both MrpA and MrpD were conserved in the complex I subunits and the detection the weak similarities between MrpC and NuoK led to the conclusion that a whole antiporter module must have been recruited (Mathiesen and Hägerhäll 2002, 2003). It seems that it is time for a yet another revision of the old scheme. We have recently demonstrated that the individual transporter proteins are not antiporters but rather single ion transporters. NuoL and its homologue MrpA are presumably Na^+ transporter proteins whereas NuoN, NuoM and MrpD conduct H^+ (Moparthi et al. 2011, Sect. 7.6.3). In addition, we know now the relative position of the proteins, with NuoL and NuoM located distally from NuoN in the P-domain of the enzyme (Holt et al. 2003; Baranova et al. 2007). Pereira and coworkers concluded that the type of electron acceptor quinone used by complex I, and thereby the amount of energy available for proton pumping, determines whether Na^+ translocation is needed to compensate for the smaller ΔE (Batista and Pereira 2011). This reasoning can be further extended considering the function of the primordial, compact complex I, that seemingly operated without an N-module or other designated partner protein. We have postulated that Q-module of the ancestral complex I acted as a versatile docking platform for several different donor and acceptor proteins and substrates with different redox potential, perhaps while simultaneously being able to operate with H_2 when possible (Moparthi and Hagerhall 2011). In such an enzyme, it would in fact be a very strong

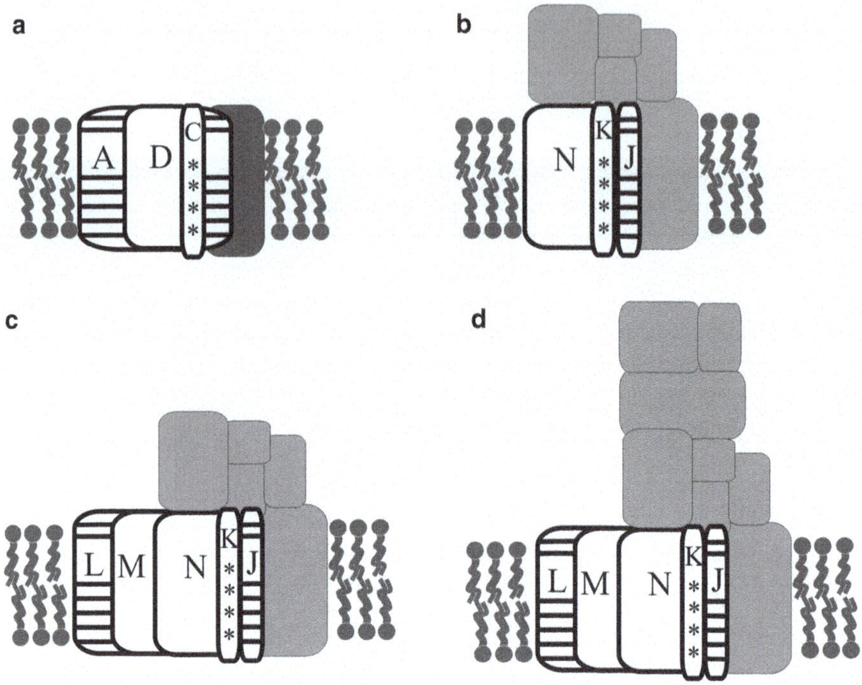

Fig. 7.3 A schematic representation of recruitment of the antiporter-module (**a**) to complex I. In the first step, an Mrp antiporter complex (**a**) is recruited to a primordial, membrane bound hydrogenase enzyme (**b**). Since this enzyme only needs a mere proton channel, the sodium transporter protein is lost. For structural reasons, the c-terminal part of MrpA is kept together with MrpC. In the second step, a whole Na^+/H^+ antiporter-module is again recruited to the ancient hydrogenase, that subsequently evolve into an enzyme similar to present day 11-subunit complex I (**c**). The N-module is finally added, to create the NADH/NAD^+-committed present-day full size complex I (**d**). The protein subunits are marked as in Fig. 7.1

advantage, perhaps even a necessity, to have means to adapt the coupling mechanism to the ΔE of different donor:acceptor redox pairs.

In the light of the novel information, the simplest and more logical evolutionary scenario is to view the recruitment of the transporter proteins as two consecutive recruitments, not involving any gene duplication at all (Fig. 7.3). First, an Mrp antiporter complex comprising MrpA(B)CD, as postulated in (Mathiesen and Hagerhall 2003), was recruited to form a primordial membrane-bound hydrogenase-type enzyme. This recruitment was presumably driven by the need for external protons for H_2 production, commonly used as an electron sink in an anaerobic world. We envision that the H^+ channel of the MrpD-like transporter subunit was leading up to the NiFe site in the NuoD-homologous protein where hydrogen was produced, thereby preventing debilitating alkalinization of the cytoplasm during H_2 production. Only a MrpD protein (i.e. a H^+ transporter) was required to perform the function needed, and therefore the Na^+ conducting part of the Mrp antiporter, the MrpA protein, could be dropped. The MrpD recruited in this first step corresponds to present day NuoN. The much later transformation of the NiFe site into a quinone binding site in the present day complex I (Brandt 2006; Kerscher et al. 2001; Darrouzet et al. 1998), sets the stage for whatever interesting energy coupling quinone chemistry your heart may desire. This discussion is beyond the scope of this chapter, but for the change of electron acceptor/donor to be feasible, a second addition to the membrane part of the enzyme was necessary. The second transporter recruitment step brought in a second antiporter module. This added both a Na^+ transporter and a second H^+ transporter to the emerging complex I enzyme (Fig. 7.3). The structural similarity of the MrpD-like protein recruited first, and the MrpD protein in the second antiporter module could in fact have directly driven such a merger. It remains to be established whether the long, membrane parallel helix structure in the NuoL C-terminal end is structurally and/or functionally related to the c-terminal extension in MrpA (see also Sects. 7.4 and 7.5). The sequence similarity in the c-terminal extensions is next to none, and the whole domain is in fact much shorter in NuoL than in MrpA. Nevertheless, one may speculate that this domain may have a role in assembly and stability of the Mrp complex, perhaps in combination with coupling the ion translocation activities of the MrpA transporter and the MrpD transporter. The recruitment of the antiporter module to complex I could therefore initially have occurred as a mere chance event, caused by a MrpA protein "hugging" the wrong partner-protein. The evolutionary advantage of this, sudden and accidental second recruitment could then lead to an immediate benefit, and an enzyme complex being able to undertake novel tasks. The emerging complex I enzyme was suddenly much more versatile and could work more efficiently in the "reverse" mode, using alternative electron donors in addition to H_2, being able to modify the output of the engine, depending on the E_m of the donors and acceptors.

In the light of the recent structural data of the complex I membrane domain (Efremov and Sazanov 2011) the NuoK of present-day complex I most likely originates from MrpC from the first Mrp complex recruitment (Fig. 7.3). Likewise, we would like to suggest that NuoJ originates from remnants of the most c-terminal end of MrpA/MrpAB that were left behind in the primordial hydrogenase complex for structural reasons, when the rest of the MrpA was lost. There is essentially no primary sequence similarity in these parts of neither proteins (Kao et al. 2003; Mathiesen and Hagerhall 2003; Kajiyama et al. 2009) to back up such a claim, but from structural point of view, the scenario still seems feasible. In the second Mrp complex recruitment step both MrpA and MrpD transporter subunits, together capable of Na^+/H^+ antiporter activity, were joined to form complex I, whereas the redundant MrpC and c-terminal MrpA/B brought in could be dropped.

7.7.2 *Implications for the Functional Mechanism of Present Day Complex I*

It was recently demonstrated that Na^+ translocation in the opposite direction compared to H^+ translocation was occurring in NADH oxidizing, present day complex I enzymes using menaquinone as electron acceptors (Batista and Pereira 2011; Batista et al. 2010). As discussed in the previous section, the addition of the whole antiporter module to an enzyme already capable of proton translocation could have added important functionality to the emerging primordial complex I enzyme. To be able to adjust the energy conserving machinery to the ΔE of a redox couple formed by different electron donors and acceptors was a strong selective advantage. In addition to the menaquinone users, there could be other present day compact complex I enzymes, those lacking the N-module, that are making use of the antiporter module in a similar way to accommodate different electron donors as well as acceptors. The use of alternative antiporter-modules in the different variants of cyanobacterial complex I (Battchikova and Aro 2007; Battchikova et al. 2011; Peng et al. 2011) is an indication that this may indeed be the case.

The *Paracoccus denitrificans* complex I uses ubiquinone as electron acceptor and show an overall close resemblance to the mammalian enzyme, albeit with only three of the so called accessory subunits (Yip et al. 2011). *E. coli* can utilize both ubiquinone and menaquinone, but in the experiments by Batista and Pereira (Batista and Pereira 2011) the *E. coli* cells were grown anaerobically, and therefore contain only menaquinone. In those experiments much less Na^+ translocation (about 10 nmol Na^+/mg protein) was seen when using inverted vesicles with *P. denitrificans* complex I compared to when using vesicles with *E. coli* complex I (about 30 nmol Na^+/mg protein) and the former activity was abolished by addition of the protonophore CCCP whereas the latter activity was further stimulated (Batista and Pereira 2011). These authors therefore concluded that only complex I in the menaquinone users were making use of an inward Na^+ transport to compensate for the smaller ΔE of NADH/ menaquinone redox reaction. Nevertheless, all three NuoL, NuoM and NuoN transporter proteins are strongly conserved in *P. denitrificans* and all other present day complex I enzymes. If many of these enzymes had no use for the Na^+ conducting NuoL, it seems highly unlikely that the primary sequence should have been conserved. On the contrary, in the membrane bound hydrogenases, degenerate forms of the transporter proteins without MrpA-type or MrpD-type primary sequence signatures, are found (see Sect. 7.3). Therefore, we would like to postulate a second, modulatory role for the NuoL+NuoM antiporter complex. We envision that after the buildup of a high ΔpH this Na^+/H^+ antiporter unit converts part of the proton gradient to a sodium gradient, while strictly maintaining the ΔΨ. This activity, while remaining electroneutral, would still facilitate further proton pumping by complex I by removing excess protons from the vicinity of the pump. This postulated function of the antiporter module would also reconcile the previous experimental results (Steuber et al. 2000; Gemperli et al. 2002; Stolpe and Friedrich 2004; Bertsova and Bogachev 2004) that gave rise to the old Na^+ or H^+ debate, since H^+ measurements are typically done using initial rates recorded on a seconds scale after a NADH or O_2 pulse, whereas the Na^+ measurements are done by recording sodium accumulation in vesicles after 5–15 min or so. In the proposed, second modulatory role, addition of CCCP is expected to abolish Na^+ translocation by the antiporter, which is exactly what was observed by Batista and Pereira in the experiment using *P. denitrificans* vesicles (Batista and Pereira 2011). The proposed role of the NuoL+NuoM antiporter module is also in agreement with the general observation that the major part of the membrane potential in mitochondria and many bacteria comprises ΔΨ, while ΔpH typically contribute only a minor part.

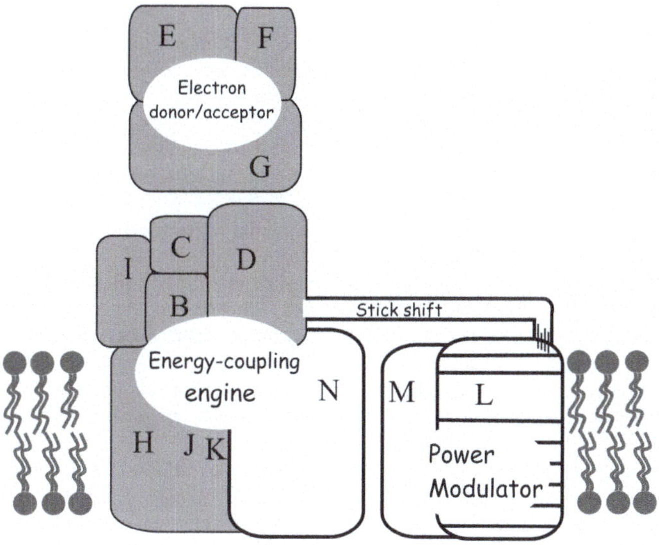

Fig. 7.4 The "machine parts" of present day complex I as envisioned in Sects. 7.1 and 7.2. The NuoN subunit H^+ transporter is build into the proton pumping engine of complex I, whereas the antiporter module, comprising the Na^+ transporter NuoL and the H^+ transporter NuoM, form a power-modulator unit that allows complex I to modify the engine output. It can be used either to modulate the H^+/e^- stoichiometry when using lower energy electron donor/acceptor pairs or to transfer ΔpH into $\Delta\Psi$ after the initial buildup of $\Delta\mu H^+$. The protein subunits are marked as in Fig. 7.1

To conclude, we propose that the transporter proteins in complex I were recruited from the Mrp antiporter complex in two consecutive steps (Fig. 7.3). In the first event, NuoN was brought in as a mere proton channel, whereas a whole Na^+/H^+ antiporter module composed of NuoL and NuoM was recruited in the second step. As outlined in Fig. 7.4, we further suggest that the proton pumping engine in complex I is composed of the Q-module, NuoH and the NuoN transporter protein providing a H^+ channel. The NuoL+NuoM antiporter-module works as a power modulator, that can modify the output of the proton pumping engine. The "coupling rod" structure should then perhaps rather be seen as a stickshift, changing the gear that the complex I machine is operating in.

References

Amarneh B, Vik SB (2003) Mutagenesis of subunit N of the *Escherichia coli* complex I. Identification of the initiation codon and the sensitivity of mutants to decylubiquinone. Biochemistry 42:4800–4808

Amarneh B, Vik SB (2010) Transmembrane topology of subunit N of complex I (NADH: ubiquinone oxidoreductase) from *Escherichia coli*. J Bioenerg Biomembr 42:511–516

Baranova EA, Morgan DJ, Sazanov LA (2007) Single particle analysis confirms distal location of subunits NuoL and NuoM in Escherichia coli complex I. J Struct Biol 159:238–242

Batista AP, Pereira MM (2011) Sodium influence on energy transduction by complexes I from *Escherichia coli* and *Paracoccus denitrificans*. Biochim Biophys Acta 1807:286–292

Batista AP, Fernandes AS, Louro RO, Steuber J, Pereira MM (2010) Energy conservation by *Rhodothermus marinus* respiratory complex I. Biochim Biophys Acta 1797:509–515
Battchikova N, Aro EM (2007) Cyanobacterial NDH-1 complexes: multiplicity in function and subunit composition. Physiol Plant 131:22–32
Battchikova N, Eisenhut M, Aro EM (2011) Cyanobacterial NDH-1 complexes: novel insights and remaining puzzles. Biochim Biophys Acta 1807:935–944
Berrisford JM, Sazanov LA (2009) Structural basis for the mechanism of respiratory complex I. J Biol Chem 284:29773–29783
Bertsova YV, Bogachev AV (2004) The origin of the sodium-dependent NADH oxidation by the respiratory chain of *Klebsiella pneumoniae*. FEBS Lett 563:207–212
Brandt U (1997) Proton-translocation by membrane-bound NADH: ubiquinone-oxidoreductase (complex I) through redox-gated ligand conduction. Biochim Biophys Acta 1318:79–91
Brandt U (2006) Energy converting NADH:quinone oxidoreductase (complex I). Annu Rev Biochem 75:69–92
Brandt U, Kerscher S, Dröse S, Zwicker K, Zickermann V (2003) Proton pumping by NADH: ubiquinone oxidoreductase. A redox driven conformational change mechanism? FEBS Lett 545:9–17
Carroll J, Fearnley IM, Skehel JM, Shannon RJ, Hirst J, Walker JE (2006) Bovine complex I is a complex of 45 different subunits. J Biol Chem 281:32724–32727
Darrouzet E, Issartel JP, Lunardi J, Dupuis A (1998) The 49-kDa subunit of NADH-ubiquinone oxidoreductase (complex I) is involved in the binding of piericidin and rotenone, two quinone-related inhibitors. FEBS Lett 431:34–38
Dutton PL, Moser CC, Sled VD, Daldal F, Ohnishi T (1998) A reductant-induced oxidation mechanism for complex I. Biochim Biophys Acta 1364:245–257
Efremov RG, Sazanov LA (2011) Structure of the membrane domain of respiratory complex I. Nature 476:414–420
Efremov RG, Baradaran R, Sazanov LA (2010) The architecture of respiratory complex I. Nature 465:441–447
Euro L, Belevich G, Verkhovsky MI, Wikström M, Verkhovskaya M (2008) Conserved lysine residues of the membrane subunit NuoM are involved in energy conversion by the proton-pumping NADH: ubiquinone oxidoreductase (complex I). Biochim Biophys Acta 1777:1166–1172
Fearnley IM, Walker JE (1992) Conservation of sequences of subunits of mitochondrial complex I and their relationships with other proteins. Biochim Biophys Acta 1140:105–134
Fisher N, Rich PR (2000) A motif for quinone binding sites in respiratory and photosynthetic systems. J Mol Biol 296:1153–1162
Friedrich T (1998) The NADH: ubiquinone oxidoreductase (complex I) from *Escherichia coli*. Biochim Biophys Acta 1364:134–146
Friedrich T (2001) Complex I: a chimaera of a redox and conformation-driven proton pump? J Bioenerg Biomembr 33:169–177
Friedrich T, Scheide D (2000) The respiratory complex I of bacteria, archaea and eukarya and its module common with membrane-bound multisubunit hydrogenases. FEBS Lett 479:1–5
Friedrich T, Weiss H (1997) Modular evolution of the respiratory NADH ubiquinone oxidoreductase and the origin of its modules. J Theor Biol 187:529–540
Friedrich T, Weidner U, Nehls U, Fecke W, Schneider R, Weiss H (1993) Attempts to define distinct parts of NADH: ubiquinone oxidoreductase (complex I). J Bioenerg Biomembr 25:331–337
Galkin A, Grivennikova VG, Vinogradov AD (1999) H^+/e^- stoichiometry in NADH-quinone reductase reactions catalyzed by bovine heart submitochondrial particles. FEBS Lett 451:157–161
Gemperli AC, Dimroth P, Steuber J (2002) The respiratory complex I (NDH I) from *Klebsiella pneumoniae*, a sodium pump. J Biol Chem 277:33811–33817
Gemperli AC, Schaffitzel C, Jakob C, Steuber J (2007) Transport of Na^+ and K^+ by an antiporter-related subunit from the *Escherichia coli* NADH dehydrogenase I produced in *Saccharomyces cerevisiae*. Arch Microbiol 188:509–521

Gong X, Xie T, Yu L, Hesterberg M, Scheide D, Friedrich T, Yu CA (2003) The ubiquinone-binding site in NADH: ubiquinone oxidoreductase from *Escherichia coli*. J Biol Chem 278:25731–25737

Gustavsson T, Trane M, Moparthi VK, Miklovyte E, Moparthi L, Gorecki K, Leiding T, Årsköld SP, Hägerhäll C (2010) A cytochrome c fusion protein domain for convenient detection, quantification, and enhanced production of membrane proteins in Escherichia coli-expression and characterization of cytochrome-tagged complex I subunits. Protein Sci 19:1445–1460

Hamamoto T, Hashimoto M, Hino M, Kitada M, Seto Y, Kudo T, Horikoshi K (1994) Characterization of a gene responsible for the Na^+/H^+ antiporter system of alkalophilic Bacillus species strain C-125. Mol Microbiol 14:939–946

Hedderich R (2004) Energy-converting [NiFe] hydrogenases from archaea and extremophiles: ancestors of complex I. J Bioenerg Biomembr 36:65–75

Hiramatsu T, Kodama K, Kuroda T, Mizushima T, Tsuchiya T (1998) A putative multisubunit Na^+/H^+ antiporter from *Staphylococcus aureus*. J Bacteriol 180:6642–6648

Hirst J (2003) The dichotomy of complex I: a sodium ion pump or a proton pump. Proc Natl Acad Sci USA 100:773–775

Holt PJ, Morgan DJ, Sazanov LA (2003) The location of NuoL and NuoM subunits in the membrane domain of the *Escherichia coli* complex I – implications for the mechanism of proton pumping. J Biol Chem 278:43114–43120

Hunte C, Screpanti E, Venturi M, Rimon A, Padan E, Michel H (2005) Structure of a Na^+/H^+ antiporter and insights into mechanism of action and regulation by pH. Nature 435:1197–1202

Hunte C, Zickermann V, Brandt U (2010) Functional modules and structural basis of conformational coupling in mitochondrial complex I. Science 329:448–451

Ito M, Guffanti AA, Oudega B, Krulwich TA (1999) Mrp, a multigene, multifunctional locus in *Bacillus subtilis* with roles in resistance to cholate and to Na^+ and in pH homeostasis. J Bacteriol 181:2394–2402

Ito M, Guffanti AA, Wang W, Krulwich TA (2000) Effects of nonpolar mutations in each of the seven *Bacillus subtilis* mrp genes suggest complex interactions among the gene products in support of Na^+ and alkali but not cholate resistance. J Bacteriol 182:5663–5670

Kajiyama Y, Otagiri M, Sekiguchi J, Kosono S, Kudo T (2007) Complex formation by the mrpABCDEFG gene products, which constitute a principal Na^+/H^+ antiporter in *Bacillus subtilis*. J Bacteriol 189:7511–7514

Kajiyama Y, Otagiri M, Sekiguchi J, Kudo T, Kosono S (2009) The MrpA, MrpB and MrpD subunits of the Mrp antiporter complex in *Bacillus subtilis* contain membrane-embedded and essential acidic residues. Microbiology 155:2137–2147

Kao MC, Di Bernardo S, Matsuno-Yagi A, Yagi T (2003) Characterization and topology of the membrane domain Nqo10 subunit of the proton-translocating NADH-quinone oxidoreductase of *Paracoccus denitrificans*. Biochemistry 42:4534–4543

Kerscher S, Kashani-Poor N, Zwicker K, Zickermann V, Brandt U (2001) Exploring the catalytic core of complex I by *Yarrowia lipolytica* yeast genetics. J Bioenerg Biomembr 33:187–196

Kitada M, Kosono S, Kudo T (2000) The Na^+/H^+ antiporter of alkaliphilic Bacillus sp. Extremophiles 4:253–258

Kosono S, Morotomi S, Kitada M, Kudo T (1999) Analyses of a *Bacillus subtilis* homologue of the Na^+/H^+ antiporter gene which is important for pH homeostasis of alkaliphilic *Bacillus* sp. C-125. Biochim Biophys Acta 1409:171–175

Krulwich TA, Hicks DB, Ito M (2009) Cation/proton antiporter complements of bacteria: why so large and diverse? Mol Microbiol 74:257–260

Kurki S, Zickermann V, Kervinen M, Hassinen I, Finel M (2000) Mutagenesis of three conserved Glu residues in a bacterial homologue of the ND1 subunit of complex I affects ubiquinone reduction kinetics but not inhibition by dicyclohexylcarbodiimide. Biochemistry 39:13496–13502

Kyte J, Doolittle RF (1982) A simple method for displaying the hydrophobic character of a protein. J Mol Biol 157:105–132

Lunardi J, Darrouzet E, Dupuis A, Issartel JP (1998) The nuoM arg(368)his mutation in NADH: ubiquinone oxidoreductase from *Rhodobacter capsulatus*: a model for the human ND4-11778 mtDNA mutation associated with Leber's hereditary optic neuropathy. Biochim Biophys Acta 1407:114–124

Magnitsky S, Toulokhonova L, Yano T, Sled VD, Hägerhäll C, Grivennikova VG, Burbaev DS, Vinogradov AD, Ohnishi T (2002) EPR characterization of ubisemiquinones and iron-sulfur cluster N2, central components of the energy coupling in the NADH-ubiquinone oxidoreductase (complex I) in situ. J Bioenerg Biomembr 34:193–208

Mathiesen C, Hägerhäll C (2002) Transmembrane topology of the NuoL, M and N subunits of NADH:quinone oxidoreductase and their homologues among membrane-bound hydrogenases and bona fide antiporters. Biochim Biophys Acta 1556:121–132

Mathiesen C, Hägerhäll C (2003) The 'antiporter module' of respiratory chain complex I includes the MrpC/NuoK subunit – a revision of the modular evolution scheme. FEBS Lett 549:7–13

Moparthi VK, Hägerhäll C (2011) The evolution of respiratory chain complex I from a smaller last common ancestor consisting of 11 protein subunits. J Mol Evol 72:484–497

Moparthi VK, Kumar B, Mathiesen C, Hägerhäll C (2011) Homologous protein subunits from *Escherichia coli* NADH:quinone oxidoreductase can functionally replace MrpA and MrpD in *Bacillus subtilis*. Biochim Biophys Acta 1807:427–436

Morino M, Natsui S, Swartz TH, Krulwich TA, Ito M (2008) Single gene deletions of mrpA to mrpG and mrpE point mutations affect activity of the Mrp Na^+/H^+ antiporter of alkaliphilic Bacillus and formation of hetero-monieric Mrp complexes. J Bacteriol 190:4162–4172

Morino M, Natsui S, Ono T, Swartz TH, Krulwich TA, Ito M (2010) Single site mutations in the hetero-oligomeric Mrp antiporter from alkaliphilic *Bacillus pseudofirmus* OF4 that affect Na^+/H^+ antiport activity, sodium exclusion, individual Mrp protein levels, or Mrp complex formation. J Biol Chem 285:30942–30950

Nakamaru-Ogiso E, Sakamoto K, Matsuno-Yagi A, Miyoshi H, Yagi T (2003) The ND5 subunit was labeled by a photoaffinity analogue of fenpyroximate in bovine mitochondrial complex I. Biochemistry 42:746–754

Nakamaru-Ogiso E, Han H, Matsuno-Yagi A, Keinan E, Sinha SC, Yagi T, Ohnishi T (2010a) The ND2 subunit is labeled by a photoaffinity analogue of asimicin, a potent complex I inhibitor. FEBS Lett 584:883–888

Nakamaru-Ogiso E, Kao MC, Chen H, Sinha SC, Yagi T, Ohnishi T (2010b) The membrane subunit NuoL(ND5) is involved in the indirect proton pumping mechanism of *Escherichia coli* complex I. J Biol Chem 285:39070–39078

Nozaki K, Inaba K, Kuroda T, Tsuda M, Tsuchiya T (1996) Cloning and sequencing of the gene for Na^+/H^+ antiporter of *Vibrio parahaemolyticus*. Biochem Biophys Res Commun 222:774–779

Ohnishi ST, Salerno JC, Ohnishi T (2010a) Possible roles of two quinone molecules in direct and indirect proton pumps of bovine heart NADH-quinone oxidoreductase (complex I). Biochim Biophys Acta 1797:1891–1893

Ohnishi T, Nakamaru-Ogiso E, Ohnishi ST (2010b) A new hypothesis on the simultaneous direct and indirect proton pump mechanisms in NADH-quinone oxidoreductase (complex I). FEBS Lett 584:4131–4137

Okun JG, Lummen P, Brandt U (1999) Three classes of inhibitors share a common binding domain in mitochondrial complex I (NADH: ubiquinone oxidoreductase). J Biol Chem 274:2625–2630

Padan E, Bibi E, Ito M, Krulwich TA (2005) Alkaline pH homeostasis in bacteria: new insights. Biochim Biophys Acta 1717:67–88

Peng L, Yamamoto H, Shikanai T (2011) Structure and biogenesis of the chloroplast NAD(P)H dehydrogenase complex. Biochim Biophys Acta 1807:945–953

Ragan CI, Hinkle PC (1975) Ion transport and respiratory control in vesicles formed from reduced nicotinamide adenine dinucleotide coenzyme Q reductase and phospholipids. J Biol Chem 250:8472–8476

Rothery RA, Workun GJ, Weiner JH (2008) The prokaryotic complex iron-sulfur molybdoenzyme family. Biochim Biophys Acta 1778:1897–1929

Saraste M (1999) Oxidative phosphorylation at the fin de siecle. Science 283:1488–1493

Sazanov LA, Hinchliffe P (2006) Structure of the hydrophilic domain of respiratory complex I from *Thermus thermophilus*. Science 311:1430–1436

Steimle S, Bajzath C, Dorner K, Schulte M, Bothe V, Friedrich T (2011) Role of subunit NuoL for proton translocation by respiratory complex I. Biochemistry 50:3386–3393

Steuber J (2001) The Na^+-translocating NADH:quinone oxidoreductase (NDH I) from *Klebsiella pneumoniae* and *Escherichia coli*: Implications for the mechanism of redox-driven cation translocation by complex I. J Bioenerg Biomembr 33:179–186

Steuber J (2003) The C-terminally truncated NuoL subunit (ND5 homologue) of the Na^+-dependent complex I from *Escherichia coli* transports Na^+. J Biol Chem 278:26817–26822

Steuber J, Schmid C, Rufibach M, Dimroth P (2000) Na^+ translocation by complex I (NADH:quinone oxidoreductase) of *Escherichia coli*. Mol Microbiol 35:428–434

Stolpe S, Friedrich T (2004) The *Escherichia coli* NADH: ubiquinone oxidoreductase (complex I) is a primary proton pump but may be capable of secondary sodium antiport. J Biol Chem 279:18377–18383

Swartz TH, Ikewada S, Ishikawa O, Ito M, Krulwich TA (2005) The Mrp system: a giant among monovalent cation/proton antiporters? Extremophiles 9:345–354

Swartz TH, Ito M, Ohira T, Natsui S, Hicks DB, Krulwich TA (2007) Catalytic properties of *Staphylococcus aureus* and *Bacillus* members of the secondary cation/proton antiporter-3 (Mrp) family are revealed by an optimized assay in an *Escherichia coli* host. J Bacteriol 189:3081–3090

Tocilescu MA, Zickermann V, Zwicker K, Brandt U (2010) Quinone binding and reduction by respiratory complex I. Biochim Biophys Acta 1797:1883–1890

Torres-Bacete J, Nakamaru-Ogiso E, Matsuno-Yagi A, Yagi T (2007) Characterization of the NuoM (ND4) subunit in *Escherichia coli* NDH-1 – conserved charged residues essential for energy-coupled activities. J Biol Chem 282:36914–36922

Torres-Bacete J, Sinha PK, Castro-Guerrero N, Matsuno-Yagi A, Yagi T (2009) Features of subunit NuoM (ND4) in *Escherichia coli* NDH-1 topology and implications of conserved Glu(144) for coupling site 1. J Biol Chem 284:33062–33069

Tuppen HAL, Blakely EL, Turnbull DM, Taylor RW (2010) Mitochondrial DNA mutations and human disease. Biochim Biophys Acta 1797:113–128

Vignais PM, Billoud B (2007) Occurrence, classification, and biological function of hydrogenases: an overview. Chem Rev 107:4206–4272

Vignais PM, Billoud B, Meyer J (2001) Classification and phylogeny of hydrogenases. FEMS Microbiol Rev 25:455–501

Wikström M (1984) Two protons are pumped from the mitochondrial matrix per electron transferred between NADH and ubiquinone. FEBS Lett 169:300–304

Yagi T, Yano T, Di Bernardo S, Matsuno-Yagi A (1998) Procaryotic complex I (NDH-1), an overview. Biochim Biophys Acta 1364:125–133

Yano T, Yagi T, Sled VD, Ohnishi T (1995) Expression and characterization of the 66-kilodalton (NQO3) iron-sulfur subunit of the proton-translocating NADH-quinone oxidoreductase of *Paracoccus denitrificans*. J Biol Chem 270:18264–18270

Yano T, Sled VD, Ohnishi T, Yagi T (1996) Expression and characterization of the flavoprotein subcomplex composed of 50-kDa (NQO1) and 25-kDa (NQO2) subunits of the proton-translocating NADH-quinone oxidoreductase of *Paracoccus denitrificans*. J Biol Chem 271:5907–5913

Yip CY, Harbour ME, Jayawardena K, Fearnley IM, Sazanov LA (2011) Evolution of respiratory complex I "supernumerary" subunits are present in the alpha-proteobacterial enzyme. J Biol Chem 286:5023–5033

Part III
Mutations in Complex I Subunits and Medical Implications

Chapter 8
Characterization of Bacterial Complex I (NDH-1) by a Genetic Engineering Approach

Takao Yagi, Jesus Torres-Bacete, Prem Kumar Sinha, Norma Castro-Guerrero, and Akemi Matsuno-Yagi

Abstract Genetic engineering has been widely adopted in many different fields of research and study on complex I is no exception. Because of current limitations in the technology when studying eukaryotes, most work has been performed using bacterial complex I (NDH-1). The bacterial system has been proven to be an excellent model in which to investigate the structure and mechanism of complex I. Sequence information, from the early days, and the availability of 3D structures, in more recent years, of NDH-1 have significantly boosted the importance of the gene engineering approach. Mutational approach has helped in identifying a number of amino acid residues that are essential not only for structural integrity but also for cofactor ligation, substrate binding and electron transfer. More recently, this approach determined residues essential for proton translocation coupled to the electron transfer from NADH to quinone. Understanding the roles of these residues, together with structural information, should lead to the elucidation of the reaction mechanism of NDH-1/complex I. This chapter is concerned with that knowledge of NDH-1 which has been advanced by a genetic engineering approach and possible applications that may extend the research to mitochondrial complex I.

Keywords NADH dehydrogenase • Complex I • NDH-1 • Genetic engineering • Energy coupling • Mitochondrial disease

T. Yagi (✉) • J. Torres-Bacete • P.K. Sinha • N. Castro-Guerrero • A. Matsuno-Yagi
The Scripps Research Institute, MEM-256, 10550 North Torrey Pines Rd.,
La Jolla, CA 92037, USA
e-mail: yagi@scripps.edu

L. Sazanov (ed.), *A Structural Perspective on Respiratory Complex I: Structure and Function of NADH:ubiquinone oxidoreductase*, DOI 10.1007/978-94-007-4138-6_8,

8.1 Introduction

Characteristics of the bacterial proton-translocating NADH-quinone (Q) oxidoreductase (NDH-1) and its counterparts are summarized in Table 8.1. NDH-1 is generally composed of 14 different subunits with two arms, peripheral and membrane (Yagi 1993; Yagi et al. 1993, 1998; Yagi and Matsuno-Yagi 2003) (see Fig. 8.1 and Table 8.1). NDH-1 houses one FMN and eight to nine iron-sulfur clusters as cofactors. Significant progress in structural studies by X-ray crystallography has led to the construction of 3D models of the peripheral segment and the transmembrane segment of NDH-1 (Hinchliffe and Sazanov 2005; Sazanov and Hinchliffe 2006; Berrisford and Sazanov 2009; Efremov et al. 2010; Efremov and Sazanov 2011) (see also Sazanov's Chap. 1).

Genetic engineering is a powerful strategy that can be employed to clarify the structure and function of multi-subunit enzyme complexes, as exemplified by oxidative phosphorylation enzyme complexes (Zannoni 2004). In the field of NDH-1/complex I, Dupuis and colleagues introduced the genetic engineering approach in 1997 by using the *Rhodobacter capsulatus* NDH-1 gene cluster. They constructed a series of gene-deletion mutants (Dupuis et al. 1997), site-directed mutants (Chevallet et al. 1997), piericidin-insensitive mutants (Darrouzet and Dupuis 1997; Dupuis et al. 1995), and a model of Leber's hereditary optic neuropathy (Lunardi et al. 1998). Since then, this strategy has widely been used to investigate the structure and function of complex I/NDH-1. We describe advantages and disadvantages of genetic engineering and its contribution to the past, current, and future research on complex I/NDH-1.

8.2 General Approach

8.2.1 NDH-1 Versus Complex I (Mitochondrial Proton-Translocating NADH-Q Oxidoreductase)

All seven ND subunits in complex I are encoded by mitochondrial DNA (mtDNA) (the nomenclature for the subunit is based on *Escherichia coli* NDH-1 and bovine complex I unless otherwise noted, see Table 8.1). The only known exceptions are subunits NuoA (ND3) and NuoK (ND4L) of *Chlamydomonas reinhardtii* (Cardol et al. 2006). One of the keys to solving the energy-transducing mechanism of NDH-1/complex I is through the understanding of the properties of hydrophobic domain subunits. However genetic manipulation of mtDNA-encoded subunits is technically difficult, making it quite challenging to conduct systematic studies of these subunits (Remacle et al. 2006). In fact there has been only a limited number of studies that involves genetic manipulation of mtDNA-encoded proteins (e.g., mutated mtDNA from patients with mitochondrial diseases). Use of the bacterial system has advantages over mitochondrial complex I, as in bacteria there are no

Table 8.1 General characteristics of the bacterial NDH-1 subunits and their homologs

	E. coli[a] / *R. cap*[b].	*T. ther.*[c] / *P. den.*[d]	Bovine/ human complex I	Cofactor	TM[e]	Homologues in other proteins: *M. barkeri* Ech hydro-genase	Homologues in other proteins: *B. subtilis* Na^+/H^+ Mrp antiporter	Ref.
Hydro-philic domain	NuoB	Nqo6	PSST/ NDUFS7	N2,[4Fe-4S]	–	EchC	–	Hinchliffe and Sazanov (2005), Sazanov and Hinchliffe (2006), Hedderich (2004), Friedrich and Scheide (2000), Hirst et al. (2003), and Flemming et al. (2003)
	NuoD[f]	Nqo4	49 kDa/ NDUFS2	–	–	EchE	–	Hedderich (2004), Friedrich and Scheide (2000), Hinchliffe and Sazanov (2005), and Sazanov and Hinchliffe (2006)
	NuoC[f]	Nqo5	30 kDa/ NDUFS3	–	–	EchD	–	Hedderich (2004), Friedrich and Scheide (2000), Hirst et al. (2003), Hinchliffe and Sazanov (2005), Sazanov and Hinchliffe (2006), and Castro-Guerrero et al. (2010)
	NuoE	Nqo2	24 kDa/ NDUFV2	N1a,[2Fe-2S]	–	–	–	Hirst et al. (2003), Zu et al. (2002), Hinchliffe and Sazanov (2005), and Sazanov and Hinchliffe (2006)
	NuoF	Nqo1	51 kDa/ NDUFV1	FMN N3,[4Fe-4S]	–	–	–	Hirst et al. (2003), Hinchliffe and Sazanov (2005), Sazanov and Hinchliffe (2006), and Velazquez et al. (2005)
	NuoG	Nqo3	75 kDa/ NDUFS1	N1b,[2Fe-2S] N4,[4Fe-4S] N5,[4Fe-4S] N7[g],[4Fe-4S]	–	–	–	Hirst et al. (2003), Hinchliffe and Sazanov (2005), Sazanov and Hinchliffe (2006), Yano et al. (2003), Nakamaru-Ogiso et al. (2005), and Nakamaru-Ogiso et al. (2008)
	NuoI	Nqo9	TYKY/ NDUFS8	N6a,[4Fe-4S] N6b,[4Fe-4S]	–	EchF	–	Hedderich (2004), Friedrich and Scheide (2000), Hirst et al. (2003), Hinchliffe and Sazanov (2005), Sazanov and Hinchliffe (2006), and Yano et al. (1999)

(continued)

Table 8.1 (continued)

	E. coli[a] / *R. cap*[b].	*T. ther.*[c] / *P. den.*[d]	Bovine/human complex I	Cofactor	TM[e]	Homologues in other proteins: *M. barkeri* Ech hydro-genase	Homologues in other proteins: *B. subtilis* Na^+/H^+ Mrp antiporter	Ref.
Hydro-phobic domain	NuoA	Nqo7	ND3	–	3	–	–	Kao et al. (2004a, b), Di Bernardo and Yagi (2001), and Di Bernardo et al. (2000)
	NuoH	Nqo8	ND1	–	8	EchB	–	Sinha et al. (2009), Hedderich (2004), Friedrich and Scheide (2000), Roth and Hagerhall (2001), Efremov et al. (2010), and Efremov and Sazanov (2011)
	NuoJ	Nqo10	ND6	–	5	–	–	Kao et al. (2003, 2005a)
	NuoK	Nqo11	ND4L	–	3	–	MrpC	Hedderich (2004) and Mathiesen and Hagerhall (2003)
	NuoL	Nqo12	ND5	–	16	EchA	MrpA/MrpD	Hedderich (2004), Mathiesen and Hagerhall (2002), Friedrich and Scheide (2000), Efremov et al. (2010), Torres-Bacete et al. (2011), and Efremov and Sazanov (2011)
	NuoM	Nqo13	ND4	–	14	EchA	MrpA/MrpD	Hedderich (2004), Mathiesen and Hagerhall (2002), Torres-Bacete et al. (2007), Friedrich and Scheide (2000), Efremov et al. (2010), Torres-Bacete et al. (2011), and Efremov and Sazanov (2011)
	NuoN	Nqo14	ND2	–	14	EchA	MrpA/MrpD	Hedderich (2004), Mathiesen and Hagerhall (2002), Friedrich and Scheide (2000), Amarneh and Vik (2003), Efremov et al. (2010), and Efremov and Sazanov (2011)

[a]*Escherichia coli*
[b]*Rhodobacter capsulatus*
[c]*Thermus thermophilus* (*T. thermophilus* has an extra subunit Nqo15 in the hydrophilic domain Hinchliffe et al. 2006)
[d]*Paracoccus denitrificans*
[e]TM: Transmembrane helixes topology
[f]NuoC and NuoD subunits are fused in *E. coli*
[g]N7 cluster is not present in higher organisms; present only in some species of bacteria (e.g., *T. thermophilus*, *E. coli*)

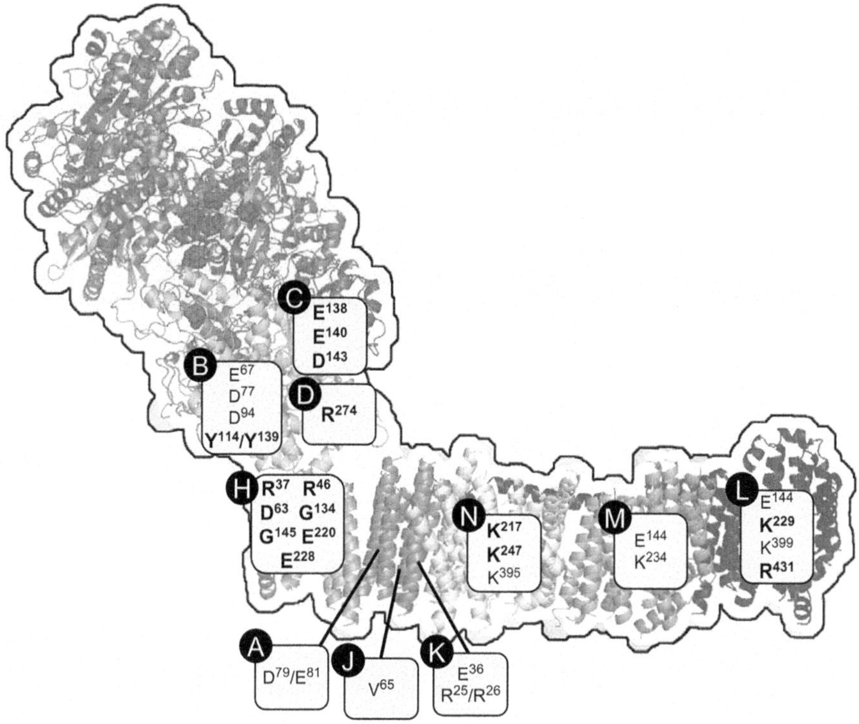

Fig. 8.1 Amino acid residues that have been reported to be essential for the energy-transducing NDH-1 activities. The letters in closed *circles* designate the *E. coli* naming of the subunits (NuoX). NuoCD is divided into the C and D regions. Residue numbers are based on *E. coli* NDH-1 subunits. Residue names in *bold* and regular letters represent the groups with and without inhibition of dNADH dehydrogenase activities, respectively

potential complications associated with import of proteins and cofactors that requires ATP and membrane potential. We have shown that *E. coli* NDH-1 works as a useful model of complex I by using systematically constructed mutants of membrane domain subunits NuoA (ND3), NuoH (ND1), NuoK (ND4L), NuoJ (ND6), NuoM (ND4), and NuoL (ND5) (Kao et al. 2004a, 2005a, b; Torres-Bacete et al. 2007, 2009, 2011; Sinha et al. 2009). In addition, mutants of NuoN (ND2) have been studied by Vik's group (Amarneh and Vik 2003). Details of these studies are presented later in this chapter.

8.2.2 *Structural Genes and URFs*

In bacterial systems, the genes encoding polypeptides that are subunits, or are involved in metabolic pathways, often construct a gene cluster or an operon. The NDH-1 gene cluster bears 14 structural genes that code for homologues of complex I subunits (Weidner et al. 1993; Dupuis et al. 1998; Yagi 1993). Dupuis and his colleagues (Dupuis et al. 1998) showed that these genes are essential by constructing gene-deletion mutants.

In the NDH-1 gene clusters from *R. capsulatus* (Dupuis et al. 1997, 1998) and *Paracoccus denitrificans* (Yagi 1993), seven and six unidentified reading frames (URFs) are present, respectively. It was unknown whether these URFs encode polypeptides and, if so, what function they have. Here, the gene disruption technique served as a powerful method. For example, Dupuis et al. (1998) constructed URF6- and URF7-deletion mutants and investigated the effect of the deletion on NDH-1 functions. It turned out that URF6 and 7 are not essential for the biosynthesis of NDH-1 in *R. capsulatus*.

8.2.3 Chromosomal Gene Manipulation Versus In Trans Complementation

The operon coding for *E. coli* NDH-1 is ~15 kb long (Yagi et al. 2004). Because this length makes it challenging to incorporate the whole operon into expression vectors, a site-specific mutation has traditionally been performed by complementation of a cassette-inserted gene carrying a site-specific mutation in the expression plasmid (designated *in trans* complementation) (Kurki et al. 2000; Amarneh and Vik 2003; Flemming et al. 2002, 2003; Chevallet et al. 1997; Lunardi et al. 1998; Kervinen et al. 2004). However, the *in trans* complementation procedure presents some problems when applied to a gene cluster. For example, a cassette inserted in chromosomal DNA might interrupt the expression of the downstream genes (Falk-Krzesinski and Wolfe 1998). This polar effect is avoided only when the mutated gene is at the last position in the operon (Flemming et al. 2003; Kervinen et al. 2004; Garofano et al. 2003). Another problem is that the mutated gene is under the control of a promoter in the expression plasmid, which often leads to varied levels of expression of the mutated subunit. In mutational studies of the NDH-1 using the *in trans* complementation approach, it has been reported that the enzymatic activities were significantly low (~20% of the wild type cells) even when the parent gene was used (Kao et al. 2004a). In addition, the NDH-1 content in host bacteria tends to vary. This variation of complex I content in cells is also often observed in eukaryotes.

An alternative method of site-directed mutation is to introduce mutations directly in chromosomal DNA (designated chromosomal DNA mutation) (Link et al. 1997; Lunardi et al. 1998; Flemming et al. 2003; Kao et al. 2004a, 2005a, b; Torres-Bacete et al. 2007, 2009, 2011; Sinha et al. 2009; Euro et al. 2008a, 2009). In this procedure, expression of all genes of the operon is regulated by authentic promoter. Although chromosomal DNA mutation is laborious and time-consuming, we adopted this technique to introduce mutations in NDH-1 subunits to minimize any complications derived from disruption of the operon. As anticipated, the NDH-1 was apparently expressed at the same level in all mutants as in the wild type. This is described in a series of our papers (Kao et al. 2005a, b; Torres-Bacete et al. 2007, 2011; Sinha et al. 2009).

8.2.4 Mutations of Amino Acids

The first step of the mutational approach is to investigate the roles of amino acid residues in the subunits of interest. It is especially important to find residues essential for the energy-transducing NDH-1 activities (dNADH oxidase and dNADH-Q reductase activities). What defines a residue to be essential? There are at least two categories by which consequences of a mutation can be assessed. The first group of mutants bring about almost complete inhibition of the energy-transducing (coupled) NDH-1 activities but scarcely affect non-coupled NADH dehydrogenase activities such as dNADH-$K_3Fe(CN)_6$ reductase and dNADH-HAR (hexammineruthenium) reductase activities. It is most likely that in that case replacing residues does not cause any major alteration in the structure of the peripheral domain. The second group of mutants display not only complete inhibition of the coupled activities but also significant suppression of non-coupled activities. In this case, mutations apparently lead to a drastic modification in the structure of NDH-1 (especially the structural arrangement of the peripheral segment). Western blotting analysis is useful to assess the presence and amount of each individual subunit. This requires specific antibodies directed to the subunits. Effects of mutations on the subunit assembly of NDH-1 can also be studied by blue native polyacrylamide gel electrophoresis (BN-PAGE). It should be noted that results obtained from BN-PAGE must be examined with caution because the procedure involves extraction of NDH-1 from the membranes by the use of detergents. It is possible that a mutation in the enzyme affects its tolerance to detergents. In fact we showed that in the case of the R25A/R26A double mutant of NuoK (ND4L), NDH-1 became extremely resistant to the detergent extraction (Kao et al. 2005b). In addition some NDH-1 mutants were much more susceptible to detergents. Therefore a combined analysis of BN-PAGE and western blotting procedures should be carried out for an assessment of subunit assembly in mutated NDH-1. The same concern regarding BN-PAGE analysis seems to be present when mutated NDH-1 is purified from the membranes.

The amino acid residues that have been identified to be essential for NDH-1 by using site-directed mutagenesis are summarized in Fig. 8.1 and Table 8.2. More details and their possible roles in the structure and function of NDH-1 will be discussed in the sections below.

8.3 Elucidation of Mechanism Using Genetic Engineering Approach

8.3.1 Electron Transfer of NDH-1

E. coli and *Thermus* NDH-1 contains 10 cofactors; one FMN, two binuclear Fe-S clusters (N1a and N1b), and seven tetranuclear Fe-S clusters (N2, N3, N5, N6a, N6b, and N7) (Ohnishi 1998; Nakamaru-Ogiso et al. 2008), all of which are housed

Table 8.2 Functionally relevant mutagenesis-based studies on the hydrophobic subunits of *E. coli* NDH-1

Subunit	Mutation	$K_3Fe(CN)_6$/HAR reductase activity	Quinone reductase activity	Comments/role	Ref.
NuoA	E51A	Not affected	30%	Mutation of the acidic residues affects the energy-transducing activity.	Kao et al. (2004a)
	D79N	Not affected	37%		
	E81A/Q	Not affected	36–50%		
	D79N/E81Q	Not affected	10		
NuoH	R37A	19%	3%	Cytoplasmic-side loops important for assembly and stability of NDH-1.	Sinha et al. (2009), Maliniemi et al. (2009), and Kervinen et al. (2006)
	R46A	47%	6%		
	D63A	18%	1%		
	G134V	54%	3%		
	G145V	50%	4%		
	R148A	49%	4%		
	V206E	67%	63%		
	R209F/K/H	70%	31–42%		
	D213E/N/A	68%	37–60%		
	E220A	49%	2%		
	E228A/K/Q/D	10–40%	8%		
	Y229H	71%	43%		
	R293M	71%	63%		
	D295A	74%	52%		
	L289C/V297P	67%	73%		
NuoJ	Y59F	Not affected	40–50%	Mutations affects energy-transducing NDH-1 activity; V65G did not show proton pump; change in ubiquinone affinity.	Patsi et al. (2008) and Kao et al. (2005a)
	G61V	Not affected	50%		
	V65G/L	Not affected	2–20%		
	M72V/C	Not affected	40–50%		

NuoK	E36A/Q	Not affected	7%	Charged transmembrane residues involved in NDH-1 activity. Role of Glu^{36} and Glu^{72} in proton translocation.	Kao et al. (2005b) and Kervinen et al. (2004)
	E72Q/A	Not affected	26–50%		
	R25A/K	Not affected	30%		
	R26A	Not affected	40%		
	R25A/R26A	Not affected	14%		
NuoM	E144A/Q	Not affected	2–10%	Charged transmembrane residues involved in NDH-1 activity. Role of Glu^{144} and Lys^{234} in proton translocation.	Torres-Bacete et al. (2007) and Euro et al. (2008a)
	E144D	Not affected	Not affected		
	K234A	Not affected	13–16%		
	K234R(Euro et al. 2008a)	70%	5%		
	K234R(Torres-Bacete et al. 2007)	Not affected	8%		
	K265A/R	Not affected	30–50%		
NuoN[a]	E133A/C/D	n.d.	70–80%	Mutations decrease NDH-1 oxidase and quinone reductase activities; possible interaction of NuoN with quinones; mutants showed reduced proton pump.	Amarneh and Vik (2003)
	K217C	No detection of NuoN	0%		
	H224K	n.d.	40%		
	K247C	Partial assembly of NDH-1	7%		
	Y300C/S	n.d	50%		
	K395C	Full assembly of NDH-1	5%		

[a] dNADH oxidase activity instead of dNADH-Quinone reductase activity

in the peripheral arm. Because of the sheer number of cofactors, it is not easy to assign them to the subunits and to determine the electron transfer pathway. Information on the primary sequence of the gene cluster of NDH-1 helped us to locate certain Fe-S cluster binding motifs (Yagi 1993; Yano et al. 1997; Dupuis et al. 1995; Weidner et al. 1993). As detailed in this section, genetic engineering techniques have contributed to this phase of research in two ways. The first is through expression of the individual subunits and the second is mutation of amino acid residues potentially involved in cluster ligation or electron transfer pathway.

Fe-S clusters of complex I/NDH-1 have been almost exclusively investigated by electron paramagnetic resonance (EPR) analysis. A major problem with a huge enzyme such as complex I/NDH-1 is that unambiguous identification of the EPR signals from the individual Fe-S clusters is nearly impossible when dealing with the whole complex. To tackle this problem, attempts were made to resolve complex I into subunits (Ohnishi et al. 1985). However the harsh conditions required for resolution of complex I resulted in destroyed or distorted EPR signals. The genetic engineering strategy to natively express individual subunits of interest has proved largely successful. It was determined that subunit NuoF (51k) carries FMN and cluster N3; NuoE (24k) houses N1a; NuoG (75k) bears N1b, N4, N5, and N7; NuoB (PSST) ligates N2. Although centers N6a and N6b are EPR-silent in complex I/NDH-1, the expressed NuoI/Nqo9 subunit exhibited EPR signals of two tetranuclear Fe-S clusters after reconstitution (Yano et al. 1999). These two Fe-S clusters *in situ* were eventually detected by using UV/vis spectrophotometry (Rasmussen et al. 2001).

The knowledge obtained by biochemical and genetic studies has immensely helped in identifying Fe-S clusters within the high resolution 3D model of the peripheral segment of *Thermus thermophilus* NDH-1 (Hinchliffe and Sazanov 2005). In this model, electron transfer along the main chain of cofactors was predicted to be FMN ↔ N3 ↔ N1b ↔ N4 ↔ N5 ↔ N6a ↔ N6b ↔ N2 ↔ Q (Hinchliffe and Sazanov 2005; Sazanov and Hinchliffe 2006). One way to verify the electron transfer pathway among cofactors is to measure real-time electron transfer in NDH-1. Once again, this is quite difficult because of overlapping EPR signals from multiple Fe-S clusters (Verkhovskaya et al. 2008; Albracht 2010a, b). An alternative approach would be to dissect the electron transfer pathway at various sites by using genetic engineering procedures. For example, one could attempt to alter the distance between electron carriers. However, there is no well-established method to systematically change the distance between carriers. A method that is technically more feasible will be to mutate the residues coordinating each cofactor. This strategy has been partially successful but, in some cases, destruction of cofactor ligation drastically affected the entire architecture of the enzyme complexes (Dupuis et al. 1998; Pohl et al. 2007). In order to circumvent this problem, a modified approach was employed to modulate the electron transfer by mutating the residues located in the proximity of cofactors but not the residues directly coordinating the cofactors. This approach might not exclude a possibility of alteration of the redox potential of cofactors. Nevertheless, with enough caution and appropriate controls, a genetic engineering approach should prove useful to clarify the electron transfer pathway of NDH-1. A successful example of this type of experiment has been reported by Euro et al. (2009) for a functional role of Glu^{95} interacting with FMN in the NuoF subunit.

The last step of the electron transfer within complex I/NDH-1is the reduction of Q. Using mutation procedures, amino acid residues involved in Q-binding and/or inhibitor-binding sites were extensively studied by the groups of Dupuis and Brandt (Dupuis et al. 1998, 2001; Tocilescu et al. 2010). Independently, Miyoshi's group investigated those binding sites by a chemical approach which utilizes photoaffinity derivatives of traditional and newly designed inhibitors/quinones (Kakutani et al. 2010; Sekiguchi et al. 2009; Murai et al. 2009; Hattori et al. 2008; Ichimaru et al. 2008). The information obtained by those two different methods provided clues to clarify the structure and function of NDH-1/complex I.

Recently, Friedrich and Koslowski (Wittekindt et al. 2009) proposed that the conserved aromatic side chains are involved in charge transfer between Fe-S clusters N3 and N2. The candidate residues are His^{327} and Phe^{328} in NuoD, Phe^{33} and Trp^{242} in NuoG, and Phe^{101} in NuoI. Using the *E. coli* NDH-1 as a model, study on the effects of mutations on the overall electron transfer rate of NDH-1 from NADH to UQ is in progress.

More recently we constructed some NuoK (ND4L) mutants that exhibited significantly suppressed energy-transducing dNADH- (exogenous) Q reductase activities, which retained dNADH oxidase activity comparable to the wild type (Torres-Bacete et al., unpublished results). We hypothesize that dissociation and/or association of exogenous Q substrate with the catalytic site of NDH-1 may take an extra pathway (possibly to endogenous Q-pool). If that is the case, these mutants will provide useful materials to investigate the mechanism of exogenous and endogenous Q reaction in NDH-1.

8.3.2 *Energy Coupling in NDH-1*

8.3.2.1 Membrane Domain Subunits

The membrane domain is considered to be involved in proton translocation and Q-binding. All seven subunits of the membrane arm are extremely hydrophobic. The first step of scrutinizing the membrane arm is to characterize the individual subunits.

Subunit NuoA (ND3) has three transmembrane helices (Di Bernardo et al. 2000) and constructs a transmembrane bundle with NuoK (ND4L) and NuoJ (ND6) (see Fig. 8.1) (Efremov et al. 2010; Efremov and Sazanov 2011). Two conserved residues, Asp^{79} and Glu^{81}, are located in the middle of the second transmembrane helix. D79A, D79N, E81A, and E81Q mutations partially suppressed the NADH-Q reductase activity. Double mutant D79N/E81Q almost completely lost the energy-transducing NDH-1 activities but did not display any loss of NADH-$K_3Fe(CN)_6$ reductase activity (Kao et al. 2004a).

NuoJ (ND6) contains five transmembrane helices. In this subunit, the only essential residue identified so far is Val^{65}, located in the third transmembrane helix which is considered to be a 'hot spot' for human mitochondrial diseases (Kao et al. 2005a).

NuoK (ND4L) is the smallest subunit of NDH-1 (Kao et al. 2005b) and harbors three transmembrane helices. The amino acid residues we targeted included two highly conserved Glu^{36} and Glu^{72} that are located in the middle of the membrane (Efremov and Sazanov 2011) and several Arg residues that are predicted to be on the cytoplasmic side. Mutations of Glu^{36} lead to almost null activity of coupled electron transfer with a concomitant loss of generation of an electrochemical gradient. Furthermore, a severe impairment of the coupled activities occurred when two vicinal Arg residues (Arg^{25} and Arg^{26}), on a short cytoplasmic loop, were simultaneously mutated.

NuoH (ND1) has eight transmembrane helices and is located near the peripheral segment (Fig. 8.1). The NuoH (ND1) subunit is one of the most conserved subunits of the membrane domain. Mutations in the NuoH (ND1) subunit from *Paracoccus* were reported to alter the affinity of NDH-1 for UQ (Kurki et al. 2000). Various inhibitors were reported to label the bovine ND1 (NuoH) subunit (Murai et al. 2007; Sekiguchi et al. 2009; Yagi and Hatefi 1988; Earley et al. 1987; Vgenopoulou et al. 2006). Subunit ND1 (NuoH) is believed to be involved in Q/inhibitor binding (Yagi and Matsuno-Yagi 2003). The recent systematic mutation experiments of *E. coli* NuoH indicate that NuoH, through its conserved residues in the cytoplasmic side loops, is involved in connecting the membrane arm and the peripheral arm, probably the NuoCD (30k + 49k) and NuoB (PSST) subunits of NDH-1 (Sinha et al. 2009).

The 3D model of the membrane domain of the *E. coli* NDH-1 contains a pattern of 14 transmembrane helices that are repeated three times, which represents subunits NuoL, M, and N (ND5, ND4, and ND2, respectively). These subunits are structurally similar to multi-subunit antiporters (Mathiesen and Hagerhall 2002; Kajiyama et al. 2009) indicating their involvement in the mechanism of proton translocation.

NuoN (ND2) contains 14 transmembrane helices and is located closest to the A/J/K bundle (see Fig. 8.1). Amarneh and Vik (2003), using site-directed mutagenesis, reported that three lysine residues, Lys^{217}, Lys^{247} and Lys^{395} were required for the energy-transducing NDH-1 activity. These lysines are all located in the transmembrane helices in the middle of membrane (Efremov and Sazanov 2011). Further analysis showed that NuoN (ND2) was absent in the membrane of mutant K217C. It seems likely that Lys^{217} is also involved in structural stability. In addition, mutant K247C had a lowered NADH dehydrogenase activity, suggestive of partial assembly of NDH-1. Mutation of Glu^{133}, which is in the membrane helix and is conserved in NuoL (Glu^{144}) and NouM (Glu^{144}), as well as antiporters, only moderately suppressed the energy-transducing NDH-1 activities.

NuoM (ND4) is the second largest subunit of the membrane domain. Initially topological experiments suggested that NuoM consisted of 14 helices and, of these, two were directed to the cytoplasmic phase (Torres-Bacete et al. 2009). However, the 3D model of the membrane arm of the *Thermus* NDH-1 based on X-ray crystallography revealed that all 14 helices are transmembrane and two of them (7 and 12) have discontinuous helix structure (Efremov et al. 2010; Efremov and Sazanov 2011).

Mutations of highly conserved Glu^{144} and Lys^{234} (Lys^{229} in NuoL and Lys^{217} in NuoN) lead to almost total elimination of energy-transducing NDH-1 activities (Torres-Bacete et al. 2007). Our results were confirmed by the Wikström's group (Euro et al. 2008a). Their NADH dehydrogenase activities were almost normal. These two residues are located in the transmembrane helices 5 and 7, respectively. NuoM was proposed to be involved in Q-binding. Photoaffinity labeling experiments suggested a possible location of this Q-binding site between Val^{184} and Asn^{206} (*E. coli* numbering) (Gong et al. 2003). However, so far no supportive evidence of the presence of Q-binding site in NuoM has been found by site-directed mutagenesis studies (Torres-Bacete et al. 2007; Euro et al. 2008a).

NuoL (ND5) is the largest subunit of the membrane arm. This subunit contains a 14 helix-bundle similar to NuoM (ND4) and NuoN (ND2). In addition, NuoL bears horizontally spanning helices through the membrane arm and one transmembrane helix located in the helix bundle of the NuoA, J, and K, suggesting that NuoL plays a role, at least, in the architecture of NDH-1 (Efremov et al. 2010; Efremov and Sazanov 2011). We have demonstrated that elimination of the last transmembrane helix (TM14) of NuoM and the TM16 (at least C-terminal seven residues) or together with the HL helix and the TM15 of the NuoL subunit lead to reduced stability of the membrane arm and therefore of the whole NDH-1 complex (Torres-Bacete et al. 2011). Comparison of this subunit among various organisms shows that it is highly conserved. Moreover its sequence is similar to the other two larger hydrophobic subunits of the membrane arm, NuoM (ND4) and NuoN (ND2), suggesting a common ancestor for the three subunits. In addition, a high homology of these subunits with certain Na^+/H^+ antiporters and energy-converting [Ni,Fe]-hydrogenase was observed (Hedderich 2004; Friedrich and Scheide 2000; Friedrich 2001). From this sequence homology, it was speculated that all the three subunits (NuoL, NuoM, and NuoN) are involved in the proton (or sodium) pump activity of NDH-1 (Steuber 2001). In fact, Steuber's group reported that a C-terminal truncated NuoL subunit from *E. coli* functions as a Na^+ translocator (Steuber 2003). Furthermore, bovine ND5 (NuoL) was labeled exclusively with a photoaffinity analog of the potent inhibitor, fenpyroximate (Nakamaru-Ogiso et al. 2003a) and, more recently, with an asimicin analog (Nakamaru-Ogiso et al. 2010a). The labeling was correlated with the inhibition of the enzyme activity. The labeling with a photoaffinity analogue of fenpyroximate was completely prevented by the pre-incubation with various complex I inhibitors including not only traditional inhibitors such as rotenone and piericidin A but also 5-(N-ethyl-N-isopropyl)-amiloride (EIPA), a specific inhibitor for Na^+/H^+ exchangers (Nakamaru-Ogiso et al. 2003a, b). This suggested that the NuoL subunit plays a role in linking between UQ-binding and proton translocation. The mutation study to clarify roles of the conserved residues in NuoL is in progress (Nakamaru-Ogiso et al. 2010b). So far, Glu^{144}, $\underline{\text{Lys}^{229}}$, Lys^{399}, and $\underline{\text{Arg}^{431}}$ are determined to be essential residues (those underlined show the second group of the essential residues). Of those four essential residues, Glu^{144}, Lys^{229}, Lys^{399} are located in the transmembrane helices and are involved in the energy-transducing NDH-1 activities but not in the NADH dehydrogenase activities.

8.3.2.2 Mechanism of Energy Coupling

The mode of coupling is debatable among direct (Dutton et al. 1998; Euro et al. 2008b), indirect (Yagi and Matsuno-Yagi 2003; Brandt 2006) and mixed (Friedrich 2001; Ohnishi et al. 2010; Berrisford and Sazanov 2009) coupling. The number and location of proton translocating sites needs to be determined. We will discuss how genetic engineering can help to obtain the answers to those questions.

Residues directly involved in proton translocation are predicted to fulfill the following conditions: (i) the residues are chargeable; (ii) the residues are located in the middle of transmembrane segment; (iii) loss of the residues almost completely inhibit the energy-transducing NDH-1 activities but scarcely affect the NADH dehydrogenase activities. As described above, the residues that meet these criteria are Glu^{36} in NuoK, Glu^{144}, Lys^{229} and Lys^{399} in NuoL, Glu^{144} and Lys^{234} in NuoM, Lys^{217}, Lys^{247} and Lys^{395} in NuoN. The conserved essential lysines are found in the middle of the membrane in putative proton-translocating half-channels (in NuoN it is Lys^{217} from discontinuous helix 7 and Lys^{395} from discontinuous helix 12) and in the connection between half-channels (Lys^{247}) (Efremov and Sazanov 2011). In addition, with regard to Glu^{144} in NuoM and Glu^{36} in NuoK, we have established the significance of the location of essential carboxyl group of Glu for proton translocation by shifting the position within membrane spanning helices (Torres-Bacete et al. unpublished results) (Torres-Bacete et al. 2009). Furthermore, it is reasonable that Glu^{144} in NuoL plays a role in the proton translocation. Together, our data suggest that NuoK, NuoL and NuoM are all involved in proton translocation. In Sazanov's hypothesis, NuoN has the same function as NuoL and NuoM (Efremov and Sazanov 2011). Although the 3D models of NuoL, NuoM, and NuoN individually contain two discontinuous transmembrane helices, which were hypothesized to be correlated with the ion translocation function by Screpani and Hunte (Screpanti and Hunte 2007), Glu^{144} in NuoM and NuoL are also conserved in Glu^{133} in NuoN. NuoN mutants E133A, E133C and E133D showed only moderately suppressed activities of the energy-transducing NDH-1 (20–30%) (Amarneh and Vik 2003). Therefore, at present, it is unclear whether NuoN is indeed a proton translocator (although effects of mutations in conserved essential lysines suggest that it might be). We can not exclude a possibility that NuoN has a bypass for proton translocation that is distinct from the Glu^{133}-involved pathway. NuoK is located in proximity of NuoB. If NuoB catalyzes proton translocation by direct (redox driven) coupling, a question arises as to whether NuoK pumps proton independent of NuoB or the two subunits cooperate in the energy transduction.

The structural work of NDH-1 by Sazanov's group (Efremov et al. 2010; Ohnishi 2010; Efremov and Sazanov 2011) may suggest the following: (i) the three antiporter-like subunits (NuoL, M, and N) function as the proton translocators; (ii) the three subunits (NuoL, M, and N) are combined by a horizontal helix of C-terminal region of NuoL; (iii) the C-terminal transmembrane helix of NuoL interacts with the helix bundle of NuoA, J and K and acts as an anchor for the horizontal helix. We have designed various truncated mutants by introducing a stop codon into the *nuoL* gene. Our preliminary results showed that removal of the end half of the anchor helix almost completely removed the energy-transducing NDH-1 activities and also suppressed NADH dehydrogenase activities by 40–50%, which is slightly higher than the activities of NuoL knock-out mutants (approximately 35%). Immunoblotting experiments showed that the membrane of these mutants contained NuoE, F and G at a level similar to the wild type. It seems likely that absence of a part of the anchoring helix of NuoL results in almost complete dysfunction of NDH-1, without drastic alteration in the overall architecture. A more detailed study on the functional role of horizontal helix and the anchor helix should provide a key to understanding the coupling mechanism of NDH-1.

8.3.3 Core of the Core Subunits

When an L-shaped structure was first reported for NDH-1/complex I by EM analyses (Hofhaus et al. 1991; Guénebaut et al. 1997, 1998), it was of interest to identify subunits involved in docking between the peripheral and the membrane segments. We carried out crosslinking experiments using *Paracoccus* NDH-1 (Di Bernardo and Yagi 2001; Kao et al. 2004b) and suggested that the NuoA (ND3) subunit interacts with the NuoB (PSST) and NuoD (49k) subunits (Kao et al. 2004b). In addition, mutation experiments of NuoH (ND1) using *E. coli* NDH-1 suggest that the cytoplasmic loops of NuoH (ND1) are associated with the peripheral segment (possibly NuoD (49k) and NuoB (PSST)) (Sinha et al. 2009). A mutation study of the NuoC (30k) subunit revealed that three highly conserved carboxyl residues (Glu^{138}, Glu^{140} and Asp^{143}), located in the third α-helix of this subunit, play a critically important role in the structural integrity of NDH-1 (Castro-Guerrero et al. 2010). These observations are quite interesting in the light of recent work on biogenesis of complex I, in which the 30k (NuoC) subunit was demonstrated to assemble with the ND1 (NuoH) at the very first stage (Perales-Clemente et al. 2010). Furthermore, according to the 3D model of NDH-1, NuoH and the membrane subunit bundle consisting of NuoA, J and K are located in proximity to the peripheral subunits NuoB and NuoD. These subunits may form a core of the core subunits. It is possible that the core serves as a support for a 'piston', postulated for the 'steam-engine' type mechanism of NDH-1/complex I. However, on the basis of our recent data, we speculated that NuoK is directly involved in the proton translocation and the recent structure indicates the presence of the putative fourth proton-translocation channel in the complex, formed mainly by NuoK with participation of NuoJ and NuoN (Efremov and Sazanov 2011). Both NuoD and NuoB subunits also participate in Q-binding and inhibitor-binding and are possibly involved in electron transfer-driven conformational change. Therefore remaining subunits are NuoA, NuoJ, NuoH and NuoC. It is of interest to investigate whether these four subunits indeed act as the prop of complex I/NDH-1 by using the genetic engineering approach. We could not exclude a possibility that a part of NuoD also participates in this role because, in certain NDH-1, NuoC and NuoD are fused.

8.4 Further Applications of Genetic Engineering

8.4.1 NDH-1 as a Model of Complex I Diseases

Another area of complex I research has been explored that takes advantage of genetic engineering by constructing a model system for mitochondrial diseases whose genetic mutations are well-defined. Using *E. coli* and *Paracoccus denitrificans* as the model systems, Hassinen's group investigated effect of mtDNA mutations in complex I diseases such as Leber's hereditary optic neuropathy (LHON) and mitochondrial encephalopathy, lactic acidosis and stroke-like episode syndrome (MELAS) (Kervinen et al. 2006; Patsi et al. 2008; Maliniemi et al. 2009).

To study the effect of MELAS mutations 3946 (E214K ND1) and 3949 (Y215H ND1), *Paracoccus* mutant (E234Q Nqo8) and *E. coli* mutant (Y229H NuoH), along with several other mutants were constructed and characterized (Kervinen et al. 2006). It turned out that the studied MELAS mutations are situated in a matrix-side loop, which appears to be highly sensitive to structural perturbations. The results provide new information on the function of the region affected by the MELAS mutations 3946 and 3949 that is not obtainable from patient samples or current eukaryote models (Kervinen et al. 2006).

To investigate the effect of LHON mutation 14484 (M64V and M64I ND6), the *E. coli* equivalents of the common LHON mutation 14484 (M64V ND6) as well as several other point mutations in its vicinity were constructed and characterized. The role of some mtDNA mutations in the development of LHON may be predicted by mutagenesis of equivalents in bacterial model systems as the current knowledge of the pathophysiology of LHON is insufficient and the delineation between pathogenic mutations and polymorphism is difficult (Patsi et al. 2008).

mtDNA is characterized by considerable sequence variation and specific combinations of polymorphisms can be used to classify mtDNA sequences into haplogroups. This variation is presumed to be neutral in effect but there is accumulating evidence against this assumption, as associations with diseases and mtDNA haplogroups have been reported. The most solid evidence of haplogroup contribution comes from the association between haplogroup J and LHON. The association is strongest with the LHON mutations 11778G>A and 14484T>C and it has been found to have functional significance (Ghelli et al. 2009). Hassinen and his colleagues have examined the functional role of 4216T>C, one of the substitutions present in mtDNA haplogroup J (Hinttala et al. 2010). Furthermore, they had previously detracted a rare 3866T>C polymorphism in the *ND1* gene of a child with skeletal muscle weakness, short stature, elevated blood lactate and an isolated complex I deficiency in the muscle and with mtDNA belonging to haplogroup J. Homologous substitutions were introduced into *E. coli* NDH-1. The NADH dehydrogenase activity, with either one or both mutations, was markedly decreased suggesting that 4216T>C and 3866T>C may have an effect on the structural integrity of complex I (Hinttala et al. 2010).

The bacterial systems seem to be useful to study, at least, mtDNA-linked complex I diseases. Recently, Tachibana et al. (2009) demonstrated that the mitochondrial genome can be efficiently replaced in mature nonhuman primate oocytes by spindle-chromosomal complex transfer from one egg to an enucleated, mitochondrial-replete egg. Nevertheless, until systematic manipulation of human mtDNA can be established, only the bacterial system will provide us with the useful information.

8.4.2 Mitochondrial Complex I Research and Genetic Engineering Approach

Structural aspect of mammalian complex I is more intricate than that of NDH-1. There are 31 "supernumerary" subunits in bovine complex I (Carroll et al. 2006). The term "supernumerary" may be misleading because some of them, such as subunit MWFE (NDUFA1), are known to be essential for the structure and function of complex I (Au et al. 1999; Yadava et al. 2002, 2007). GRIM-19 (Huang et al. 2004; Fearnley et al. 2001) and acyl carrier protein (NDUFAB1) (Schneider et al. 1995) give complex I additional functions which do not seem to be present in bacterial enzyme. Facilities for genetic manipulation, though somewhat limited, have become available in fungi and utilized in complex I research (Videira and Duarte 2002; Zickermann et al. 2008). However gene engineering techniques for mammals, to carry out systematic manipulation of nuclear DNA and mitochondrial DNA, are still lagging. Recent progress includes the construction of knock-out mutants of GRIM-19 and NDUFAB1 (Huang et al. 2004; Fearnley et al. 2001; Schneider et al. 1995). Also, knock-out mouse mutant of FDUFS4 (18k, AQDQ) was produced in body (Kruse et al. 2008) and brain (Quintana et al. 2010) and was characterized. The development in technology would eventually make it possible to apply a genetic engineering approach to a comprehensive study of mammalian complex I.

References

Albracht SP (2010a) The reaction of NADPH with bovine mitochondrial NADH:ubiquinone oxidoreductase revisited: I. Proposed consequences for electron transfer in the enzyme. J Bioenerg Biomembr 42:261–278

Albracht SP (2010b) The reaction of NADPH with bovine mitochondrial NADH:ubiquinone oxidoreductase revisited: II. Comparison of the proposed working hypothesis with literature data. J Bioenerg Biomembr 42:279–292

Amarneh B, Vik SB (2003) Mutagenesis of subunit N of the *Escherichia coli* complex I. Identification of the initiation codon and the sensitivity of mutants to decylubiquinone. Biochemistry 42:4800–4808

Au HC, Seo BB, Matsuno-Yagi A, Yagi T, Scheffler IE (1999) The *NDUFA1* gene product (MWFE protein) is essential for activity of complex I in mammalian mitochondria. Proc Natl Acad Sci USA 96:4354–4359

Berrisford JM, Sazanov LA (2009) Structural basis for the mechanism of respiratory complex I. J Biol Chem 284:29773–29783

Brandt U (2006) Energy converting NADH:quinone oxidoreductase (complex I). Annu Rev Biochem 75:69–92

Cardol P, Lapaille M, Minet P, Franck F, Matagne RF, Remacle C (2006) ND3 and ND4L subunits of mitochondrial complex I, both nucleus encoded in Chlamydomonas reinhardtii, are required for activity and assembly of the enzyme. Eukaryot Cell 5:1460–1467

Carroll J, Fearnley IM, Skehel JM, Shannon RJ, Hirst J, Walker JE (2006) Bovine complex I is a complex of forty-five different subunits. J Biol Chem 281:32724–32727

Castro-Guerrero N, Sinha PK, Torres-Bacete J, Matsuno-Yagi A, Yagi T (2010) Pivotal roles of three conserved carboxyl residues of the NuoC (30 k) segment in the structural integrity of proton-translocating NADH-quinone oxidoreductase from Escherichia coli. Biochemistry 49:10072–10080

Chevallet M, Dupuis A, Lunardi J, Van Belzen R, Albracht SPJ, Issartel JP (1997) The NuoI subunit of the *Rhodobacter capsulatus* respiratory complex I (equivalent to the bovine TYKY subunit) is required for proper assembly of the membraneous and peripheral domains of the enzyme. Eur J Biochem 250:451–458

Darrouzet E, Dupuis A (1997) Genetic evidence for the existence of two quinone related inhibitor binding sites in NADH-CoQ reductase. Biochim Biophys Acta 1319:1–4

Di Bernardo S, Yagi T (2001) Direct interaction between a membrane domain subunit and a connector subunit in the H^+-translocating NADH-quinone oxidoreductase. FEBS Lett 508:385–388

Di Bernardo S, Yano T, Yagi T (2000) Exploring the membrane domain of the reduced nicotinamide adenine dinucleotide-quinone oxidoreductase of *Paracoccus denitrificans*: characterization of the NQO7 subunit. Biochemistry 39:9411–9418

Dupuis A, Peinnequin A, Chevallet M, Lunardi J, Darrouzet E, Pierrard B, Procaccio V, Issartel JP (1995) Identification of five *Rhodobacter capsulatus* genes encoding the equivalent of ND subunits of the mitochondrial NADH- ubiquinone oxidoreductase. Gene 167:99–104

Dupuis A, Peinnequin A, Darrouzet E, Lunardi J (1997) Genetic disruption of the respiratory NADH-ubiquinone reductase of *Rhodobacter capsulatus* leads to an unexpected photosynthesis- negative phenotype. FEMS Microbiol Lett 148:107–114

Dupuis A, Chevallet M, Darrouzet E, Duborjal H, Lunardi J, Issartel JP (1998) The complex I from *Rhodobacter capsulatus*. Biochim Biophys Acta 1364:147–165

Dupuis A, Prieur I, Lunardi J (2001) Toward a characterization of the connecting module of complex I. J Bioenerg Biomembr 33:159–168

Dutton PL, Moser CC, Sled VD, Daldal F, Ohnishi T (1998) A reductant-induced oxidation mechanism for complex I. Biochim Biophys Acta 1364:245–257

Earley FGP, Patel SD, Ragan CI, Attardi G (1987) Photolabelling of a mitochondrially encoded subunit of NADH dehydrogenase with [^{3}H]dihydrorotenone. FEBS Lett 219:108–113

Efremov RG, Sazanov LA (2011) Structure of the membrane domain of respiratory complex I. Nature 476:414–420

Efremov RG, Baradaran R, Sazanov LA (2010) The architecture of respiratory complex I. Nature 465:441–445

Euro L, Belevich G, Verkhovsky MI, Wikstrom M, Verkhovskaya M (2008a) Conserved lysine residues of the membrane subunit NuoM are involved in energy conversion by the proton-pumping NADH:ubiquinone oxidoreductase (complex I). Biochim Biophys Acta 1777:1166–1172

Euro L, Bloch DA, Wikstrom M, Verkhovsky MI, Verkhovskaya M (2008b) Electrostatic interactions between FeS clusters in NADH:ubiquinone oxidoreductase (complex I) from Escherichia coli. Biochemistry 47:3185–3193

Euro L, Belevich G, Bloch DA, Verkhovsky MI, Wikstrom M, Verkhovskaya M (2009) The role of the invariant glutamate 95 in the catalytic site of complex I from Escherichia coli. Biochim Biophys Acta 1787:68–73

Falk-Krzesinski H, Wolfe AJ (1998) Genetic analysis of the *nuo* locus, which encodes the proton-translocating NADH dehydrogenase in *Escherichia coli*. J Bacteriol 180:1174–1184

Fearnley IM, Carroll J, Shannon RJ, Runswick MJ, Walker JE, Hirst J (2001) GRIM-19, a cell death regulatory gene product, is a subunit of bovine mitochondrial NADH:ubiquinone oxidoreductase (complex I). J Biol Chem 276:38345–38348

Flemming D, Hellwig P, Friedrich T (2002) Involvement of tyrosines 114 and 139 of subunit NuoB in the proton pathway around cluster N2 in *Escherichia coli* NADH:ubiquinone oxidoreductase. J Biol Chem 278:3055–3062

Flemming D, Schlitt A, Spehr V, Bischof T, Friedrich T (2003) Iron-sulfur cluster N2 of the *Escherichia coli* NADH:ubiquinone oxidoreductase (complex I) is located on subunit NuoB. J Biol Chem 278:47602–47609

Friedrich T (2001) Complex I: a chimaera of a redox and conformation-driven proton pump? J Bioenerg Biomembr 33:169–177

Friedrich T, Scheide D (2000) The respiratory complex I of bacteria, archaea and eukarya and its module common with membrane-bound multisubunit hydrogenases. FEBS Lett 479:1–5

Garofano A, Zwicker K, Kerscher S, Okun P, Brandt U (2003) Two aspartic acid residues in the PSST-homologous NUKM subunit of complex I from *Yarrowia lipolytica* are essential for catalytic activity. J Biol Chem 278:42435–42440

Ghelli A, Porcelli AM, Zanna C, Vidoni S, Mattioli S, Barbieri A, Iommarini L, Pala M, Achilli A, Torroni A, Rugolo M, Carelli V (2009) The background of mitochondrial DNA haplogroup J increases the sensitivity of Leber's hereditary optic neuropathy cells to 2,5-hexanedione toxicity. PLoS One 4:e7922

Gong X, Xie T, Yu L, Hesterberg M, Scheide D, Friedrich T, Yu CA (2003) The ubiquinone-binding site in NADH:ubiquinone oxidoreductase from *Escherichia coli*. J Biol Chem 278:25731–25737

Guénebaut V, Vincentelli R, Mills D, Weiss H, Leonard KR (1997) Three-dimensional structure of NADH-dehydrogenase from *Neurospora crassa* by electron microscopy and conical tilt reconstruction. J Mol Biol 265:409–418

Guénebaut V, Schlitt A, Weiss H, Leonard K, Friedrich T (1998) Consistent structure between bacterial and mitochondrial NADH:ubiquinone oxidoreductase (complex I). J Mol Biol 276:105–112

Hattori Y, Furuhata SI, Okajima M, Konno H, Abe M, Miyoshi H, Goto T, Makabe H (2008) Synthesis of pyranicin and its inhibitory action with bovine heart mitochondrial complex I. Org Lett 10:717–720

Hedderich R (2004) Energy-converting [NiFe] hydrogenases from archaea and extremophiles: ancestors of complex I. J Bioenerg Biomembr 36:65–75

Hinchliffe P, Sazanov LA (2005) Organization of iron-sulfur clusters in respiratory complex I. Science 309:771–774

Hinchliffe P, Carroll J, Sazanov LA (2006) Identification of a novel subunit of respiratory complex I from *Thermus thermophilus*. Biochemistry 45:4413–4420

Hinttala R, Kervinen M, Uusimaa J, Maliniemi P, Finnila S, Rantala H, Remes AM, Hassinen IE, Majamaa K (2010) Analysis of functional consequences of haplogroup J polymorphisms m.4216 T>C and m.3866 T>C in human MT-ND1 Mutagenesis of homologous positions in Escherichia coli. Mitochondrion 10:358–361

Hirst J, Carroll J, Fearnley IM, Shannon RJ, Walker JE (2003) The nuclear encoded subunits of complex I from bovine heart mitochondria. Biochim Biophys Acta 1604:135–150

Hofhaus G, Weiss H, Leonard K (1991) Electron microscopic analysis of the peripheral and membrane parts of mitochondrial NADH dehydrogenase (complex I). J Mol Biol 221:1027–1043

Huang G, Lu H, Hao A, Ng DC, Ponniah S, Guo K, Lufei C, Zeng Q, Cao X (2004) GRIM-19, a cell death regulatory protein, is essential for assembly and function of mitochondrial complex I. Mol Cell Biol 24:8447–8456

Ichimaru N, Murai M, Kakutani N, Kako J, Ishihara A, Nakagawa Y, Nishioka T, Yagi T, Miyoshi H (2008) Synthesis and characterization of new piperazine-type inhibitors for mitochondrial NADH-ubiquinone oxidoreductase (complex I). Biochemistry 47:10816–10826

Kajiyama Y, Otagiri M, Sekiguchi J, Kudo T, Kosono S (2009) The MrpA, MrpB and MrpD subunits of the Mrp antiporter complex in Bacillus subtilis contain membrane-embedded and essential acidic residues. Microbiology 155:2137–2147

Kakutani N, Murai M, Sakiyama N, Miyoshi H (2010) Exploring the binding site of deltalac-acetogenin in bovine heart mitochondrial NADH-ubiquinone oxidoreductase. Biochemistry 49:4794–4803

Kao MC, Di Bernardo S, Matsuno-Yagi A, Yagi T (2003) Characterization and topology of the membrane domain Nqo10 subunit of the proton-translocating NADH-quinone oxidoreductase of *Paracoccus denitrificans*. Biochemistry 42:4534–4543

Kao MC, Di Bernardo S, Perego M, Nakamaru-Ogiso E, Matsuno-Yagi A, Yagi T (2004a) Functional roles of four conserved charged residues in the membrane domain subunit NuoA of the proton-translocating NADH-quinone oxidoreductase from *Escherichia coli*. J Biol Chem 279:32360–32366

Kao MC, Matsuno-Yagi A, Yagi T (2004b) Subunit proximity in the H^+-translocating NADH-quinone oxidoreductase probed by zero-length cross-linking. Biochemistry 43:3750–3755

Kao MC, Di Bernardo S, Nakamaru-Ogiso E, Miyoshi H, Matsuno-Yagi A, Yagi T (2005a) Characterization of the membrane domain subunit NuoJ (ND6) of the NADH-quinone oxidoreductase from *Escherichia coli* by chromosomal DNA manipulation. Biochemistry 44:3562–3571

Kao MC, Nakamaru-Ogiso E, Matsuno-Yagi A, Yagi T (2005b) Characterization of the membrane domain subunit NuoK (ND4L) of the NADH-quinone oxidoreductase from *Escherichia coli*. Biochemistry 44:9545–9554

Kervinen M, Patsi J, Finel M, Hassinen IE (2004) A pair of membrane-embedded acidic residues in the NuoK subunit of *Escherichia coli* NDH-1, a counterpart of the ND4L subunit of the mitochondrial complex I, are required for high ubiquinone reductase activity. Biochemistry 43:773–781

Kervinen M, Hinttala R, Helander HM, Kurki S, Uusimaa J, Finel M, Majamaa K, Hassinen IE (2006) The MELAS mutations 3946 and 3949 perturb the critical structure in a conserved loop of the ND1 subunit of mitochondrial complex I. Hum Mol Genet 15:2543–2552

Kruse SE, Watt WC, Marcinek DJ, Kapur RP, Schenkman KA, Palmiter RD (2008) Mice with mitochondrial complex I deficiency develop a fatal encephalomyopathy. Cell Metab 7:312–320

Kurki S, Zickermann V, Kervinen M, Hassinen I, Finel M (2000) Mutagenesis of three conserved Glu residues in a bacterial homologue of the ND1 subunit of complex I affects ubiquinone reduction kinetics but not inhibition by dicyclohexylcarbodiimide. Biochemistry 39:13496–13502

Link AJ, Phillips D, Church GM (1997) Methods for generating precise deletions and insertions in the genome of wild-type *Escherichia coli*: application to open reading frame characterization. J Bacteriol 179:6228–6237

Lunardi J, Darrouzet E, Dupuis A, Issartel JP (1998) The *nuoM arg368his* mutation in NADH: ubiquinone oxidoreductase from *Rhodobacter capsulatus*: a model for the human *nd4*-11778 mtDNA mutation associated with Leber's hereditary optic neuropathy. Biochim Biophys Acta 1407:114–124

Maliniemi P, Kervinen M, Hassinen IE (2009) Modeling of human pathogenic mutations in Escherichia coli complex I reveals a sensitive region in the fourth inside loop of NUOH. Mitochondrion 9:394–401

Mathiesen C, Hagerhall C (2002) Transmembrane topology of the NuoL, M and N subunits of NADH:quinone oxidoreductase and their homologues among membrane-bound hydrogenases and bona fide antiporters. Biochim Biophys Acta 1556:121–132

Mathiesen C, Hagerhall C (2003) The 'antiporter module' of respiratory chain Complex I includes the MrpC/NuoK subunit – a revision of the modular evolution scheme. FEBS Lett 549:7–13

Murai M, Ishihara A, Nishioka T, Yagi T, Miyoshi H (2007) The ND1 subunit constructs the inhibitor binding domain in bovine heart mitochondrial complex I. Biochemistry 46:6409–6416

Murai M, Sekiguchi K, Nishioka T, Miyoshi H (2009) Characterization of the inhibitor binding site in mitochondrial NADH-ubiquinone oxidoreductase by photoaffinity labeling using a quinazoline-type inhibitor (dagger). Biochemistry 48:688–698

Nakamaru-Ogiso E, Sakamoto K, Matsuno-Yagi A, Miyoshi H, Yagi T (2003a) The ND5 subunit was labeled by a photoaffinity analogue of fenpyroximate in bovine mitochondrial complex I. Biochemistry 42:746–754

Nakamaru-Ogiso E, Seo BB, Yagi T, Matsuno-Yagi A (2003b) Amiloride inhibition of the proton-translocating NADH-quinone oxidoreductase of mammals and bacteria. FEBS Lett 549:43–46

Nakamaru-Ogiso E, Yano T, Yagi T, Ohnishi T (2005) Characterization of the iron-sulfur cluster N7(N1c) in the subunit NuoG of the proton-translocating NADH-quinone oxidoreductase from *Escherichia coli*. J Biol Chem 280:301–307

Nakamaru-Ogiso E, Matsuno-Yagi A, Yoshikawa S, Yagi T, Ohnishi T (2008) Iron-sulfur cluster N5 is coordinated by a HXXXCXXCXXXXXC motif in the nuog subunit of *E. coli* NADH:quinone oxidoreductase (complex I). J Biol Chem 283:25979–25987

Nakamaru-Ogiso E, Han H, Matsuno-Yagi A, Keinan E, Sinha SC, Yagi T, Ohnishi T (2010a) The ND2 subunit was labeled by a photoaffinity analogue of asimicin, a potent complex I inhibitor. FEBS Lett 584:883–888

Nakamaru-Ogiso E, Kao MC, Chen H, Sinha SC, Yagi T, Ohnishi T (2010b) The membrane subunit NuoL(ND5) is involved in the indirect proton pumping mechanism of *E. coli* complex I. J Biol Chem 285:39070–39078

Ohnishi T (1998) Iron-sulfur clusters semiquinones in complex I. Biochim Biophys Acta 1364:186–206

Ohnishi T (2010) Structural biology: piston drives a proton pump. Nature 465:428–429

Ohnishi T, Ragan CI, Hatefi Y (1985) EPR studies of iron-sulfur cluster in isolated subunits and subfractions of NADH-ubiquinone oxidoreductase. J Biol Chem 260:2782–2788

Ohnishi ST, Salerno JC, Ohnishi T (2010) Possible roles of two quinone molecules in direct and indirect proton pumps of bovine heart NADH-quinone oxidoreductase (complex I). Biochim Biophys Acta 1797:1891–1893

Patsi J, Kervinen M, Finel M, Hassinen IE (2008) Leber hereditary optic neuropathy mutations in the ND6 subunit of mitochondrial complex I affect ubiquinone reduction kinetics in a bacterial model of the enzyme. Biochem J 409:129–137

Perales-Clemente E, Fernandez-Vizarra E, Acin-Perez R, Movilla N, Bayona-Bafaluy MP, Moreno-Loshuertos R, Perez-Martos A, Fernandez-Silva P, Enriquez JA (2010) Five entry points of the mitochondrially encoded subunits in mammalian complex I assembly. Mol Cell Biol 30:3038–3047

Pohl T, Bauer T, Dorner K, Stolpe S, Sell P, Zocher G, Friedrich T (2007) Iron-sulfur cluster N7 of the NADH:ubiquinone oxidoreductase (complex I) is essential for stability but not involved in electron transfer. Biochemistry 46:6588–6596

Quintana A, Kruse SE, Kapur RP, Sanz E, Palmiter RD (2010) From the cover: complex I deficiency due to loss of Ndufs4 in the brain results in progressive encephalopathy resembling Leigh syndrome. Proc Natl Acad Sci USA 107:10996–11001

Rasmussen T, Scheide D, Brors B, Kintscher L, Weiss H, Friedrich T (2001) Identification of two tetranuclear FeS clusters on the ferredoxin-type subunit of NADH:ubiquinone oxidoreductase (complex I). Biochemistry 40:6124–6131

Remacle C, Cardol P, Coosemans N, Gaisne M, Bonnefoy N (2006) High-efficiency biolistic transformation of Chlamydomonas mitochondria can be used to insert mutations in complex I genes. Proc Natl Acad Sci USA 103:4771–4776

Roth R, Hagerhall C (2001) Transmembrane orientation and topology of the NADH:quinone oxidoreductase putative quinone binding subunit NuoH. Biochim Biophys Acta 1504:352–362

Sazanov LA, Hinchliffe P (2006) Structure of the hydrophilic domain of respiratory complex I from *Thermus thermophilus*. Science 311:1430–1436

Schneider R, Massow M, Lisowsky T, Weiss H (1995) Different respiratory defective phenotypes of *Neurospora crassa* and *Saccharomyces cerevisiae* after inactivation of the gene encoding the mitochondrial acyl carrier protein. Curr Genet 29:10–17

Screpanti E, Hunte C (2007) Discontinuous membrane helices in transport proteins and their correlation with function. J Struct Biol 159:261–267

Sekiguchi K, Murai M, Miyoshi H (2009) Exploring the binding site of acetogenin in the ND1 subunit of bovine mitochondrial complex I. Biochim Biophys Acta 1787:1106–1111

Sinha PK, Torres-Bacete J, Nakamaru-Ogiso E, Castro-Guerrero N, Matsuno-Yagi A, Yagi T (2009) Critical roles of subunit NuoH (ND1) in the assembly of peripheral subunits with the membrane domain of *Escherichia coli* NDH-1. J Biol Chem 284:9814–9823

Steuber J (2001) Na^+ translocation by bacterial NADH:quinone oxidoreductases: an extension to the complex-I family of primary redox pumps. Biochim Biophys Acta 1505:45–56

Steuber J (2003) The C-terminally Truncated NuoL Subunit (ND5 Homologue) of the Na^+-dependent complex I from *Escherichia coli* Transports Na^+. J Biol Chem 278:26817–26822

Tachibana M, Sparman M, Sritanaudomchai H, Ma H, Clepper L, Woodward J, Li Y, Ramsey C, Kolotushkina O, Mitalipov S (2009) Mitochondrial gene replacement in primate offspring and embryonic stem cells. Nature 461:367–372

Tocilescu MA, Zickermann V, Zwicker K, Brandt U (2010) Quinone binding and reduction by respiratory complex I. Biochim Biophys Acta 1797:1883–1890

Torres-Bacete J, Nakamaru-Ogiso E, Matsuno-Yagi A, Yagi T (2007) Characterization of the NuoM (ND4) subunit in *Escherichia coli* NDH-1: conserved charged residues essential for energy-coupled activities. J Biol Chem 282:36914–36922

Torres-Bacete J, Sinha PK, Castro-Guerrero N, Matsuno-Yagi A, Yagi T (2009) Features of subunit NuoM (ND4) in *Escherichia coli* NDH-1: topology and implication of conserved Glu144 for coupling site 1. J Biol Chem 284:33062–33069

Torres-Bacete J, Sinha PK, Matsuno-Yagi A, Yagi T (2011) Structural contribution of C-terminal segments of NuoL (ND5) and NuoM (ND4) subunits of complex I from Escherichia coli. J Biol Chem 286:34007–34014

Velazquez I, Nakamaru-Ogiso E, Yano T, Ohnishi T, Yagi T (2005) Amino acid residues associated with cluster N3 in the NuoF subunit of the proton-translocating NADH-quinone oxidoreductase from *Escherichia coli*. FEBS Lett 579:3164–3168

Verkhovskaya ML, Belevich N, Euro L, Wikstrom M, Verkhovsky MI (2008) Real-time electron transfer in respiratory complex I. Proc Natl Acad Sci USA 105:3763–3767

Vgenopoulou I, Gemperli AC, Steuber J (2006) Specific modification of a Na^+ binding site in NADH:quinone oxidoreductase from *Klebsiella pneumoniae* with dicyclohexylcarbodiimide. J Bacteriol 188:3264–3272

Videira A, Duarte M (2002) From NADH to ubiquinone in Neurospora mitochondria. Biochim Biophys Acta 1555:187–191

Weidner U, Geier S, Ptock A, Friedrich T, Leif H, Weiss H (1993) The gene locus of the proton-translocating NADH:ubiquinone oxidoreductase in *Escherichia coli*. Organization of the 14 genes and relationship between the derived proteins and subunits of mitochondrial complex I. J Mol Biol 233:109–122

Wittekindt C, Schwarz M, Friedrich T, Koslowski T (2009) Aromatic amino acids as stepping stones in charge transfer in respiratory complex I: an unusual mechanism deduced from atomistic theory and bioinformatics. J Am Chem Soc 131:8134–8140

Yadava N, Potluri P, Smith EN, Bisevac A, Scheffler IE (2002) Species-specific and mutant MWFE proteins: their effect on the assembly of a functional mammalian mitochondrial complex I. J Biol Chem 277:21221–21230

Yadava N, Potluri P, Scheffler IE (2007) Investigations of the potential effects of phosphorylation of the MWFE and ESSS subunits on complex I activity and assembly. Int J Biochem Cell Biol 40:447–460

Yagi T (1993) The bacterial energy-transducing NADH-quinone oxidoreductases. Biochim Biophys Acta 1141:1–17

Yagi T, Di Bernardo S, Nakamaru-Ogiso E, Kao MC, Seo BB, Matsuno-Yagi A (2004) NADH dehydrogenase (NADH-quinone oxidoreductase). In: Zannoni D (ed) Respiration in archaea and bacteria, vol 15. Kluwer Academic Publishers, Dordrecht, pp 15–40. Advances in photosynthesis and respiration. Govindjee

Ref Type: Serial (Book, Monograph)

Yagi T, Hatefi Y (1988) Identification of the DCCD-binding subunit of NADH-ubiquinone oxidoreductase (complex I). J Biol Chem 263:16150–16155
Yagi T, Matsuno-Yagi A (2003) The proton-translocating NADH-quinone oxidoreductase in the respiratory chain: the secret unlocked. Biochemistry 42:2266–2274
Yagi T, Yano T, Matsuno-Yagi A (1993) Characteristics of the energy-transducing NADH-quinone oxidoreductase of *Paracoccus denitrificans* as revealed by biochemical, biophysical, and molecular biological approaches. J Bioenerg Biomembr 25:339–345
Yagi T, Yano T, Di Bernardo S, Matsuno-Yagi A (1998) Procaryotic complex I (NDH-1), an overview. Biochim Biophys Acta 1364:125–133
Yano T, Chu SS, Sled' VD, Ohnishi T, Yagi T (1997) The proton-translocating NADH-quinone oxidoreductase (NDH-1) of thermophilic bacterium *Thermus thermophilus* HB-8: complete DNA sequence of the gene cluster and thermostable properties of the expressed NQO2 subunit. J Biol Chem 272:4201–4211
Yano T, Magnitsky S, Sled' VD, Ohnishi T, Yagi T (1999) Characterization of the putative 2x[4Fe-4S] binding NQO9 subunit of the proton-translocating NADH-quinone oxidoreductase (NDH-1) of *Paracoccus denitrificans*: expression, reconstitution, and EPR characterization. J Biol Chem 274:28598–28605
Yano T, Sklar J, Nakamaru-Ogiso E, Takahashi Y, Yagi T, Ohnishi T (2003) Characterization of cluster N5 as a fast-relaxing [4Fe-4S] cluster in the Nqo3 subunit of the proton-translocating NADH-ubiquinone oxidoreductase from *Paracoccus denitrificans*. J Biol Chem 278:15514–15522
Zannoni D (2004) Respiration in archaea and bacteria: diversity of prokaryotic electron transport carriers. Kluwer Academic Publishers, Dordrecht

Ref Type: Edited Book

Zickermann V, Drose S, Tocilescu MA, Zwicker K, Kerscher S, Brandt U (2008) Challenges in elucidating structure and mechanism of proton pumping NADH:ubiquinone oxidoreductase (complex I). J Bioenerg Biomembr 40:475–483
Zu Y, Di Bernardo S, Yagi T, Hirst J (2002) Redox properties of the [2Fe-2S] center in the 24 kDa (NQO2) subunit of NADH:ubiquinone oxidoreductase (complex I). Biochemistry 41:10056–10069

Chapter 9
Cellular Consequences of mtDNA-Encoded Mutations in NADH:Ubiquinone Oxidoreductase

Mina Pellegrini, Jan A.M. Smeitink, Peter H.G.M. Willems, and Werner J.H. Koopman

Abstract Accumulation of mitochondrial DNA (mtDNA) mutations appears to be causatively linked to mitochondrial disorders. However, given the complex genetics guiding mtDNA inheritance, the link between a specific mtDNA mutation and the ensuing clinical phenotype is not always clear. In cells, mitochondrial and cellular functions are intricately linked at the level of (energy) metabolism, signal transduction and apoptosis induction. Although mtDNA mutations generally induce malfunctioning of the oxidative phosphorylation (OXPHOS) system, it is currently unclear how mtDNA mutations induce cell dysfunction. Information about the latter is important to complement the biochemical characterization of mtDNA-linked diseases and thereby aid for the rational design of therapeutic strategies. Complex I (NADH:ubiquinone oxidoreductase) is the first complex of the OXPHOS system. This chapter first provides a brief summary of mtDNA

M. Pellegrini • P.H.G.M. Willems (✉)
Department of Biochemistry, Nijmegen Centre for Molecular Life Sciences, Radboud University Nijmegen Medical Centre, P.O. Box 9101, NL-6500 HB Nijmegen, The Netherlands

Department of Pediatrics, Nijmegen Centre of Mitochondrial Disorders, Radboud University Nijmegen Medical Centre, Nijmegen, The Netherlands
e-mail: p.willems@ncmls.ru.nl

J.A.M. Smeitink
Department of Pediatrics, Nijmegen Centre of Mitochondrial Disorders, Radboud University Nijmegen Medical Centre, Nijmegen, The Netherlands

W.J.H. Koopman
Department of Biochemistry, Nijmegen Centre for Molecular Life Sciences, Radboud University Nijmegen Medical Centre, P.O. Box 9101 NL-6500 HB Nijmegen, The Netherlands

L. Sazanov (ed.), *A Structural Perspective on Respiratory Complex I: Structure and Function of NADH:ubiquinone oxidoreductase*, DOI 10.1007/978-94-007-4138-6_9,

genetics in mammals, mtDNA mutations and their clinical phenotype, complex I structure and function and mtDNA-encoded mutations in complex I. Then, an overview is given of the current knowledge concerning the cell biological consequences of pathological mtDNA-encoded mutations in complex I.

Keywords OXPHOS • mtDNA mutation • LHON • Cell pathology • ATP • Reactive oxygen species

Abbreviations

$\Delta\psi$	mitochondrial membrane potential
ETC	electron transport chain
LHON	Leber Hereditary Optic Neuropathy
MELAS	Mitochondrial Encephalomyopathy, Lactic Acidosis, and Stroke-like episodes
MIM	mitochondrial inner membrane
MOM	mitochondrial outer membrane
MTND genes	mtDNA genes encoding structural subunits of CI
MPT	mitochondrial permeability transition pore
mtDNA	mitochondrial DNA
nDNA	nuclear DNA
OXPHOS	oxidative phosphorylation
PMF	proton-motive force
ROS	reactive oxygen species
TCA	tricarboxylic acid

9.1 Introduction

Mitochondria are organelles present in the cytoplasm of most eukaryotic cells. Each mitochondrion consists of a double membrane system in which a matrix compartment is surrounded by the mitochondrial inner membrane (MIM) and the mitochondrial outer membrane (MOM). Mitochondria are central to the aerobic synthesis of cellular ATP, a process that is driven by the four MIM-embedded protein complexes (CI-CIV) of the electron transport chain (ETC). Within the ETC, electrons are extracted from NADH (at CI) and $FADH_2$ (at CII), transported to CIII and CIV, and finally donated to molecular oxygen (O_2). The energy generated by the ETC is used to sustain an inward-negative proton gradient across the MIM. This creates an inward proton-motive force (PMF), which is used by the F_0F_1-ATPase (CV) for ATP production. Together with the ETC complexes, CV constitutes the oxidative phosphorylation (OXPHOS) system (Smeitink et al. 2001; Koopman et al. 2010).

Mitochondria contain their own genome in the form of multiple copies of a double-stranded DNA (mtDNA), which encodes for only a small set of genes that are crucial for proper biological functioning of the organelle. Out of a total of ~90 different proteins that make up the OXPHOS system, 13 are encoded by the mtDNA and the reminder by the nuclear DNA (nDNA). All the *trans*-acting factors directly involved in mitochondrial gene expression are also encoded by the nDNA, therefore mitochondrial biogenesis and proper OXPHOS function require a coordinated expression of mtDNA- and nDNA-encoded subunits in space and time (Scarpulla 2008).

In addition to aerobic ATP synthesis, mitochondria are important players in many other metabolic and signaling pathways including the tricarboxylic acid (TCA) cycle, fatty acid oxidation, calcium (Ca^{2+}) homeostasis, redox signaling and apoptosis induction (Walsh et al. 2009; Kowaltowski et al. 2009; Pradelli et al. 2010). Since a functional ETC is essential to maintaining these functions, it is not surprising that mutations in ETC proteins are linked to a broad spectrum of heterogeneous and often multisystemic diseases (Schapira 2006).

The first human mitochondrial disorder was described in the early 1960s (Luft et al. 1962) and almost simultaneously it was discovered that mitochondria contain their own DNA (Nass and Nass 1963). However, it took more than 20 years before the first pathogenic mtDNA mutations were identified (Holt et al. 1988). Since then, over 250 different pathogenic point mutations and a large number of mtDNA rearrangements (mainly deletions and duplications) have been linked to human pathologies (see: MITOMAP Human Mitochondrial Genome Database: http://www.mitomap.org). Because mitochondrial functioning strongly co-depends on proper expression of nDNA-encoded proteins, mutations in these proteins are also associated with mitochondrial disorders. Considering both nDNA and mtDNA mutations, the incidence of mitochondrial disorders in humans is at least one in 5,000 live births (Dimauro and Davidzon 2005). This makes these disorders far more common than previously anticipated. Patients with mutations in nDNA-encoded CI subunits generally present with Leigh syndrome, although genetic heterogeneity and different onset of the disease has been observed (Finsterer 2008). The cellular consequences of CI deficiency due to nDNA mutations, in patient-derived fibroblasts, have been reviewed in details elsewhere (Willems et al. 2008, 2009; Valsecchi et al. 2009; Distelmaier et al. 2009). Below we briefly introduce some basic concepts in mammalian mtDNA genetics. Then, the relationship between the nature of the mtDNA mutations and the clinical phenotype is discussed. Finally, an overview is presented of the current cell biological aspects associated with pathogenic mutations in mtDNA-encoded CI subunits.

9.2 Mammalian mtDNA Genetics

In mammalian cells, each mitochondrion contains between 2 and 10 copies of mtDNA (Shuster et al. 1988; Wiesner et al. 1992). Moreover, the number of mitochondria varies between cells, possibly due to the metabolic state of the cell and its tissue of origin. For a given individual, all mtDNA copies are supposed to be identical, a condition known as '***homoplasmy***'. However, in case of a mtDNA mutation: (a) the mutation can arise in one or multiple mtDNA copies; (b) the mutation can be maintained or amplified at different levels, and (c) the mutation can coexist with wild-type mtDNA (‘***heteroplasmy***’).

During cell division, mitochondria and their genomes are randomly distributed between cells (‘***mitotic segregation***’) and hence, starting from a certain heteroplasmy percentage, an unequal partition of mutated mtDNA molecules between cells will lead to different cellular heteroplasmy levels. This results in genetic drift towards either purely mutant or wild type mtDNA.

The ratio between mutant and wild type mtDNA in a heteroplasmic cell can also rapidly change due to ‘***relaxed replication***’ of the mtDNA. This phenomenon is characterized by mtDNA replication occurring independently from nDNA replication throughout the cell cycle. As a consequence, a certain mtDNA molecule could be replicated many times or not at all during subsequent cell divisions, leading to different levels of mutated mtDNA in different cells.

The above factors might explain the ‘***threshold effect***’ in mtDNA-linked diseases, which requires a mutation to reach a certain percentage, usually higher than 60–90%, in order to manifest its pathological effects (Rossignol et al. 2003; Tuppen et al. 2010).

It has been recognized that mammalian mtDNA is only transmitted via the female germ line. Despite a single described case of paternal mtDNA transmission (Schwartz and Vissing 2002), it is today well accepted that sperm mitochondria are lost during early embryogenesis due to a ubiquitin-dependent mechanism, although they are transferred to the oocyte during fertilization (Sutovsky et al. 1999). During oogenesis the ‘***bottleneck phenomenon***’ occurs, by which only a small subset of the maternally inherited mtDNA molecules are amplified and then transmitted to the offspring. This phenomenon contributes to the rapid segregation of new mtDNA variants, which are either lost during transmission, or reach high levels and affect the progeny. If the bottleneck phenomenon is responsible for the individual load of mutated mtDNA, the ‘***purifying selection***’ during oocyte development ensures that deleterious mtDNA mutations, such as those affecting protein-coding genes, are selected against (Stewart et al. 2008). It is still unclear whether the bottleneck mechanism and purifying selection are linked processes, but it is likely that both have a profound impact on the transmission of mtDNA mutations in human diseases.

9.3 Mutations in mtDNA and Their Clinical Phenotype

Diseases due to mtDNA mutations are well defined and their genetic background is generally known. However, the relationship between the mtDNA defect and the clinical phenotype is usually unclear because the nature of the protein alteration due to a specific mutation does not suffice to explain the observed pathology. In general, tissues and organs that heavily depend on oxidative phosphorylation for their energy supply are most severely affected. Among these organs, brain malfunctioning is generally observed, leading to neurological defects, but also muscular, endocrine and ophthalmologic dysfunctions are well documented (Chinnery et al. 1999; Chinnery and Schon 2003). Although the prime cause of these dysfunctions is obviously an OXPHOS defect, the cellular context might be very important for the clinical features of the diseases, thereby contributing to their heterogeneity. This is because mitochondria not merely function in ATP generation, but they are involved in many other cell metabolic pathways that can be differently regulated in different cell types. Moreover, genetic and environmental factors, such as the mitochondrial haplogroup of the patient, could also influence the phenotype and progression of the disease in a patient-specific way, by a mechanism that is still poorly understood (Gomez-Duran et al. 2010). Although the precise threshold value for the biochemical manifestation of the deficiency varies from mutation to mutation, the percentage of mutated mtDNA in the affected tissues generally correlates with the severity of the disease (Chinnery et al. 1997). Quite often the percentage of mutated mtDNA in individual tissues can change during development, potentially influencing the individual phenotype. A possible explanation resides in the fact that mtDNA replication occurs continuously and independently from nDNA replication. In theory this mechanism could lead to changes in the proportion of mutated mtDNA of a patient, providing also a possible explanation for the late onset and progression of some mtDNA disorders. In a dividing tissue, the proportion of mutated mtDNA may also change because of the mitotic segregation of the mtDNA during cell division, thereby affecting the clinical phenotype of the patients as they grow older. Furthermore, the bottleneck phenomenon can explain the clinical variability of a mitochondrial disorder among siblings, while heteroplasmy and the threshold effect likely explicate why patients harboring pathogenic mtDNA defects often have a mixture of mutated and wild type mtDNA molecules.

In principle, all the above factors can influence the onset of the symptoms, creating the variability in the observed phenotype and in the penetrance of different mitochondrial disorders. But still some unknown mechanisms play a role in the manifestation of mtDNA disorders. For instance, mtDNA mutations will not be transmitted in a purely Mendelian way because of their maternal inheritance, in contrast to nDNA defects. However, the inheritance pattern of mtDNA-related diseases can vary from case to case: some patients appear to be sporadic cases, while others are clearly familiar (Chinnery and Schon 2003). It is also puzzling that many mitochondrial disorders affect multiple organs, whereas others display an organ-specific phenotype as for Leber's hereditary optic neuropathy (LHON) (Tonska et al. 2010).

9.4 Complex I Structure and Function

Within the family of mitochondrial disorders, CI deficiency is the most frequently observed defect (Distelmaier et al. 2009). Mammalian CI or NADH:ubiquinone oxidoreductase (EC 1.6.5.3) is an L-shaped protein complex comprising a peripheral arm that protrudes into the mitochondrial matrix, and a second, MIM-embedded arm. The CI holocomplex consists of 45 subunits, seven of which are encoded by the mtDNA. Detailed information about CI biogenesis can be found elsewhere (Koopman et al. 2010). Evolutionary analysis revealed that the CI catalytic function requires only a minimal set of 14 'core' subunits, encoded by both the nDNA (NDUFV1, NDUFV2, NDUFS1, NDUFS2, NDUFS3, NDUFS7, NDUFS8) and the mtDNA (ND1, ND2, ND3, ND4, ND4L, ND5, ND6). Functionally, CI extracts electrons from NADH and donates them to ubiquinone. From thereon, electrons are transported further through the ETC. The energy released by the electron transport in CI is indirectly coupled to protons (H^+) pumping across the MIM. In fact, the X-ray crystal structure of the complete *Y. lipolytica* CI suggests that an helical transmission element is needed to bridge together the conformational changes occurring in the peripheral arm of CI due to the electrons transfer, with the proton-pumping machinery located in the membrane arm of CI (Hunte et al. 2010). Combining X-ray structural analysis of the MIM-embedded part of *E. coli* CI, and of the entire CI from *T. thermophilus*, it was recently reported that the proton-pumping machinery has an unusual 'piston' mechanism involving the NuoL (*i.e.* mammalian ND5), NuoM (ND4) and NuoN (ND2) subunits that controls the H^+ movement across the MIM (Efremov et al. 2010). In this way CI, together with CIII and CIV contributes to maintain the inward PMF and inside negative membrane potential ($\Delta\psi$) across the MIM.

9.5 Mutations in mtDNA-Encoded Genes of Complex I

In the case of CI, pathological mutations in nDNA- and mtDNA-encoded subunits can be associated with a similar clinical phenotype of Leigh syndrome. In this sense, Leigh syndrome-associated mutations were found in: (a) the nDNA-encoded subunits NDUFS1 (Martin et al. 2005); NDUFS3 (Benit et al. 2004); NDUFS7 (Lebon et al. 2007); NDUFS8 (Loeffen et al. 1998) and NDUFA2 (Hoefs et al. 2008), and (b) the mtDNA-encoded subunits ND3 (Sarzi et al. 2007); ND5 (Taylor et al. 2002) and ND6 (Kirby et al. 2000).

On the other hand, the same mutation can cause very different clinical phenotypes. In this respect the G14459A mutation in the mtDNA gene encoding the ND6 subunit of CI can be associated with classical LHON and dystonia (Jun et al. 1994), but also with Leigh syndrome (Kirby et al. 2000). As explained above, mtDNA mutations are usually maternally inherited, and their clinical expression depends on the mutant load in the affected tissue, the energy demand of the same and the site of the mutation. In general, mtDNA mutations result in diseases with a milder phenotype than those caused by mutations in nDNA genes, often with a late onset, although some severe and infantile-onset mtDNA diseases have been reported (Sue et al. 1999; Wong et al. 2001). Below, we will focus on mutations in mtDNA genes encoding structural subunits of CI (MTND genes). An overview of the several mtDNA mutations causing human diseases can be found elsewhere (Tuppen et al. 2010).

Although MTND gene mutations can be associated with variable clinical phenotypes, the most commonly observed disorder is LHON. This is a maternally transmitted disease characterized by acute or subacute bilateral visual loss, affecting mainly

Table 9.1 Confirmed pathological mutations in MTND genes

Affected gene	Mutation and nucleotide position	Clinical phenotype	Reference first describing mutation
MTND1	**G3460A**	**LHON**	Howell et al. (1991) and Huoponen et al. (1991)
	G3635A	LHON	Brown et al. (2001)
	G3697A	MELAS/LS	Kirby et al. (2004)
MTND3	T10158C	Leigh disease	Lebon et al. (2003)
	T10191C	ESOC/Leigh-like disease/LS	Taylor et al. (2001)
	G10197A	LS/dystonia/stroke	Sarzi et al. (2006)
MTND4	C11777A	Leigh disease	Komaki et al. (2003) and Deschauer et al. (2003)
	G11778A	**LHON**	Wallace et al. (1988)
	G11778A	Progressive dystonia	Meire et al. (1995)
MTND5	T12706C	Leigh disease	Taylor et al. (2002)
	G13513A	MELAS/Leigh disease	Santorelli et al. (1997)
	A13514G	MELAS/Leigh disease	Corona et al. (2001)
MTND6	G14459A	LDYT/Leigh disease	Jun et al. (1994)
	C14482G	LHON	Valentino et al. (2004)
	T14484C	**LHON**	Johns et al. (1992) and Mackey and Howell (1992)
	T14487C	Dystonia/Leigh disease/OA/Ataxia	Ugalde et al. (2003) and Solano et al. (2003)
	C14568T	LHON	Wissinger et al. (1997)

This table was composed using Mitomap (www.mitomap.org). Confirmed status indicates that at least two or more independent laboratories have published reports on the pathogenicity of a specific mutation. The mutations reported in bold are considered the three primary LHON mutations
Abbreviations: *ESOC* Epilepsy, Strokes, Optic atrophy, and Cognitive decline, *LHON* Leber Hereditary Optic Neuropathy, *LDYT* LHON and DYsTonia, *LS* Leigh Syndrome, *MELAS* Mitochondrial myopathy, Encephalopathy, Lactic Acidosis, and Stroke-like episodes, *OA* OligoAsthenoteratozoospermia

young men. The first report of a relationship between a mtDNA mutation and LHON was published in 1988 by Wallace et al. (1988), and since then several other mutations in the mtDNA were reported as being linked to LHON (Wong 2007). Quantitatively, about half of the pathological mutations found in the MTND genes are LHON mutations, which were identified in several unrelated families and are generally homoplasmic (source: MITOMAP). The large majority of the LHON cases (*i.e.* 95%) are caused by the three primary mtDNA mutations: (a) G11778A in the ND4 gene (Wallace et al. 1988), (b) G3460A in the ND1 gene (Huoponen et al. 1991; Howell et al. 1991), and (c) T14484C in the ND6 gene (Johns et al. 1992; Mackey and Howell 1992). Among these mutations, the G3460A is associated with the most severe phenotype, while the T14484C mutation is the mildest (Yen et al. 2006). Other rare LHON mutations have mainly been found in the two LHON mutation hotspot genes ND1 and ND6 (Yen et al. 2006). The remainder of the MTND gene mutations are connected to Leigh and MELAS syndromes (Wong 2007) or even to overlapping LHON/MELAS (Blakely et al. 2005), MELAS/LHON/Leigh (Liolitsa et al. 2003) and MELAS/Leigh (Santorelli et al. 1997) (Table 9.1).

9.6 Cellular Consequences of Mutations in mtDNA-Encoded CI Subunits

Given the central position of mitochondria in cell functioning and survival, it is evident that the study of the consequences of mutations in mtDNA-encoded CI subunits, in the context of living cells, represents an advantage compared to merely *in vitro* systems. At the cellular level, mtDNA mutations are generally analyzed using the 'cybrid' technology. Cybrids, or cytoplasmic hybrid cells, are generated by introducing the mutated mtDNA into cells depleted of their endogenous mtDNA (ρ0 cell), so that the effect of the mtDNA mutation can be studied in the controlled nuclear background of the recipient cells (Wiseman and Attardi 1978). The first created cybrid cell line was used to gain insight into the mechanisms of a known disease-associated mtDNA mutation, by transferring the mtDNA from myoblasts of patients harboring such a mutation into mtDNA-depleted osteosarcoma cells (Chomyn et al. 1991). However, the nuclear background of the ρ0 cell line used to generate the cybrids can lead to different results. For instance, Yoneda and colleagues (Yoneda et al. 1992) reported a shift towards mutated mtDNA in several heteroplasmic cybrid clones (derived from 143B osteosarcoma cells) containing mtDNA from MELAS patients. When the same mutated mtDNA was introduced into a different ρ0 cell line (derived from A549 cells), the shift observed was towards pure wild type mtDNA rather than mutant (Dunbar et al. 1995). This is not unexpected, since it is known that mitochondrial physiology varies between different cells, tissues, organs and organisms. It is also important to realize that most of the ρ0 cell lines are derived from tumor cells, which are aneuploid, a condition that might affect the nDNA-mtDNA stoichiometry of the ETC subunits. Moreover, tumor cells are generally highly glycolytic, in contrast to the highly aerobic tissues that are mostly affected in patients with mitochondrial diseases. This suggests that using cybrid cell lines with a more aerobic metabolism might be useful (Swerdlow 2007).

Nevertheless, the cybrid technology has been an invaluable tool to investigate human diseases linked to mtDNA mutations. In this sense it has been applied in complementation studies of heteroplasmic mtDNA mutations, to investigate the pathogenic effects of unknown mtDNA mutations (Trounce and Pinkert 2007) and mtDNA integrity in common neurodegenerative disorders (Swerdlow 2007).

Below we provide an overview of the current knowledge concerning the cellular consequences of mutations in mtDNA-encoded CI subunits with respect to: ATP generation, oxygen consumption, reactive oxygen species (ROS) production, mitochondrial membrane potential, permeability transition pore (MPT) opening, calcium (Ca^{2+}) homeostasis and apoptosis.

9.6.1 ATP Generation and Oxygen Consumption

Since mtDNA encodes for core subunits of CI, CIII, CIV and CV of the OXPHOS system, it is expected that mutations in such genes will affect mitochondrial bioenergetics and ATP production in particular. The biochemical consequences of the three primary LHON mutations have been described for different cybrid models (Brown et al. 2000; Baracca et al. 2005), but also for isolated mitochondria from lymphocytes (Majander et al. 1991), platelets (Howell et al. 1991) and fibroblasts (Cock et al. 1999; Chevrollier et al. 2008) of LHON patients. Given the fact that these mutations are all present in mtDNA-encoded CI subunits, one would expect the induction of similar bioenergetic defects. Although the reported results might be affected by the nuclear background of the used system, in general only modest or slight changes in mitochondrial bioenergetics have been detected for those LHON mutations (Brown et al. 2000), whereas no differences were observed between cells derived from LHON patients and cells from unaffected LHON carriers (Baracca et al. 2005). The G3460A and G11778A mutations seem to cause the most severe impairments, with a 79% CI enzymatic activity reduction, relative to controls, for the G3460A mutation and 25% reduction for the G11778A mutation. No differences have been detected for the T14484C mutation compared to the controls (Brown et al. 2000; Baracca et al. 2005).

Polarographic measurements of CI-dependent oxygen consumption in isolated mitochondria, showed a 10–15% reduction for the milder T14484C mutation, while a 30–36% decrease was found for the G11778A mutation, and a 20–28% reduction for the most severe G3640A, all compared to the controls (Brown et al. 2000). These data nicely correlate with the clinical phenotype of the three primary LHON mutations, since the T14484C mutation is associated with the highest spontaneous visual recovery rate among all the LHON mutations (Johns et al. 1993). Moreover this same mutation is linked with a specific mtDNA background, European haplogroup J, unlike the more severe G3460A mutation (Brown et al. 1997) and this might explain the different biochemical profile of the T14484C mutation.

Other studies tested the sensitivity of CI to different specific CI inhibitors leading to the conclusion that all three primary LHON mutations affect the interaction of CI with ubiquinone (Carelli et al. 1997, 1999). As a consequence, one would expect a defective electron flow and proton pumping through CI, causing eventually a decrease in ATP production. Indeed, this was the case when CI-driven ATP synthesis was measured in isolated mitochondria from cybrid cell lines carrying the three primary LHON mutations, with the G11778A one presenting the most severe ATP reduction, compared to cybrid controls (Baracca et al. 2005). A defective ATP synthesis was also confirmed in mitochondria from peripheral blood mononuclear cells (PBMC) of several LHON carriers with normal vision and visually impaired LHON patients (Korsten et al. 2010), as well as *in vivo* using ^{31}P magnetic resonance spectroscopy (MRS) of skeletal muscle and brain of LHON patients (Lodi et al. 1997, 2002). However, the total cellular ATP content of the LHON cybrid cell lines, and that of the PBMC, was similar to controls. This suggests the action of compensatory mechanisms, probably through an up-regulation of the glycolytic and CII/glycerol 3-phosphate dehydrogenase pathways (Baracca et al. 2005), or a compensatory increase of mitochondrial density and/or mass (Korsten et al. 2010).

Although the above results at first seem contradictory, they probably reflect different energy demands and modes of ATP production (*i.e.* OXPHOS *vs.* glycolysis) in the investigated systems: CI-driven ATP measurements in isolated mitochondria are done in the presence of an excess of ADP, resembling the state 3 of mitochondrial respiration (*i.e.* a state of high energy demand). The mtDNA mutation thus would only enhance the defective phenotype when ATP production is stimulated above basal levels, a situation probably not achieved in the whole cell experiments.

9.6.2 ROS Production

Mitochondrial energy metabolism constitutes an important source of ROS (Murphy 2009; Koopman et al. 2010), with the oxygen superoxide anion ($O_2^{\bullet-}$) being the primary mitochondrial ROS that is generated mainly at the level of CIII and CI (Kowaltowski et al. 2009). ROS can not only damage biomolecules like proteins, lipids and DNA (oxidative stress), but also participate in many signaling pathways. Therefore, whether a cell will suffer from oxidative stress, depends on the balance between the increase in ROS generation and their removal by endogenous antioxidant defense systems.

In principle, mtDNA mutations could lead to increased ROS production (Liu et al. 2009; Wu et al. 2010) and a role for ROS in the pathogenesis of LHON was demonstrated (Degli Esposti et al. 1994; Wong et al. 2002; Carelli et al. 2002; Beretta et al. 2004; Gonzalo et al. 2005). Quantification of specific types of ROS in the living cells is far from straightforward. However, chemical and protein-based reporter molecules can be combined with live cell microscopy imaging to measure ROS (see: Forkink et al. 2010 and the references therein). Using 2′–7′-dichlorofluorescin diacetate (DCFDA), a chemical reporter of cellular oxidant levels, differentiated NT2-LHON cybrid cell lines displayed two-fold higher ROS levels than control neurons (Wong et al. 2002). Another chemical ROS reporter (dihydroethidium, DHE) was oxidized faster in LHON neurons carrying the more severe G3460A mutation when compared to the milder G11778A mutation (Wong et al. 2002). In osteosarcoma cybrid cell lines carrying the three primary LHON mutations, quantification of DCFDA-oxidation (Carelli et al. 2002) and dihydrorhodamine-123 (RH-123) oxidation (Beretta et al. 2004), revealed a positive correlation between ROS levels and the severity of the mutations in patients.

Interestingly, also a decrease in endogenous antioxidants has been reported for the cybrid cell lines harboring the three primary LHON mutations (Floreani et al. 2005). However, the ability of these cells to grow normally in a glucose-containing medium suggests that this decrease is not lethal under these conditions (Floreani et al. 2005). When the same cells were grown in galactose medium, a further decrease in antioxidant levels was observed (Floreani et al. 2005), together with a time-dependent depletion of cellular ATP content (Zanna et al. 2003) and apoptotic cell death (Ghelli et al. 2003). These findings suggest involvement of ROS in the pathogenesis of LHON. Endogenous and/or exogenous factors might trigger an already unbalanced situation between ROS production and removal, rendering the cells more prone to oxidative stress and eventually death by apoptosis (see below).

9.6.3 *Mitochondrial Membrane Potential and MPT Opening*

Given the role of CI in maintaining mitochondrial membrane potential ($\Delta\psi$), mtDNA mutations in CI subunits might induce $\Delta\psi$ depolarization. Indeed, a mitochondrial coupling defect has been reported for primary fibroblasts from several LHON patients, which in turn causes $\Delta\psi$ depolarization (Chevrollier et al. 2008). Similarly, $\Delta\psi$ depolarization has been demonstrated in digitonin-permeabilized PBMC from LHON patients (Korsten et al. 2010). However, NT2-LHON cybrids expressing the G11778A and the G3640A LHON mutation did not show $\Delta\psi$ alterations (Wong et al. 2002). Another study did reveal $\Delta\psi$ depolarization in a cybrid cell line harboring both the T14484C and the G14279A LHON mutations, but only when these cells were treated with rotenone and oligomycin (Porcelli et al. 2009). This depolarization was prevented by cyclosporine A (CsA), a well-known inhibitor of mitochondrial permeability transition pore (MPT) opening (Porcelli et al. 2009). Additional evidence for involvement of MPT opening in mtDNA diseases was obtained in cybrids harboring non-LHON mtDNA mutations (Wong and Cortopassi 1997).

MPT opening occurs when the extent of $\Delta\psi$ depolarization reaches a certain threshold value (Bernardi 1992). This value is, for instance, affected by (pathological) increases in ROS or ionic calcium concentration ($[Ca^{2+}]$) (Petronilli et al. 1993). It was proposed that the rotenone-induced $\Delta\psi$ depolarization in LHON cybrids results from ATP depletion, leading to insufficient fueling of CV 'reverse mode' action (Porcelli et al. 2009). Elevated ROS levels and Ca^{2+} overload (see next section) can sensitize MPT opening, thereby possibly stimulating cell death (Wong and Cortopassi 1997).

9.6.4 *Calcium Handling*

Ionic calcium (Ca^{2+}) regulates many cell functions ranging from contraction to division. Mitochondria not only take up and release Ca^{2+} but also generate ATP required for active Ca^{2+} extrusion from the cytosol by Ca^{2+} ATPases. In this way, mitochondria actively participate in cellular Ca^{2+} homeostasis.

In principle, $\Delta\psi$ depolarization due to CI malfunction might affect mitochondrial uniporter-mediated Ca^{2+} uptake and/or mitochondrial Ca^{2+} release by the Na^{+}/Ca^{2+}-exchanger. Additionally, CI malfunction might hamper the fueling with mitochondrial ATP of Ca^{2+} ATPases that pump Ca^{2+} into the endoplasmic reticulum (ER) or across the plasmamembrane. In case of nDNA-encoded mutations in CI, evidence was provided that $\Delta\psi$ depolarization was associated with a reduced ER Ca^{2+} content in resting patient-derived skin fibroblasts (Willems et al. 2008). When these cells were stimulated with the hormone bradykinin (Bk) Ca^{2+} was released from the ER, leading to a transient increase in cytosolic $[Ca^{2+}]$ ($[Ca^{2+}]_c$). In the patient cells, both the amplitude and the rate of decay of the $[Ca^{2+}]_c$ transient were significantly lower than in healthy fibroblasts. The reduced ER Ca^{2+} content, lower $[Ca^{2+}]_c$ peak amplitude and slower $[Ca^{2+}]_c$ decay were fully normalized by treatment of the patient cells with the benzothiazepine CGP37157, an inhibitor of the mitochondrial Na^+/Ca^{2+}-exchanger. Analysis of the Bk-induced Ca^{2+}-stimulated increase in mitochondrial ATP level ($[ATP]_m$)revealed a reduced peak $[ATP]_m$ in patient fibroblasts. Also this reduction was fully restored to control levels by CGP37157 (Visch et al. 2004).

Concerning the effect of mtDNA-encoded mutations in CI subunits on mitochondrial and cellular Ca^{2+} handling, experimental information is very limited. Cellular Ca^{2+} deregulation has been observed in LHON cybrids upon hormone stimulation (Haroon et al. 2007). In these cells, cellular Ca^{2+} concentration remained elevated for longer time than in control cells. This effect was alleviated when MPT opening was blocked by cyclosporine A (CsA) treatment (Haroon et al. 2007). Interestingly, cybrid cell lines harboring different mtDNA mutations (including LHON mutations), were protected from oxidative stress when the Ca^{2+} was depleted from the extracellular medium (Wong and Cortopassi 1997). It was proposed that the increase in ROS induced by mitochondrial dysfunction triggers a rise in the intracellular Ca^{2+}, which eventually triggers MPT opening. This model is consistent with previous evidence linking excessive mitochondrial Ca^{2+} accumulation to mitochondrial oxidative stress and opening of the MPT, followed by release of cytochrome *c* and apoptosis induction (Kowaltowski et al. 2001).

9.6.5 Apoptosis Induction

Mitochondria play a central role in the regulation of active cell death (apoptosis) in response to cellular dysfunctioning. Under conditions of Ca^{2+} overload, elevated ROS levels and/or low ATP, MPT opening can occur. The latter allows diffusion of low MW solutes across the MIM. This leads to mitochondrial swelling, release of inter-membrane space proteins into the cytosol, such as cytochrome *c*, a key activator of programmed cell death (Caroppi et al. 2009).

An increased sensitivity to apoptotic cell death has been demonstrated in lymphoblasts of LHON patients (Battisti et al. 2004) and LHON cybrids grown in galactose medium (Ghelli et al. 2003). The latter medium was used to block glycolysis and stimulate OXPHOS-mediated ATP generation. In the LHON cybrids, which harbored the three primary LHON mutations, apoptotic features included the typical apoptotic nuclear morphology, DNA fragmentation and cytochrome *c* release (Ghelli et al. 2003). More detailed studies revealed that apoptosis of the LHON cells was independent of caspase activation and mediated by mitochondrial release of cytochrome *c*, AIF and Endo G (Zanna et al. 2003, 2005). It was proposed that apoptosis becomes activated because LHON cells are unable to maintain their ATP content during the early stages of galactose incubation. Although still an artificial condition, the LHON cybrid-galactose model likely resembles the natural circumstances

Table 9.2 Summary of the main cell biological consequences of the three primary LHON mutations

mtDNA mutation	Complex I enzymatic activity	Mitochondrial membrane potential (*)	Mitochondrial respiration	Mitochondrial ATP generation	Mitochondrial ROS production	Ca^{2+} handling	Apoptosis (#)
LHON **G3460A**	79% ↓	Depolarized	20–28% ↓	↓	↑	Impaired	Yes
LHON **G11778A**	25% ↓	Depolarized	30–36% ↓	↓	↑	Impaired	Yes
LHON **T14484C**	No difference compared to controls	Depolarized	10–15% ↓	↓	↑	Impaired	Yes

Abbreviations: ↓ decrease, ↑ increase, (*) general result, see text for details, (#) general result, see text for details associated with mitochondrial bioenergetic failure in LHON patients. It has been reported that environmental toxins, such as tobacco smoke (Sadun et al. 2003), represent a risk factor for the onset of LHON. Therefore external factors might codetermine the onset of bioenergetic failure and ensuing ATP depletion, leading to triggering of the apoptotic pathway and disease onset.

9.7 Concluding Remarks

The diagnosis and development of rational treatment strategies for mtDNA disorders is hampered by the lack of a clear genotype-phenotype correlation and limited insight into their pathogenic mechanism (Mancuso et al. 2009). However, strategies to ameliorate (some of) the cellular consequences of mitochondrial disorders in general (Wallace et al. 2010) and mtDNA mutations in particular were recently presented elsewhere (Koene and Smeitink 2010). In this sense, promising approaches include the pronuclear transfer between zygotes (Craven et al. 2010) and the metaphase II spindle transfer (Tachibana et al. 2010). In case of mtDNA-encoded mutations in CI subunits, the limited cell evidence that is currently available indicates that the most common mutations in mtDNA-encoded CI subunits induce a reduction in CI activity and $\Delta\psi$ depolarization (Table 9.2). This is associated with a reduced mitochondrial respiration, hampered mitochondrial ATP generation, increased mitochondrial ROS production, aberrant Ca^{2+} homeostasis and apoptosis induction. Clearly, a proper understanding of CI deficiency due to mtDNA-mutations requires additional cell biological analysis of appropriate cell and animal models.

References

Baracca A, Solaini G, Sgarbi G, Lenaz G, Baruzzi A, Schapira AH, Martinuzzi A, Carelli V (2005) Severe impairment of complex I-driven adenosine triphosphate synthesis in Leber hereditary optic neuropathy cybrids. Arch Neurol 62:730–736

Battisti C, Formichi P, Cardaioli E, Bianchi S, Mangiavacchi P, Tripodi SA, Tosi P, Federico A (2004) Cell response to oxidative stress induced apoptosis in patients with Leber's hereditary optic neuropathy. J Neurol Neurosurg Psychiatry 75:1731–1736

Benit P, Slama A, Cartault F, Giurgea I, Chretien D, Lebon S, Marsac C, Munnich A, Rotig A, Rustin P (2004) Mutant NDUFS3 subunit of mitochondrial complex I causes Leigh syndrome. J Med Genet 41:14–17

Beretta S, Mattavelli L, Sala G, Tremolizzo L, Schapira AH, Martinuzzi A, Carelli V, Ferrarese C (2004) Leber hereditary optic neuropathy mtDNA mutations disrupt glutamate transport in cybrid cell lines. Brain 127:2183–2192

Bernardi P (1992) Modulation of the mitochondrial cyclosporin A-sensitive permeability transition pore by the proton electrochemical gradient. Evidence that the pore can be opened by membrane depolarization. J Biol Chem 267:8834–8839

Blakely EL, De Silva R, King A, Schwarzer V, Harrower T, Dawidek G, Turnbull DM, Taylor RW (2005) LHON/MELAS overlap syndrome associated with a mitochondrial MTND1 gene mutation. Eur J Hum Genet 13:623–627

Brown MD, Sun F, Wallace DC (1997) Clustering of Caucasian Leber hereditary optic neuropathy patients containing the 11778 or 14484 mutations on an mtDNA lineage. Am J Hum Genet 60:381–387

Brown MD, Trounce IA, Jun AS, Allen JC, Wallace DC (2000) Functional analysis of lymphoblast and cybrid mitochondria containing the 3460, 11778, or 14484 Leber's hereditary optic neuropathy mitochondrial DNA mutation. J Biol Chem 275:39831–39836

Brown MD, Zhadanov S, Allen JC, Hosseini S, Newman NJ, Atamonov VV, Mikhailovskaya IE, Sukernik RI, Wallace DC (2001) Novel mtDNA mutations and oxidative phosphorylation dysfunction in Russian LHON families. Hum Genet 109:33–39

Carelli V, Ghelli A, Ratta M, Bacchilega E, Sangiorgi S, Mancini R, Leuzzi V, Cortelli P, Montagna P, Lugaresi E, Degli Esposti M (1997) Leber's hereditary optic neuropathy: biochemical effect of 11778/ND4 and 3460/ND1 mutations and correlation with the mitochondrial genotype. Neurology 48:1623–1632

Carelli V, Ghelli A, Bucchi L, Montagna P, De Negri A, Leuzzi V, Carducci C, Lenaz G, Lugaresi E, Degli Esposti M (1999) Biochemical features of mtDNA 14484 (ND6/M64V) point mutation associated with Leber's hereditary optic neuropathy. Ann Neurol 45:320–328

Carelli V, Napoli E, Valente L, Valentino L, Martinuzzi A (2002) ROS production in cybrids carrying the three primary mutations associated with Leber's hereditary optic neuropathy. Neurology 58:A507–A507

Caroppi P, Sinibaldi F, Fiorucci L, Santucci R (2009) Apoptosis and human diseases: mitochondrion damage and lethal role of released cytochrome C as proapoptotic protein. Curr Med Chem 16:4058–4065

Chevrollier A, Guillet V, Loiseau D, Gueguen N, De Crescenzo MA, Verny C, Ferre M, Dollfus H, Odent S, Milea D, Goizet C, Amati-Bonneau P, Procaccio V, Bonneau D, Reynier P (2008) Hereditary optic neuropathies share a common mitochondrial coupling defect. Ann Neurol 63:794–798

Chinnery PF, Schon EA (2003) Mitochondria. J Neurol Neurosurg Psychiatry 74:1188–1199

Chinnery PF, Howell N, Lightowlers RN, Turnbull DM (1997) Molecular pathology of MELAS and MERRF. The relationship between mutation load and clinical phenotypes. Brain 120:1713–1721

Chinnery PF, Howell N, Andrews RM, Turnbull DM (1999) Clinical mitochondrial genetics. J Med Genet 36:425–436

Chomyn A, Meola G, Bresolin N, Lai ST, Scarlato G, Attardi G (1991) In vitro genetic transfer of protein synthesis and respiration defects to mitochondrial DNA-less cells with myopathy-patient mitochondria. Mol Cell Biol 11:2236–2244

Cock HR, Cooper JM, Schapira AH (1999) Functional consequences of the 3460-bp mitochondrial DNA mutation associated with Leber's hereditary optic neuropathy. J Neurol Sci 165:10–17

Corona P, Antozzi C, Carrara F, D'Incerti L, Lamantea E, Tiranti V, Zeviani M (2001) A novel mtDNA mutation in the ND5 subunit of complex I in two MELAS patients. Ann Neurol 49: 106–110

Craven L, Tuppen HA, Greggains GD, Harbottle SJ, Murphy JL, Cree LM, Murdoch AP, Chinnery PF, Taylor RW, Lightowlers RN, Herbert M, Turnbull DM (2010) Pronuclear transfer in human embryos to prevent transmission of mitochondrial DNA disease. Nature 465:82–85

Degli Esposti M, Carelli V, Ghelli A, Ratta M, Crimi M, Sangiorgi S, Montagna P, Lenaz G, Lugaresi E, Cortelli P (1994) Functional alterations of the mitochondrially encoded ND4 subunit associated with Leber's hereditary optic neuropathy. FEBS Lett 352:375–379

Deschauer M, Bamberg C, Claus D, Zierz S, Turnbull DM, Taylor RW (2003) Late-onset encephalopathy associated with a C11777A mutation of mitochondrial DNA. Neurology 60:1357–1359

Dimauro S, Davidzon G (2005) Mitochondrial DNA and disease. Ann Med 37:222–232

Distelmaier F, Koopman WJ, Van Den Heuvel LP, Rodenburg RJ, Mayatepek E, Willems PH, Smeitink JA (2009) Mitochondrial complex I deficiency: from organelle dysfunction to clinical disease. Brain 132:833–842

Dunbar DR, Moonie PA, Jacobs HT, Holt IJ (1995) Different cellular backgrounds confer a marked advantage to either mutant or wild-type mitochondrial genomes. Proc Natl Acad Sci U S A 92:6562–6566

Efremov RG, Baradaran R, Sazanov LA (2010) The architecture of respiratory complex I. Nature 465:441–445
Finsterer J (2008) Leigh and Leigh-like syndrome in children and adults. Pediatr Neurol 39: 223–235
Floreani M, Napoli E, Martinuzzi A, Pantano G, De Riva V, Trevisan R, Bisetto E, Valente L, Carelli V, Dabbeni-Sala F (2005) Antioxidant defences in cybrids harboring mtDNA mutations associated with Leber's hereditary optic neuropathy. FEBS J 272:1124–1135
Forkink M, Smeitink JA, Brock R, Willems PH, Koopman WJ (2010) Detection and manipulation of mitochondrial reactive oxygen species in mammalian cells. Biochim Biophys Acta 1797: 1034–1044
Ghelli A, Zanna C, Porcelli AM, Schapira AH, Martinuzzi A, Carelli V, Rugolo M (2003) Leber's hereditary optic neuropathy (LHON) pathogenic mutations induce mitochondrial-dependent apoptotic death in transmitochondrial cells incubated with galactose medium. J Biol Chem 278:4145–4150
Gomez-Duran A, Pacheu-Grau D, Lopez-Gallardo E, Diez-Sanchez C, Montoya J, Lopez-Perez MJ, Ruiz-Pesini E (2010) Unmasking the causes of multifactorial disorders: OXPHOS differences between mitochondrial haplogroups. Hum Mol Genet 19:3343–3353
Gonzalo R, Garcia-Arumi E, Llige D, Marti R, Solano A, Montoya J, Arenas J, Andreu AL (2005) Free radicals-mediated damage in transmitochondrial cells harboring the T14487C mutation in the ND6 gene of mtDNA. FEBS Lett 579:6909–6913
Haroon MF, Fatima A, Scholer S, Gieseler A, Horn TF, Kirches E, Wolf G, Kreutzmann P (2007) Minocycline, a possible neuroprotective agent in Leber's hereditary optic neuropathy (LHON): studies of cybrid cells bearing 11,778 mutation. Neurobiol Dis 28:237–250
Hoefs SJ, Dieteren CE, Distelmaier F, Janssen RJ, Epplen A, Swarts HG, Forkink M, Rodenburg RJ, Nijtmans LG, Willems PH, Smeitink JA, Van Den Heuvel LP (2008) NDUFA2 complex I mutation leads to Leigh disease. Am J Hum Genet 82:1306–1315
Holt IJ, Harding AE, Morgan-Hughes JA (1988) Deletions of muscle mitochondrial DNA in patients with mitochondrial myopathies. Nature 331:717–719
Howell N, Bindoff LA, Mccullough DA, Kubacka I, Poulton J, Mackey D, Taylor L, Turnbull DM (1991) Leber hereditary optic neuropathy: identification of the same mitochondrial ND1 mutation in six pedigrees. Am J Hum Genet 49:939–950
Hunte C, Zickermann V, Brandt U (2010) Functional modules and structural basis of conformational coupling in mitochondrial complex I. Science 329:448–451
Huoponen K, Vilkki J, Aula P, Nikoskelainen EK, Savontaus ML (1991) A new mtDNA mutation associated with Leber hereditary optic neuroretinopathy. Am J Hum Genet 48:1147–1153
Johns DR, Neufeld MJ, Park RD (1992) An ND-6 mitochondrial DNA mutation associated with Leber hereditary optic neuropathy. Biochem Biophys Res Commun 187:1551–1557
Johns DR, Heher KL, Miller NR, Smith KH (1993) Leber's hereditary optic neuropathy. Clinical manifestations of the 14484 mutation. Arch Ophthalmol 111:495–498
Jun AS, Brown MD, Wallace DC (1994) A mitochondrial DNA mutation at nucleotide pair 14459 of the NADH dehydrogenase subunit 6 gene associated with maternally inherited Leber hereditary optic neuropathy and dystonia. Proc Natl Acad Sci U S A 91:6206–6210
Kirby DM, Kahler SG, Freckmann ML, Reddihough D, Thorburn DR (2000) Leigh disease caused by the mitochondrial DNA G14459A mutation in unrelated families. Ann Neurol 48:102–104
Kirby DM, Mcfarland R, Ohtake A, Dunning C, Ryan MT, Wilson C, Ketteridge D, Turnbull DM, Thorburn DR, Taylor RW (2004) Mutations of the mitochondrial ND1 gene as a cause of MELAS. J Med Genet 41:784–789
Koene S, Smeitink J (2010) Metabolic manipulators: a well founded strategy to combat mitochondrial dysfunction. J Inherit Metab Dis 34:315–325
Komaki H, Akanuma J, Iwata H, Takahashi T, Mashima Y, Nonaka I, Goto Y (2003) A novel mtDNA C11777A mutation in Leigh syndrome. Mitochondrion 2:293–304
Koopman WJ, Nijtmans LG, Dieteren CE, Roestenberg P, Valsecchi F, Smeitink JA, Willems PH (2010) Mammalian mitochondrial complex I: biogenesis, regulation, and reactive oxygen species generation. Antioxid Redox Signal 12:1431–1470

Korsten A, De Coo IF, Spruijt L, De Wit LE, Smeets HJ, Sluiter W (2010) Patients with Leber hereditary optic neuropathy fail to compensate impaired oxidative phosphorylation. Biochim Biophys Acta 1797:197–203
Kowaltowski AJ, Castilho RF, Vercesi AE (2001) Mitochondrial permeability transition and oxidative stress. FEBS Lett 495:12–15
Kowaltowski AJ, De Souza-Pinto NC, Castilho RF, Vercesi AE (2009) Mitochondria and reactive oxygen species. Free Radic Biol Med 47:333–343
Lebon S, Chol M, Benit P, Mugnier C, Chretien D, Giurgea I, Kern I, Girardin E, Hertz-Pannier L, De Lonlay P, Rotig A, Rustin P, Munnich A (2003) Recurrent de novo mitochondrial DNA mutations in respiratory chain deficiency. J Med Genet 40:896–899
Lebon S, Rodriguez D, Bridoux D, Zerrad A, Rotig A, Munnich A, Legrand A, Slama A (2007) A novel mutation in the human complex I NDUFS7 subunit associated with Leigh syndrome. Mol Genet Metab 90:379–382
Liolitsa D, Rahman S, Benton S, Carr LJ, Hanna MG (2003) Is the mitochondrial complex I ND5 gene a hot-spot for MELAS causing mutations? Ann Neurol 53:128–132
Liu CY, Lee CF, Wei YH (2009) Role of reactive oxygen species-elicited apoptosis in the pathophysiology of mitochondrial and neurodegenerative diseases associated with mitochondrial DNA mutations. J Formos Med Assoc 108:599–611
Lodi R, Taylor DJ, Tabrizi SJ, Kumar S, Sweeney M, Wood NW, Styles P, Radda GK, Schapira AH (1997) In vivo skeletal muscle mitochondrial function in Leber's hereditary optic neuropathy assessed by 31P magnetic resonance spectroscopy. Ann Neurol 42:573–579
Lodi R, Carelli V, Cortelli P, Iotti S, Valentino ML, Barboni P, Pallotti F, Montagna P, Barbiroli B (2002) Phosphorus MR spectroscopy shows a tissue specific in vivo distribution of biochemical expression of the G3460A mutation in Leber's hereditary optic neuropathy. J Neurol Neurosurg Psychiatry 72:805–807
Loeffen J, Smeitink J, Triepels R, Smeets R, Schuelke M, Sengers R, Trijbels F, Hamel B, Mullaart R, Van Den Heuvel L (1998) The first nuclear-encoded complex I mutation in a patient with Leigh syndrome. Am J Hum Genet 63:1598–1608
Luft R, Ikkos D, Palmieri G, Ernster L, Afzelius B (1962) A case of severe hypermetabolism of nonthyroid origin with a defect in the maintenance of mitochondrial respiratory control: a correlated clinical, biochemical, and morphological study. J Clin Invest 41:1776–1804
Mackey D, Howell N (1992) A variant of Leber hereditary optic neuropathy characterized by recovery of vision and by an unusual mitochondrial genetic etiology. Am J Hum Genet 51:1218–1228
Majander A, Huoponen K, Savontaus ML, Nikoskelainen E, Wikstrom M (1991) Electron transfer properties of NADH:ubiquinone reductase in the ND1/3460 and the ND4/11778 mutations of the Leber hereditary optic neuroretinopathy (LHON). FEBS Lett 292:289–292
Mancuso M, Orsucci D, Coppede F, Nesti C, Choub A, Siciliano G (2009) Diagnostic approach to mitochondrial disorders: the need for a reliable biomarker. Curr Mol Med 9:1095–1107
Martin MA, Blazquez A, Gutierrez-Solana LG, Fernandez-Moreira D, Briones P, Andreu AL, Garesse R, Campos Y, Arenas J (2005) Leigh syndrome associated with mitochondrial complex I deficiency due to a novel mutation in the NDUFS1 gene. Arch Neurol 62:659–661
Meire FM, Van Coster R, Cochaux P, Obermaier-Kusser B, Candaele C, Martin JJ (1995) Neurological disorders in members of families with Leber's hereditary optic neuropathy (LHON) caused by different mitochondrial mutations. Ophthalmic Genet 16:119–126
Murphy MP (2009) How mitochondria produce reactive oxygen species. Biochem J 417:1–13
Nass MM, Nass S (1963) Intramitochondrial fibers with DNA characteristics. I. Fixation and electron staining reactions. J Cell Biol 19:593–611
Petronilli V, Cola C, Massari S, Colonna R, Bernardi P (1993) Physiological effectors modify voltage sensing by the cyclosporin A-sensitive permeability transition pore of mitochondria. J Biol Chem 268:21939–21945
Porcelli AM, Angelin A, Ghelli A, Mariani E, Martinuzzi A, Carelli V, Petronilli V, Bernardi P, Rugolo M (2009) Respiratory complex I dysfunction due to mitochondrial DNA mutations shifts the voltage threshold for opening of the permeability transition pore toward resting levels. J Biol Chem 284:2045–2052

Pradelli LA, Beneteau M, Ricci JE (2010) Mitochondrial control of caspase-dependent and -independent cell death. Cell Mol Life Sci 67:1589–1597

Rossignol R, Faustin B, Rocher C, Malgat M, Mazat JP, Letellier T (2003) Mitochondrial threshold effects. Biochem J 370:751–762

Sadun AA, Carelli V, Salomao SR, Berezovsky A, Quiros PA, Sadun F, Denegri AM, Andrade R, Moraes M, Passos A, Kjaer P, Pereira J, Valentino ML, Schein S, Belfort R (2003) Extensive investigation of a large Brazilian pedigree of 11778/haplogroup J Leber hereditary optic neuropathy. Am J Ophthalmol 136:231–238

Santorelli FM, Tanji K, Kulikova R, Shanske S, Vilarinho L, Hays AP, Dimauro S (1997) Identification of a novel mutation in the mtDNA ND5 gene associated with MELAS. Biochem Biophys Res Commun 238:326–328

Sarzi E, Brown M, Lebon S, Chretien D, Munnich A, Rotig A, Procaccio V (2006) A novel recurrent mtDNA mutation in ND3 gene causing Leigh syndrome and dystonia. Am J Med Genet 143A:33–41

Sarzi E, Brown MD, Lebon S, Chretien D, Munnich A, Rotig A, Procaccio V (2007) A novel recurrent mitochondrial DNA mutation in ND3 gene is associated with isolated complex I deficiency causing Leigh syndrome and dystonia. Am J Med Genet A 143:33–41

Scarpulla RC (2008) Transcriptional paradigms in mammalian mitochondrial biogenesis and function. Physiol Rev 88:611–638

Schapira AH (2006) Mitochondrial disease. Lancet 368:70–82

Schwartz M, Vissing J (2002) Paternal inheritance of mitochondrial DNA. N Engl J Med 347:576–580

Shuster RC, Rubenstein AJ, Wallace DC (1988) Mitochondrial DNA in anucleate human blood cells. Biochem Biophys Res Commun 155:1360–1365

Smeitink J, Van Den Heuvel L, Dimauro S (2001) The genetics and pathology of oxidative phosphorylation. Nat Rev Genet 2:342–352

Solano A, Roig M, Vives-Bauza C, Hernandez-Pena J, Garcia-Arumi E, Playan A, Lopez-Perez MJ, Andreu AL, Montoya J (2003) Bilateral striatal necrosis associated with a novel mutation in the mitochondrial ND6 gene. Ann Neurol 54:527–530

Stewart JB, Freyer C, Elson JL, Wredenberg A, Cansu Z, Trifunovic A, Larsson NG (2008) Strong purifying selection in transmission of mammalian mitochondrial DNA. PLoS Biol 6:e10

Sue CM, Bruno C, Andreu AL, Cargan A, Mendell JR, Tsao CY, Luquette M, Paolicchi J, Shanske S, Dimauro S, De Vivo DC (1999) Infantile encephalopathy associated with the MELAS A3243G mutation. J Pediatr 134:696–700

Sutovsky P, Moreno RD, Ramalho-Santos J, Dominko T, Simerly C, Schatten G (1999) Ubiquitin tag for sperm mitochondria. Nature 402:371–372

Swerdlow RH (2007) Mitochondria in cybrids containing mtDNA from persons with mitochondriopathies. J Neurosci Res 85:3416–3428

Tachibana M, Sparman M, Mitalipov S (2010) Chromosome transfer in mature oocytes. Nat Protoc 5:1138–1147

Taylor RW, Singh-Kler R, Hayes CM, Smith PE, Turnbull DM (2001) Progressive mitochondrial disease resulting from a novel missense mutation in the mitochondrial DNA ND3 gene. Ann Neurol 50:104–107

Taylor RW, Morris AA, Hutchinson M, Turnbull DM (2002) Leigh disease associated with a novel mitochondrial DNA ND5 mutation. Eur J Hum Genet 10:141–144

Tonska K, Kodron A, Bartnik E (2010) Genotype-phenotype correlations in Leber hereditary optic neuropathy. Biochim Biophys Acta 1797:1119–1123

Trounce IA, Pinkert CA (2007) Cybrid models of mtDNA disease and transmission, from cells to mice. Curr Top Dev Biol 77:157–183

Tuppen HA, Blakely EL, Turnbull DM, Taylor RW (2010) Mitochondrial DNA mutations and human disease. Biochim Biophys Acta 1797:113–128

Ugalde C, Triepels RH, Coenen MJ, Van Den Heuvel LP, Smeets R, Uusimaa J, Briones P, Campistol J, Majamaa K, Smeitink JA, Nijtmans LG (2003) Impaired complex I assembly in a Leigh syndrome patient with a novel missense mutation in the ND6 gene. Ann Neurol 54:665–669

Valentino ML, Barboni P, Ghelli A, Bucchi L, Rengo C, Achilli A, Torroni A, Lugaresi A, Lodi R, Barbiroli B, Dotti M, Federico A, Baruzzi A, Carelli V (2004) The ND1 gene of complex I is a mutational hot spot for Leber's hereditary optic neuropathy. Ann Neurol 56:631–641
Valsecchi F, Esseling JJ, Koopman WJ, Willems PH (2009) Calcium and ATP handling in human NADH:ubiquinone oxidoreductase deficiency. Biochim Biophys Acta 1792:1130–1137
Visch HJ, Rutter GA, Koopman WJH, Koenderink JB, Verkaart S, De Groot T, Varadi A, Mitchell KJ, Van Den Heuvel LP, Smeitink JAM, Willems PHGM (2004) Inhibition of mitochondrial Na^+-Ca^{2+} exchange restores agonist-induced ATP production and Ca^{2+} handling in human complex I deficiency. J Biol Chem 279:40328–40336
Wallace DC, Singh G, Lott MT, Hodge JA, Schurr TG, Lezza AM, Elsas LJ 2nd, Nikoskelainen EK (1988) Mitochondrial DNA mutation associated with Leber's hereditary optic neuropathy. Science 242:1427–1430
Wallace DC, Fan W, Procaccio V (2010) Mitochondrial energetics and therapeutics. Annu Rev Pathol 5:297–348
Walsh C, Barrow S, Voronina S, Chvanov M, Petersen OH, Tepikin A (2009) Modulation of calcium signalling by mitochondria. Biochim Biophys Acta 1787:1374–1382
Wiesner RJ, Ruegg JC, Morano I (1992) Counting target molecules by exponential polymerase chain reaction: copy number of mitochondrial DNA in rat tissues. Biochem Biophys Res Commun 183:553–559
Willems PH, Valsecchi F, Distelmaier F, Verkaart S, Visch HJ, Smeitink JA, Koopman WJ (2008) Mitochondrial Ca^{2+} homeostasis in human NADH:ubiquinone oxidoreductase deficiency. Cell Calcium 44:123–133
Willems PH, Smeitink JA, Koopman WJ (2009) Mitochondrial dynamics in human NADH:ubiquinone oxidoreductase deficiency. Int J Biochem Cell Biol 41:1773–1782
Wiseman A, Attardi G (1978) Reversible tenfod reduction in mitochondria DNA content of human cells treated with ethidium bromide. Mol Gen Genet 167:51–63
Wissinger B, Besch D, Baumann B, Fauser S, Christ-Adler M, Jurklies B, Zrenner E, Leo-Kottler B (1997) Mutation analysis of the ND6 gene in patients with Lebers hereditary optic neuropathy. Biochem Biophys Res Commun 234:511–515
Wong LJ (2007) Pathogenic mitochondrial DNA mutations in protein-coding genes. Muscle Nerve 36:279–293
Wong A, Cortopassi G (1997) mtDNA mutations confer cellular sensitivity to oxidant stress that is partially rescued by calcium depletion and cyclosporin A. Biochem Biophys Res Commun 239:139–145
Wong LJ, Dai P, Tan D, Lipson M, Grix A, Sifry-Platt M, Gropman A, Chen TJ (2001) Severe lactic acidosis caused by a novel frame-shift mutation in mitochondrial-encoded cytochrome c oxidase subunit II. Am J Med Genet 102:95–99
Wong A, Cavelier L, Collins-Schramm HE, Seldin MF, Mcgrogan M, Savontaus ML, Cortopassi GA (2002) Differentiation-specific effects of LHON mutations introduced into neuronal NT2 cells. Hum Mol Genet 11:431–438
Wu SB, Ma YS, Wu YT, Chen YC, Wei YH (2010) Mitochondrial DNA mutation-elicited oxidative stress, oxidative damage, and altered gene expression in cultured cells of patients with MERRF syndrome. Mol Neurobiol 41:256–266
Yen MY, Wang AG, Wei YH (2006) Leber's hereditary optic neuropathy: a multifactorial disease. Prog Retin Eye Res 25:381–396
Yoneda M, Chomyn A, Martinuzzi A, Hurko O, Attardi G (1992) Marked replicative advantage of human mtDNA carrying a point mutation that causes the MELAS encephalomyopathy. Proc Natl Acad Sci U S A 89:11164–11168
Zanna C, Ghelli A, Porcelli AM, Carelli V, Martinuzzi A, Rugolo M (2003) Apoptotic cell death of cybrid cells bearing Leber's hereditary optic neuropathy mutations is caspase independent. Ann N Y Acad Sci 1010:213–217
Zanna C, Ghelli A, Porcelli AM, Martinuzzi A, Carelli V, Rugolo M (2005) Caspase-independent death of Leber's hereditary optic neuropathy cybrids is driven by energetic failure and mediated by AIF and Endonuclease G. Apoptosis 10:997–1007

Part IV
Subunit Composition and Assembly of Mitochondrial Complex I

Chapter 10
The Assembly of Human Complex I

Jessica Nouws, Maria Antonietta Calvaruso, and Leo Nijtmans

Abstract Human mitochondrial complex I is one of the biggest membrane bound enzymes found in man. Its assembly is a complicated process and requires the combination of 45 subunits, encoded by the nuclear and mitochondrial genomes. It has become clear that the assembly of complex I takes place in different steps in which some parts are assembled separately and connected to form a functional enzyme. Remarkably this stepwise process follows the assembly of the distinct functional and evolutionary conserved modules, the hydrogenase part, the membrane transporter part and the NADH dehydrogenase part. Assembly of these parts and further combination of these modules requires the help of specialized proteins called chaperones. A number of these chaperones have been identified and they play a role in the different stages of the assembly. Although the precise molecular mechanisms concerning complex I assembly are still unknown, the overall picture of this process is getting shape as discussed in this chapter.

Keywords Mitochondria • Oxidative Phosphorylation • Complex I • Assembly • Human • Subunits • Assembly chaperones • Supercomplexes

10.1 Introduction

Mitochondrial complex I, NADH:ubiquinone oxidoreductase (EC1.6.5.3), is a central enzyme in the regulation of cell metabolism. As major entry point of electrons to the respiratory chain it couples the transfer of electrons from NADH

J. Nouws • M.A. Calvaruso • L. Nijtmans (✉)
Department of Pediatrics, Nijmegen Center for Mitochondrial Disorders, Radboud University Nijmegen Medical Centre, PO Box 9101, 6500 HB Nijmegen, The Netherlands
e-mail: j.nouws@cukz.umcn.nl; mary.calvaruso@virgilio.it; l.nijtmans@cukz.umcn.nl1

L. Sazanov (ed.), *A Structural Perspective on Respiratory Complex I: Structure and Function of NADH:ubiquinone oxidoreductase*, DOI 10.1007/978-94-007-4138-6_10,

to ubiquinone with the translocation of protons across the mitochondrial inner membrane (Brandt 2006). In this fashion it contributes to the generation of a membrane potential which is used by complex V to generate ATP. Thus by controlling cellular NADH, ATP levels and the mitochondrial membrane potential, complex I is an important regulator of metabolic pathways. Beside its function in oxidative phosphorylation, complex I has been implied to be the main generator of reactive oxygen species (ROS) (Kussmaul and Hirst 2006; Lambert and Brand 2009). These ROS are known to be toxic for the cell by damaging proteins, membranes and DNA, but also function as molecular signaling molecules (D'Autreaux and Toledano 2007). Moreover dysfunction of complex I is believed to play an important role in mitochondrial triggered programmed cell death (Perier et al. 2005). With such a crucial biochemical role it is expected that defects in this enzyme give rise to a large number of clinical heterogeneous mitochondrial disorders.

To understand how complex I activities influence the function of a cell it is important to know how this complex is assembled and how this assembly is regulated. Although in recent years excellent progression has been made in revealing the structure (Efremov et al. 2010; Hinchliffe et al. 2006; Hunte et al. 2010), the evolution and function of complex I (Gabaldon et al. 2005), the molecular mechanisms involved in its biogenesis are still poorly understood. This might not come as a surprise when one realizes that complex I is one of the largest membrane bound complexes in the cell, its components are encoded by mitochondrial and nuclear genes, it contains a large number of prosthetic groups, it is part of respiratory supercomplexes and its assembly requires specialized proteins, called assembly chaperones.

In this chapter we will describe our current knowledge on how this large membrane bound complex is assembled. Firstly, to know how complex I assembly takes place it is essential to know what the components are and where they are located in the complex. Secondly, because complex I is believed to originate from different enzymes which have been combined into one complex, the assembly of complex I might be reflected by its evolutionary route. Our understanding of the 45 subunit-containing mammalian complex I assembly will therefore benefit from the studies which were performed to investigate how simpler forms of complex I assemble in lower organisms, like in *Escherichia coli* and *Neurospora crassa*. Thirdly, it is not difficult to imagine that the assembly and regulation of complex I assembly requires additional factors. These factors, the so-called assembly chaperones, facilitate certain steps in the assembly, however they are not part of the holo-complex. Finally, because the mitochondrial respiratory chain is believed to be organized in supercomplexes it is very well possible that the assembly of complex I does not stand alone but also depends on the coordinate assembly of the other members of the respiratory chain.

10.2 Subunits

To understand complex I assembly one should first have knowledge of its components. Most complete experimental subunit determination of mammalian complex I has been performed by the group of Walker who studied the composition of complex I in bovine heart. This study identified 45 different subunits (Carroll et al. 2006). Based on sequence comparison and experimental data, the complex I subunit content of many other organisms, including *Homo sapiens*, has been established (see Table 10.1).

When comparing the complex I subunit content of the different species there are quite some differences. Beside a basic set of subunits which are conserved throughout all organisms from bacteria to man, also several subunits were acquired later in evolution and these are specific for different species (Gabaldon et al. 2005). The minimal set of subunits required to fulfill the enzymatic function of complex I are the 14 subunits found in most bacteria (two of them are fused in *E. coli* (Leif et al. 1993)), which are considered to be the core subunits. These include the seven mitochondrially encoded subunits (ND1-ND6, ND4L), subunits containing prosthetic groups (NDUFV1, NDUFV2, NDUFS1, NDUFS7 and NDUFS8) and a subunit involved in ubiquinone binding (NDUFS2). These subunits are distributed over three evolutionary conserved modules, the membrane arm (ND1-ND6, ND4L), the hydrogenase part (NDUFS2, NDUFS3, NDUFS7 and NDUFS8) and the NADH-dehydrogenase part (NDUFS1, NDUFV1 and NDUFV2). These evolutionary modules will be discussed below.

Beside these 14 core subunits, mammalian complex I has gained another 31 subunits, which are called accessory or supernumerary subunits (Table 10.1). Why human complex I has acquired these accessory subunits, which are not required for its basic enzymatic function, is not known. They might play a role in the regulation of the enzyme as suggested by the fact that some of these subunits can be phosphorylated (Chen et al. 2004; Palmisano et al. 2007; Schilling et al. 2005; Ugalde et al. 2004b). Or they can be associated with other enzymatic functions as is suggested by the fact that: NDUFA13 (Grim 19), has been shown to play a role in apoptosis (Fearnley et al. 2001); NDUFAB1 (SDAP), is proposed to play a role in lipoic acid biosynthesis (Cronan et al. 2005; Runswick et al. 1991) and NDUFS9 which contains a NADPH binding site (Schulte et al. 1999; Yamaguchi et al. 2000). They might also play a role in the regulation of the assembly. Suggestive for such a role in regulating assembly of complex I is the fact that some subunits are homologous to proteins which are involved in the mitochondrial biogenesis like for instance: NDUFA2 and NDUFA10, which are homologues of mitochondrial ribosomal proteins L43 and S25; NDUFA11, which has homology with Tim17/22 family of proteins involved in mitochondrial protein import and NDUFA12, which is paralogue of complex I assembly factor NDUFAF2 (Gabaldon et al. 2005). In addition it has been proposed that these accessory subunits might be involved in the stability and protection against ROS or proteolytic degradation. Nevertheless, experimental evidence for any of such functions is still lacking.

Table 10.1 Complex I subunit nomenclature in different organisms

Escherichia coli	*Thermus thermophilus (NDH-1)*	*Neurospora crassa*	*Yarrowia lypolitica*	*Chlamydomonas reinhardtii*	*Bos taurus*	*Homo sapiens*
NuoH	NQO8	ND1	NU1M	ND1	ND1	ND1
NuoN	NQO14	ND2	NU2M	ND2	ND2	ND2
NuoA	NQO7	ND3	NU3M	NUO3	ND3	ND3
NuoM	NQO13	ND4	NU4M	ND4	ND4	ND4
NuoK	NQO11	ND4L	NULM	NUO11	ND4L	ND4L
NuoL	NQO12	ND5	NU5M	ND5	ND5	ND5
NuoJ	NQO10	ND6	NU6M	ND6	ND6	ND6
NuoF	NQO1	51 kDa	NUBM	NUO6	51 kDa	NDUFV1
NuoE	NQO2	24 kDa	NUHM	NUO5	24 kDa	NDUFV2
–		–	–	–	10 kDa	NDUFV3
NuoG	NQO3	78 kDa	NUAM	NUOS1	75 kDa	NDUFS1
NuoD (together with NuoC)	NQO4	49 kDa	NUCM	NUO7	49 kDa	NDUFS2
NuoC (together with NuoD)	NQO5	30.4 kDa	NUGM	NUO9	30 kDa	NDUFS3
–		21 kDa	NUYM	NUOS4	18 kDa (AQDQ)	NDUFS4
–		11,5 kDa	NIPM	15 kDa	15 kDa (PFFD)	NDUFS5
–		28.7 kDa	NUMM	NUOS6	13 kDa	NDUFS6
NuoB	NQO6	19.3 kDa	NUKM	NUO10	PSST	NDUFS7
NuoI	NQO9	21.3c kDa	NUIM	NUO8	TYKY	NDUFS8
–		9.8 kDa	NIMM	7.5 kDa	MWFE	NDUFA1
–		10.5 kDa	NI8M	NUOB8	B8	NDUFA2
–		9.3 kDa	NI9M	–	B9	NDUFA3
–		–	–	–	MLRQ	NDUFA4
–		29.9 kDa	NUFM	NUOB13	B13	NDUFA5
–		14.8 kDa	NB4M	NUOB14	B14	NDUFA6
–		–	–	–	B14.5a	NDUFA7
–		20.8 kDa	NUPM	NUO8A	PGIV	NDUFA8
–		40 kDa	NUEM	NUOA9	39 kDa	NDUFA9
–		–	–	–	42 kDa	NDUFA10

–		21.3b kDa	NUJM	23 kDa	B14.7	NDUFA11
–		13.4 kDa	N7BM	NUO13	B17.2	NDUFA12
–		13,5 kDa	NB6M	NUOB16	B16.6	NDUFA13/Grim-19
–		–	–	–	MNLL	NDUFB1
–		–	–	–	AGGG	NDUFB2
–		–	NB2M	6.5 kDa	B12	NDUFB3
–		7 kDa	NUVM	NUO2P	B15	NDUFB4
–		–	–	–	SGHD	NDUFB5
–		–	–	–	B17	NDUFB6
–		89.7 kDa	NB8M	NUOB18	B18	NDUFB7
–		19 kDa	NIAM	–	ASHI	NDUFB8
–		18.3 kDa	NI2M	NUOB22	B22	NDUFB9
–		12.3 kDa	NIDM	NUOB10	PDSW	NDUFB10
–		11,7	NUWM	NUO17	ESSS	NDUFB11
–		9.6 kDa ACP	ACPM	ACP1	SDAP	NDUFAB1
–		–	–	–	KFYI	NDUFC1
–		–	–	–	B14.5b	NDUFC2
		21.3 kDa				
		21.3a kDa	NUZM	–	–	–
		20.9 kDa	NUXM	NUO21		
		17.8 kDa	–	–	–	–
	NQO15					
				NUOP1		
				NUOP3		
				FBP1		
				FBP2		
				CAH9		
				NUOP4		
				NUOP5		
				NUOP6		

10.3 Topology and Evolution

The subunit topology will put constrains on complex I assembly. Recent crystal structures of complex I of *Thermus thermophilus* (Hinchliffe et al. 2006), *Escherichia coli* (Efremov et al. 2010), and *Yarrowia lypolytica* (Hunte et al. 2010) have given much insight into the structure and how subunits are arranged within these complexes. In higher eukaryotes the most detailed topological structure of subunits within complex I is obtained from experiments using mild chaiotropic salts to break up complex I into subcomplexes. First studies performed by Hatefi and colleagues enabled to fractionate complex I into a flavoprotein part, a hydrogenase part and a membrane part (Galante and Hatefi 1978). This approach was later refined by the group of Walker which was able to define Iα, Iβ, Iλ, Iγ fragments and resolve the subunit content of these parts by mass spectrometry (Carroll et al. 2003; Sazanov et al. 2000). This has led to a crude topological view of where the subunits are localized in complex I (Fig. 10.1).

With mildly chaotropic detergents, intact bovine complex I has been resolved into various subcomplexes that have subsequently been subjected to extensive analysis (Carroll et al. 2003). Subcomplex Iλ represents the peripheral arm of the complex, subcomplex Iα consists of subcomplex Iλ plus part of the membrane arm and subcomplex Iβ forms the major part of the membrane arm. Subunits that do not associate

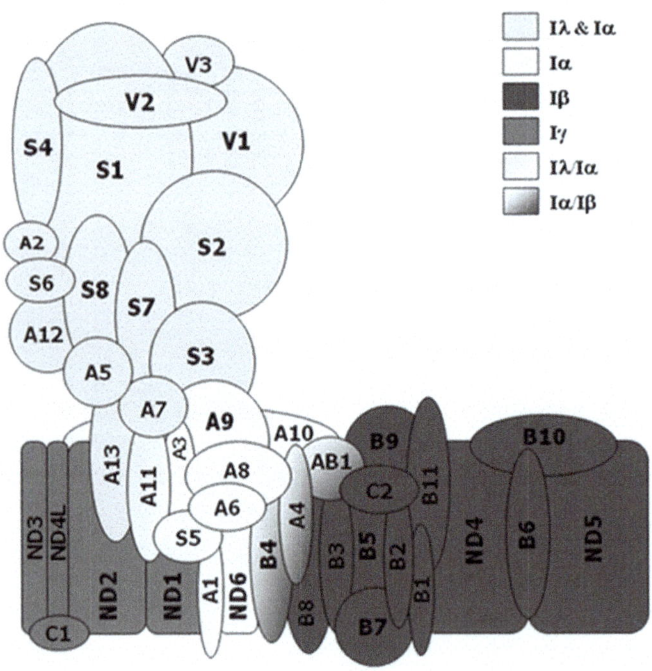

Fig. 10.1 A model for the topology of the subunits of mammalian complex I

with the Iα or Iβ subcomplexes are grouped as the Iγ 'subcomplex', a fraction for which it is not clear whether it represents a true fragment of complex I. Some subunits associate with two subcomplexes (Iα/Iβ), for others it is not exactly clear which subcomplex they associate with (Iλ/Iα). The nuclear DNA-encoded subunits are named according to the human nomenclature and preceded by the prefix 'NDUF'.

Given the fact that complex I has evolved from a 14 subunit enzyme into a 45 subunit complex, evolution might provide some insights into how this increasing complexity has developed and how this is reflected in its assembly. Of particular interest is the modular evolution of complex I proposed by Friedrich and Weiss (Friedrich and Weiss 1997). Based on homologies with bacterial enzymes, this theory postulates that complex I originates from fusion of distinct pre-existing protein assemblies, which have combined their activities in one complex. The hydrogenase has evolved from the small and large subunit of the bacterial nickel-Hydrogenases, which are conserved in subunit NDUFS7 and NDUFS2 respectively (Albracht 1993). This hydrogenase has acquired a ferredoxin (NDUFS8), a membrane bound quinone reduction site (ND1), a transport protein (ND4) and a protein of unknown function (NDUFS3). Further upon evolution the transporter module has expanded by the recruitment of ND5, a gene duplication of ND4 to form ND2 and the incorporation of ND3, ND6 and ND4L (Mathiesen and Hagerhall 2003). This enzyme was equipped with the NDUFS1, NDUFV1-V2 NADH-dehydrogenase module. Further gain of (accessory) subunits in the evolutionary lineage has occurred in all stages and in all fractions of the complex (Gabaldon et al. 2005).

10.4 Complex I Assembly

10.4.1 Complex I Assembly in Escherichia coli

The 14 subunit containing bacterial NDH-1 complex is considered to be the structural minimal form of complex I (Leif et al. 1993; Weidner et al. 1993). Because all NDH-1 subunits are homologues to human complex I subunits (see Table 10.1), assembly of complex I in this organism might serve as an example for the assembly of the core of human complex I. The genes encoding the *Escherichia coli* complex I subunits are called nuoA to nuoN, including nuoC and nuoD which have been fused to one gene (Leif et al. 1995) (see Table 10.1). Six of these genes code for the seven subunits (NuoB-G and NuoI) of the peripheral hydrophilic part including the subunits harboring the flavin-mononucleotide (FMN), iron sulfur (Fe-S) cluster, the site binding NADH (NuoF) (Velazquez et al. 2005), and the site binding quinone (NuoB and NuoD) (Prieur et al. 2001). The other seven subunits (NuoA, H, J-N) are mostly hydrophobic proteins which are incorporated into the membrane. Recently, Sazanov and colleagues have isolated and crystallized the membrane domain of complex I from *E. coli*. In the crystal structure NuoL (ND5) is localized at the distal part of the complex followed by NuoM (ND4), NuoN (ND2) and closer to the peripheral arm of complex I NuoA (ND3), J (ND6) and K (ND4) (Efremov et al. 2010) are positioned, whereas NuoH (ND1) is not present in the crystal structure since is it readily dissociated from the complex.

Thanks to the availability of *E. coli* strains with disruption of each of the complex I genes (Baba et al. 2006) a systematic investigation to the effect caused by the lack of each subunit on complex I assembly could be performed (Schneider et al. 2008). In all mutants the specific NADH oxidase activity was absent indicating that all complex I genes are required for proper complex I function. Further investigations revealed most mutants still form partially assembled or degraded subassemblies. In deletion mutants for NuoA-D (ND3, NDUFS7, NDUFS3 and NDUFS2) and NuoH-N (ND1-ND6, NDUFS8) a subassembly containing NuoEFG (NDUFV1, NDUFV2, NDUFS1) including Fe-S clusters N1a and N1b could be recovered from the cytoplasm. Consistent with the presence of iron sulfur clusters this subassembly still had activity for NADH oxidation (Schneider et al. 2008). This NuoEFG subassembly forms the NADH dehydrogenase module and it was proposed to be an assembly intermediate of complex I, which is in line with the modular evolution scheme. In fact, it seems that the NADH dehydrogenase module is the most recent acquisition of the precursor of complex I. However, joint overexpression of the genes encoding the subunits NuoE, F, G leads to the overproduction of these subunits which are not assembled into any subcomplex (Braun et al. 1998). This could be possibly due to a lack of cofactors involved in this part or to the absence of a stabilizing complex I preassembly. In mutant nuoL (ND5) a membrane-bound complex I fragment, which only lacks this subunit, without a detectable NADH oxidase activity was found (Schneider et al. 2008). In mutant nuoH (ND1) the membrane arm is severely disturbed (Kervinen et al. 2006). Disruption of nuoJ (ND6) gene did not result in hampering of assembly, indicating that this protein is not essential to pursue the assembly, or it might be added in a rather late stage (Kao et al. 2005). NuoL (ND5) and nuoM (ND4) are localized in the distal part so it is supposed that these subunits are also added late stages in the assembly (Baranova et al. 2007; Holt et al. 2003).

In summary, the hydrogenase part, containing NuoCD-B-I (NDUFS3, NDUFS2, NDUFS7, NDUFS8), is added to NuoEFG (dehydrogenase part) (NDUFV2, NDUFV1, NDUFS1) and part of the membrane arm (NuoK, J, A and N; ND4L, ND6, ND3 and ND2). At a later stage Nuo M-L (ND4 and ND5) are added to this subassembly, this corresponds to the crystal structure in which these subunits are localized at the end of the membrane domain.

10.4.2 Complex I Assembly in Neurospora crassa

In eukaryotes the level of complexity of complex I and its assembly increases. Firstly, the number of subunits rises significantly from 14 up to 45 in man. Secondly, the complex has become under dual genetic control harboring subunits encoded by the nuclear DNA and mitochondrial DNA. The new nuclear-encoded subunits are assembled in coordination with the mitochondrially encoded ones and this intricate process requires regulated signaling between nucleus and mitochondrion (Ryan and Hoogenraad 2007). Thirdly, complex I is located in the mitochondria requiring coordinated transportation and import of the nuclear encoded subunits in this cellular compartment.

The fungus *Neurospora crassa* has been used as one of the main model organisms to investigate complex I for different reasons: first the fungal and the mammalian enzymes are very similar; second, this fungus grows quite rapidly and can be easily genetically manipulated; and finally, since the genetics and biochemistry of this organism are well understood, it is possible to perform a wide range of experiments.

At the end of the 1980s, the group of Weiss described the presence of two different sub-assemblies in *N. crassa* that can be formed independently from each other. When mitochondrial protein synthesis was inhibited by using chloramphenicol, an accumulation of the peripheral arm of complex I constituted by nuclear encoded subunits with NADH:Q reductase activity could be observed (Friedrich et al. 1989); whereas in cells grown in medium lacking manganese, membrane arm accumulation was observed (Schmidt et al. 1992). The same group followed the flux of radioactivity in complex I polypeptides, which demonstrated that the peripheral arm and membrane arm of complex I undergo independent assembly (Tuschen et al. 1990).

The first gene inactivated in *N. crassa* was 21,3b (NDUFA11) which encodes a subunit situated in the membrane arm. The authors found that complex I accumulated in two intermediates: a larger intermediate containing four mitochondrially encoded subunits and six nuclear encoded subunits; and a smaller intermediate containing two mitochondrially encoded and two nuclear-encoded subunits. The larger intermediate was found connected to two CIA-proteins (Complex I intermediate-associated): CIA30 and CIA84 (Kuffner et al. 1998; Nehls et al. 1992). These proteins are not part of the fully assembled complex I and have been shown to cycle between bound and unbound state (Kuffner et al. 1998). Subsequently, other subunits of the membrane arm were disrupted leading to the accumulation of the peripheral arm and the formation of two membrane intermediates or to the formation of an incomplete membrane arm like in mutants nuo20.9 (Schulte and Weiss 1995), nuo12,3(NDUFB10) (Duarte et al. 1995; Schulte et al. 1995), nuo14 (NDUFA13) (Marques et al. 2005) and nuo20,8 (NDUFA8) (da Silva et al. 1996). In mutant nuo9,8 (NDUFA1) (Marques et al. 2003) both membrane and peripheral arm fail to assemble. Instead, when the 11.5 kDa subunit (NDUFS5) was disrupted the peripheral arm was not detected, nevertheless the accumulation of two membrane arm intermediates was observed (Marques et al. 2005). In addition this gene was mutated in four conserved cysteines, and for all the four mutants a complete and inactive complex I was detected and an accumulation of the peripheral arm with a molecular weight of ~500 kDa was observed. This suggests that the peripheral arm accumulated due to the inhibited assembly of the membrane part (Marques et al. 2007).

So far, just one mutant, the E35 stopper mutant, was described to cause depletion in mtDNA subunits (ND2/ND3), in this mutant only a structure resembling the peripheral arm was formed (Alves and Videira 1998). Different subunits of the peripheral arm were disrupted and in some cases an almost complete complex I was found such as in mutants nuo21 (NDUFS4) (Ferreirinha et al. 1999), nuo51 (NDUFV1) (Duarte et al. 2005; Fecke et al. 1994), nuo21,3a (Alves and Videira 1994), nuo24 (NDUFV2) (Almeida et al. 1999) and nuo40 (NDUFA9) (Schulte et al. 1999). In other cases, the absence of one subunit of the peripheral arm led to an accumulation of the membrane arm with the absence of assembly of the peripheral part like in nuo19,3 (NDUFS7) (Duarte et al. 2002, 2005), nuo29,9 (NDUFA5) (Duarte et al. 1995), nuo49 (NDUFS2) (Schulte et al. 1995), nuo21,3c (NDUFS8) (Duarte and Videira 2000; Duarte et al. 2005), nuo30,4 (NDUFS3) (Duarte et al. 1998) and nuo78 (NDUFS1) (Harkness et al. 1995). An exception was in the mutant for ACP (NDUFAB1) protein (acyl-carrier protein) which was not only unable to assemble the peripheral arm but it also failed to assemble the membrane part (Schneider et al. 2008).

To put all these findings together, complex I assembly in *N. crassa* is a process divided in different steps in which the different modules are assembled independently from each other. Subsequently, they are joined together to form holo-complex I. Mutations in subunits of the peripheral arm lead to the accumulation of the membrane part and *vice versa*. Further, none of the manipulated mutations in subunits of the NADH dehydrogenase module cause a disturbance of the peripheral arm meaning that this module is added at a later stage of complex I assembly. In total four different intermediates were described in complex I assembly in *N. crassa*: a small and a large membrane arm intermediate, the complete membrane arm, and a peripheral arm. First the small and large membrane parts are joined together with the formation of the membrane arm intermediate then also the peripheral arm is connected to the membrane part to form a complete complex I.

10.4.3 Mammalian Complex I Assembly

The most complicated structure of NADH:oxidoreductase is the 45 subunit-containing mammalian complex I. Since isolated complex I deficiency is the most common form of defective oxidative phosphorylation, complex I function and assembly has been frequently studied using samples of patients with mutations in one of the subunits or assembly chaperones (Janssen et al. 2006). Complex I assembly will be stalled if one of the structural subunits or assembly factors is mutated. In patient fibroblasts, complex I assembly can be checked by two dimensional blue native/SDS electrophoresis (2D-BN/SDS PAGE) to determine where the assembly is disturbed. Hereafter we will discuss experimental data gathered by *in vitro* systems and by analysis of patient material harboring mutations in subunits or assembly factors to show where they are required in the assembly pathway. These mutations will be discussed in the context of the earlier introduced evolutionary modules.

10.4.3.1 Assembly of Membrane Arm (Proton Translocating Unit)

The membrane arm is known to contain the mitochondrially encoded subunits. Over the years many mutations in mitochondrially encoded subunits have been described. In general, such mutations lead to a defective enzyme activity, and more precisely to a disturbed membrane arm assembly. An important step forward was the development of cybrid cell line technology: cell lines depleted of mitochondrial DNA can be repopulated with mutant DNA allowing researchers to study specific mutations in a control background. In an early study it was discovered that in the cybrid cell line lacking the ND4 subunit, the membrane arm was not assembled (Hofhaus and Attardi 1993). Moreover, it was found that some of the nuclear subunits were still assembled as suggested by presence of NADH:Fe$(CN)_6$ activity. Further investigations demonstrated that the absence of either ND4 or ND5 led to a loss of the NADH:ubiquinone oxidoreductase and the NADH:NTB activity, and to the formation of a subassembly containing NDUFS2, NDUFS3 and NDUFS8 (Bourges et al. 2004). These three subunits are supposed to be part of the connective complex I module between the membrane arm and the peripheral arm (Bourges et al. 2004). ND5 was shown to be more important for the complex I stability rather than for the assembly (Bai and Attardi 1998; Hofhaus and Attardi 1995). By tagging NDUFS3 with GFP in combination with 2D-BN/SDS PAGE it was shown that the assembly of different peripheral arm subcomplexes occurred prior to the incorporation of membrane subunit ND1 (Vogel et al. 2007a).

Recently, Enriquez and colleagues exploited their collection of ND mutated cybrids to perform a thorough investigation of the assembly of membrane arm ND subunits (Perales-Clemente et al. 2010). The authors identified five entry points of ND subunits in the complex I assembly process. The first entry point was for ND1 that anchors peripheral arm subunits to the inner mitochondrial membrane, forming an intermediate of ~400 kDa. The second entry point was the assembly of an ND2, ND3 and ND4L containing intermediate of ~460, which was independent from the ND1 intermediate. The third entry point was the incorporation of ND4. The incorporation of ND6 defined the fourth entry point and seems to facilitate bringing together the two subcomplexes containing ND1 and ND2. ND5 is incorporated at a later stage and it is the last entry point of the mitochondrially encoded subunits (Perales-Clemente et al. 2010). ND5 does not seem to be essential for complex I assembly, supported by the fact that in mutant ND5 cybrids all mitochondrially encoded subunits, except ND5, could be detected. Studies on patient cell lines harboring mutations in either ND1 or ND6 showed that these subunits are important for the assembly of complex I (Kirby et al. 2004a, b). The lack of ND2 resulted in an impairment of assembly displayed by an accumulation of all NDUFS3 containing subcomplexes (Antonicka et al. 2003; Ugalde et al. 2007).

In summary, ND1 assembles into a subcomplex, which solely contains nuclear encoded subunits. This part is added to a separate ND2 containing subassembly which has acquired also subunits ND4L, ND3 and ND6. In a next stage transporter module ND4 is possibly added followed by the incorporation of ND5 (Fig. 10.2a).

10.4.3.2 Assembly of the Hydrogenase Part

In the recent years several models for the assembly have been proposed (Antonicka et al. 2003; Lazarou et al. 2007; Ugalde et al. 2004b; Vogel et al. 2007a). Although the models show some differences, almost all agree that the hydrogenase part is made independently, including NDUFS3 as an essential component. Vogel and colleagues have shown that the assembly of the hydrogenase module starts at least with the association of NDUFS3 and NDUFS2 (Vogel et al. 2007a). These subunits are part of three subcomplexes of which the last two are associated with the mitochondrial inner membrane (Fig. 10.2b). Further studies using blue native electrophoresis in combination with mass spectrometry revealed that the membrane bound subcomplexes contained the subunits NDUFS2, NDUFS3, NDUFS7, and NDUFS8 (Wessels et al. 2009). This suggests that the hydrogenase part preassembles and binds to the membrane prior to the incorporation of subunit ND1. Consistent with this finding is the fact that patients with mutations in the subunits NDUFS2, NDUFS7 or NDUFS8 lack the occurrence of these subcomplexes whereas mutations in other subunits, which enter the assembly pathway at a later stage, such as NDUFS4, still allow assembly of the hydrogenase part (Ugalde et al. 2004a).

In brief, the hydrogenase part is assembled by the association of NDUFS3 and NDUFS2, of note these are the subunits which are fused in *E. coli*. This matrix assembly of NDUFS3 and NDUFS2, binds to NDUFS7 and NDUFS8 and this subcomplex tethers to the membrane. This basically completes the assembly of the hydrogenase part which goes further in the assembly pathway by acquiring, amongst others, ND1 (Fig. 10.2b).

10.4.3.3 Assembly of the NADH Dehydrogenase Part

The NADH dehydrogenase part contains at least NDUFV1, NDUFV2, NDUFV3 and NDUFS1. Mutations in NDUFV1 do allow complex I assembly to proceed to a subassembly of about 830 kDa (Ogilvie et al. 2005; Vogel et al. 2007c). This subassembly does not show any NADH reactivity anymore and lacks at least subunits NDUFV1 and NDUFS4 (Ogilvie et al. 2005). This suggests that the 830 kDa subassembly seems to lack the NADH-dehydrogenase part, indicating that this part is assembled independently and added at a later stage. In line with this idea is the finding of a similar subassembly of 830 kDa in patients with mutations in the NADH dehydrogenase subunit NDUFS1 (Hoefs et al. 2010). Furthermore, this 830 kDa subassembly is found in patients with mutations in NDUFS4 (Ogilvie et al. 2005; Vogel et al. 2007c) and NDUFS6 (Kirby et al. 2004b), meaning that this is a more

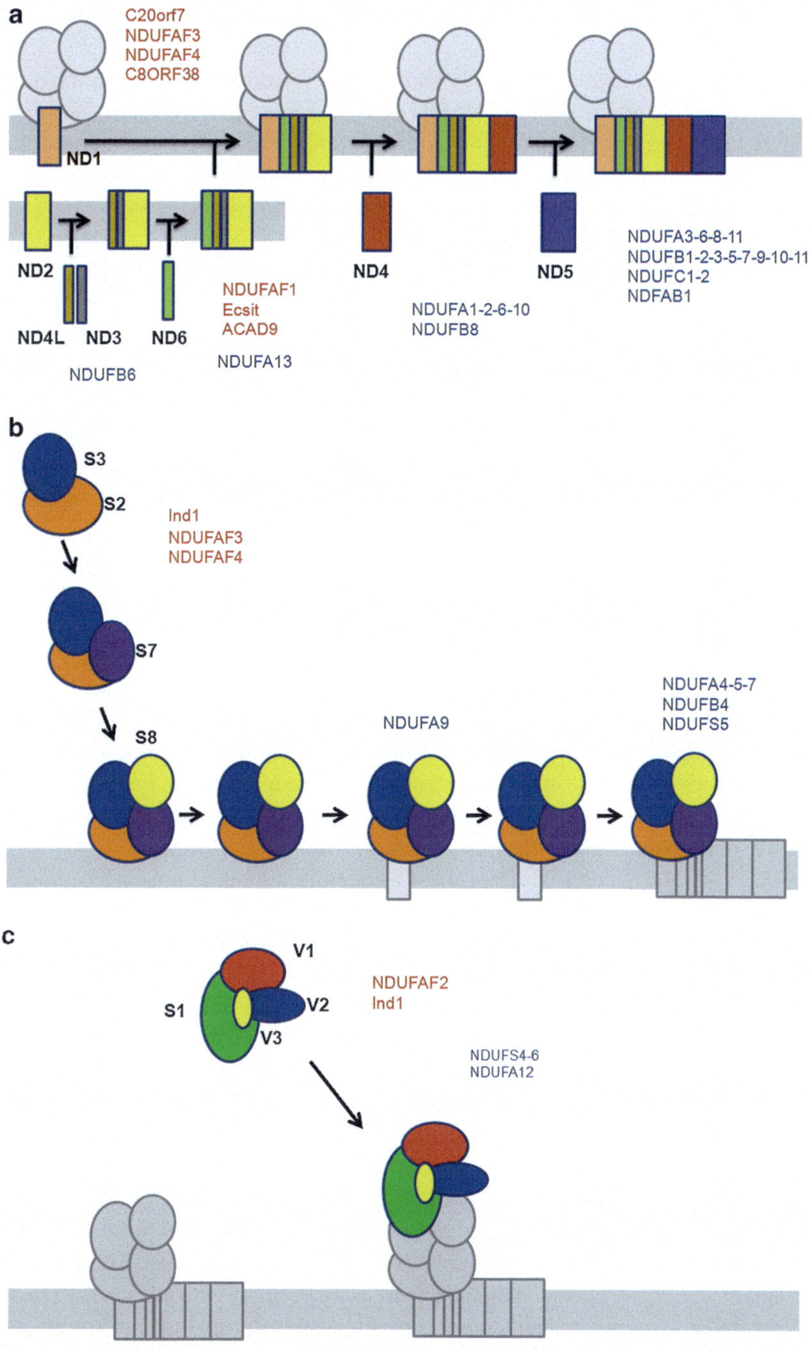

Fig. 10.2 Schematic overview of the assembly of the different modules of complex I. This overview includes the assembly of (**a**) the membrane arm, (**b**) the hydrogenase part and (**c**) the NADH dehydrogenase part. For clarity the core subunits are represented by *colored symbols*. Accessory subunits (*blue text*) and assembly factors (*red text*) are indicated

frequently impaired stage of assembly and that the subunits NDUFS4 and NDUFS6 are involved in the addition of the NADH dehydrogenase part to an almost finished complex I subassembly. In an *in vitro* assembly study, the import and incorporation into complex I subcomplexes of radiolabeled subunits in isolated mitochondria was studied. This method showed that in mitochondria isolated from patient cells containing an NDUFS4 mutation, subunits NDUFS2, NDUFS7, NDUFA9 and NDUFA10 were present in an 830 kDa subassembly. By contrast, the subunits NDUFV1, NDUFV2 and NDUFS1 were incorporated only in lower but similar molecular weight subcomplexes, possibly suggesting the preassembly of the NADH dehydrogenase part (Lazarou et al. 2007).

These findings suggest that the NADH dehydrogenase part preassembles and enters the complex I assembly at a late stage to complete complex I (Fig. 10.2c). Nevertheless, whether this NADH-dehydrogenase part is assembled independently as shown in *E. coli* (see above) or if it requires other complex I subassemblies still remains to be established.

10.4.3.4 Assembly of Accessory Subunits

Most mammalian complex I subunits are not homologous to components of *E. coli* complex I, the so-called core subunits. The extra subunits are referred to as the accessory subunits. Although these subunits are not supposed to be required for complex I activity, mutations in these subunits causing complex I deficiency have been found. These subunits enter the assembly pathway at different stages and probably play a role in the assembly or stabilization of intermediates.

NDUFS4 is the subunit which is most frequently mutated in isolated complex I deficiencies. NDUFS4 mutations cause an accumulation of NDUFS3 containing subcomplexes and subunits NDUFS5 and NDUFA9 accumulate just before entering into the holocomplex (Ugalde et al. 2004b). In three patients with homozygous NDUFS4 mutations an accumulation of NDUFA5 in a subcomplex of approximately 800 kDa was observed (Scacco et al. 2003). These data are consistent with the previously mentioned findings of Lazarou and coworkers who showed that in an NDUFS4 patient NDUFV1, NDUFV2, NDUFV3, NDUFS1 and NDUFS6 could not assemble into complex I, while subunits NDUFS2, NDUFS7, NDUFA9 and NDUFA10 could integrate into intermediates of complex I (Lazarou et al. 2007).

NDUFA1 mutations can cause a decrease of NDUFA9 and NDUFS3 containing subcomplexes (Fernandez-Moreira et al. 2007). In the patient described by Fernandez-Moreira and coworkers the mutant NDUFA1 protein itself is still detectable, albeit at a lower level. This causes no accumulation but rather a decrease of several NDUFS3 containing subcomplexes, suggestive of rapid incorporation of the remaining NDUFA1 protein. Another NDUFA1 deficient patient harboring a missense mutation (Potluri et al. 2009) shows a decreased expression level of NDUFS2, NDUFS 6, NDUFA13, NDUFB8, all located in different modules of complex I. In the mouse cell lines used to demonstrate the entry points of the mitochondrially encoded ND subunits it was shown using antibodies against membrane subunits NDUFB6 and NDUFB8, that NDUFB6 already assembles in an ND2 containing subcomplex whereas NDUFB8 integration occurred at a late stage when almost all ND subunits where assembled (Perales-Clemente et al. 2010). The mutation causing the exon skipping in NDUFA2, leads to a mild accumulation of NDUFS3 containing subcomplexes, but no accumulation of NDUFB6 in any subcomplex could be detected (Hoefs et al. 2008). Mutations in NDUFA11 have been reported, although its effect on the assembly is unclear (Berger et al. 2008). These data indicate that the accessory subunits enter the assembly at different stages and mutations cause a decrease in amount of fully assembled complex I accompanied with the accumulation of subassemblies. Based on the accumulation of different subassemblies it can be deduced where in the assembly pathway some of these subunits might come into play (Fig. 10.2). Nevertheless it remains an intriguing question what the exact role of these subunits in assembly is.

10.5 Complex I Chaperones

Beside all the nuclear and mitochondrially encoded subunits and prosthetic groups, complex I requires additional proteins to assemble. These proteins, called assembly chaperones, aid the assembly of complex I but they are not part of the holocomplex. To date, a number of these chaperones has been described. The firstly discovered mammalian chaperone was found by a homology search of the known assembly factor in *Neurospora crassa*, CIA30 (NDUFAF1) (Janssen et al. 2002; Vogel et al. 2005). The second assembly chaperone, NDUFAF2, was a protein which appeared to be a paralogue of complex I subunit NDUFA12 (Ogilvie et al. 2005). In the last 3 years several approaches like; homozygosity mapping, comparative genomic screens and co-purifications of subassemblies or other chaperones, have led to the identification of new chaperones. These methods showed that NDUFAF3, C20orf7, C8orf38, Ecsit, ACAD9, NDUFAF4 and Ind1 are involved in complex I biogenesis and further analysis showed their requirement for complex I assembly (Sheftel et al. 2009; Pagliarini et al. 2008; Saada et al. 2008, 2009; Sugiana et al. 2008; Vogel et al. 2007b; Nouws et al. 2010). The molecular working mechanism of the assembly chaperones is still unclear. In analogy of the better known complex IV assembly proteins, they are suggested to be involved in the incorporation and maturation of prosthetic groups, the insertion of hydrophobic mitochondrial membrane proteins and assembly or stabilisation of subcomplexes.

Despite our lack of knowledge on the precise working mechanisms of the assembly chaperones, it is possible to pinpoint in what stage of the assembly process some of these chaperones play a role (McKenzie and Ryan 2010).

10.5.1 Chaperones for the Membrane Unit

The assembly of the membrane unit involves translation and coordinated insertion of the mitochondrial encoded subunits. From the assembly of complex IV it is known that similar steps in the biogenesis of this complex involve chaperones which aid this process. In complex I assembly the chaperone C20orf7 is a likely candidate to play a role in the ND1 insertion into the already preassembled hydrogenase core. In patients with mutations in C20orf7, subunit ND1 was absent, suggesting a decreased translation or an increased instability due to the lack of assembly (Sugiana et al. 2008). Two other assembly chaperones, NDUFAF3 and NDUFAF4, were both associated with a number of assembly intermediates starting with the hydrogenase subassembly prior to the incorporation of ND1 (Saada et al. 2009). This suggests that they might be involved in the insertion of mitochondrially translated subunits. Both proteins were found to bind to the mitochondrial inner membrane and knock down of either of these chaperones by RNA interference shows that their stability is interdependent. Remarkably, after inhibition of mitochondrial translation, NDUFAF4 stays bound to NDUFS3 containing subcomplex, which solely consists of nuclear encoded subunits, whereas NDUFAF3 interactions seem to disappear under these conditions. This suggests that dynamic interactions of these proteins play a role in the regulation of this assembly step. Gene order conservation analysis of NDUFAF3 shows significant conservation with the SecF/SecD/YajC gene cluster, indicative of a function in membrane protein insertion (Saada et al. 2009). This correlates with the observed association of NDUFAF3 with higher molecular weight complexes in the presence of mitochondrial DNA encoded subunits, so these subunits are probably helped by NDUFAF3 to enter the subcomplexes. Interestingly, assembly chaperone C8orf38 was also shown in the gene order conservation analysis to be associated with NDUFAF3. This might suggest that this chaperone is also involved in this stage of the complex I assembly process, however up till now no further experimental data exist which support this.

Also for the incorporation of ND2 into a membrane subcomplex, the involvement of chaperones is suggested. In a patient with a mutation in chaperone NDUFAF1 it appears that the ND2 translation is impaired or that the stability of ND2 is decreased (Dunning et al. 2007). The NDUFAF1 protein has been shown to associate with at least two other proteins, Ecsit (Vogel et al. 2007b) and ACAD9 (Nouws et al. 2010). Down regulation of any of these proteins leads to a destabilisation of the other chaperones, suggesting that their stability depends on each other. Also knock down of any of these proteins still allows assembly of the ND1 subcomplexes to proceed to the stage where the incorporation of ND2 takes place, consistent with the fact that they play a role in this stage of the assembly. Remarkably, the heterogeneity of these proteins is large. Ecsit was already proposed to play a role in the toll-like pathway and the bone morphogenetic pathway, both essential processes which take place in the cytosol. Whereas ACAD9 structurally appears to be a mitochondrial fatty acid oxidation enzyme.

Nevertheless these proteins combined with NDUFAF1, are required for complex I assembly. Although the three proteins are co-dependent and co-migrate on two-dimensional gels, no associations of ACAD9 and Ecsit with complex I subunits have been found (Nouws et al. 2010).

10.5.2 Chaperones for the Core Hydrogenase Part

An important step in the assembly of the core hydrogenase part is the incorporation of iron-sulfur clusters, as two subunits of this part contain these prosthetic groups (NDUFS7 and NDUFS8). Although maturation of iron-sulfur clusters depends on many proteins and a general defect in iron-sulfur cluster maturation would also affect a number of other proteins, it is quite conceivable that the specific transfer of iron-sulfur clusters in complex I might require specific proteins. Recently Ind1 was shown to bind an Fe/S cluster via a conserved CXXC motive and knock down of this protein led to a strong complex I decrease (Sheftel et al. 2009). Analysis of the complex I subunit and subcomplex content in HeLa cells lacking Ind1 demonstrated a significant decrease of peripheral arm subunits NDUFS1, NDUFS3 and NDUFA13, and an accumulation of a 450 kDa membrane arm subcomplex. The decrease of hydrogenase subunit NDUFS3 indicates that this subunit is broken-down rapidly most likely because the subunit cannot continue its assembly route. This is consistent with the idea that Ind1 might play a role in the incorporation of Fe/S clusters in NDUFS7 and NDUFS8 and co-ordinately leading to the prevention of assembly of this hydrogenase core.

The previously mentioned chaperones NDUFAF3 and NDUFAF4 associate with this subassembly and might play a role in the membrane tethering or stabilisation of this part of complex I prior to the incorporation of ND1 (Saada et al. 2009). Interestingly a recently described putative methyl transferase MidA (also called C2orf56 or PRO1853) was shown to play a role in complex I assembly (Carilla-Latorre et al. 2010). Because of its interaction with subunit NDUFS2 one could suggest that it plays a role in the assembly of this part of the complex.

10.5.3 Chaperones for the NADH Dehydrogenase Part

Because the NADH dehydrogenase part also contains subunits which contain Fe-S clusters it could be assumed that Ind1 might play a role in the assembly of this subcomplex. Indeed the observed severely reduced levels of the Fe-S containing subunits NDUFV1 and NDUFS1 are in line with this assumption (Sheftel et al. 2009). Nevertheless, it is possible that the absence of the subassemblies containing the hydrogenase part may destabilize the assembly of the NADH dehydrogenase part therefore this observation might also be an indirect effect.

The assembly of the NADH dehydrogenase part into the complex I core takes place at a late stage. In patients with a mutation in NDUFS4, NDUFV1 or NDUFS6, assembly can proceed up to an 830 kDa subassembly. Further analysis has shown that this subassembly specifically lacks the subunits NDUFV1-3 and NDUFS1, suggesting the absence of the NADH dehydrogenase part. Interestingly, this 830 kDa subassembly also contains assembly chaperone NDUFAF2, which indicates that this chaperone might specifically play a role in the acquisition of the NADH-dehydrogenase part (Ogilvie et al. 2005). The total lack of the NDUFAF2 protein leads to the accumulation of five NDUFS2 and NDUFS3 containing subcomplexes (Janssen et al. 2009). This confirms that in the absence of this chaperone complex I assembly can proceed to the stage in which the NADH-dehydrogenase part is incorporated.

10.5.4 Other Chaperones

Pagliarini and co-workers searched for mitochondrial proteins without annotated function and essential for complex I (Pagliarini et al. 2008). To accomplish this mitochondrial protein compendium they performed mass spectrometry analysis on mitochondria derived from 14 mouse organs and they used several verification methods, furthermore they analyzed whether the proteins are mitochondrially localized by tagging them with GFP. Some candidates were also checked by means of RNA interference, and 19 candidates survived the selection. To date four proteins of this list are established as genuine complex I assembly chaperones, C8orf38, C20orf7, PRO1853 (C2orf56, MidA) and C3orf60 (NDUFAF3) (Carilla-Latorre et al. 2010; Pagliarini et al. 2008; Saada et al. 2009; Sugiana et al. 2008). In addition three proteins, FOXRED1, LYRM5 and LACT8, were shown to reduce complex I activity upon knock down of these proteins. Nevertheless, whether the role of these proteins can also be found in complex I assembly is unknown. The other proteins on the list also need to be confirmed.

10.6 Complex I in Supercomplexes

Initially, the organization of the respiratory complexes was described as randomly dispersed enzymes functionally connected by diffusion of smaller redox components (Coenzyme Q and cytochrome *c*). Only recently, the supercomplex concept is considered as a physiological condition and not just an experimental artifact. By using mild extraction procedures, Schagger and coworkers demonstrated by native electrophoresis the existence of supercomplexes and the re-introduced the idea of a respirosome to define the supramolecular organization of the respiratory chain complexes (Schagger 2001; Schagger and Pfeiffer 2001; Wittig et al. 2006). The study about the position of complex I in the supercomplexes was performed in bovine mitochondria and the authors suggested a model in which complex III-complex IV are both associated with the membrane arm of complex I (Schafer et al. 2007). Based on this study, the positions of the coenzyme Q binding sites of complex I and III and the cytochrome *c* binding sites of complex III and IV, are in close proximity in the supercomplex structure. This supports the hypothesis that the supercomplex is a more efficient organization, since their existence reduces the diffusion distance of the substrates and improves the electron transfer. Another possible role of the supercomplex organization is to decrease ROS formation. Although this is plausible, so far, no direct studies were performed to verify this hypothesis.

Moreover, mutations in cytochrome *b*, causing a destabilization of complex III, cause a secondary complex I deficiency (Acin-Perez et al. 2004; Schafer et al. 2007); the same holds true for mutations in complex IV (Diaz et al. 2006; Li et al. 2007). Interestingly, patient cell lines with complex III decreased activity but no disturbance in assembly, did not show any secondary complex I deficiency (Acin-Perez et al. 2004). These data imply that supercomplexes also have another role, namely to stabilize complex I by the physical presence of complex III or complex IV. Though in the model organism, *N. Crassa*, this is not the case, in fact disruption of complex III does not interfere with the stabilization of complex I or complex IV, maybe because an alternative oxidase is present (Duarte et al. 2009). Further, in patient cell line with a mutation in NDUFS4, a supercomplex complex I-complex III was formed, in which complex I is not the holocomplex but a subassembly of 830 kDa, meaning that a supercomplex can be formed even if the complex I assembly process is not finished (Ogilvie et al. 2005). This situation might represent a pathological condition which is not present in healthy cells, however it can be hypothesized that complex III assists complex I during the late stage of the assembly process.

10.7 Concluding Remarks

Complex I assembly and its regulation are studied extensively nowadays, however, it seems like only the basic part of the assembly has been uncovered. Hitherto it has been shown that complex I assembles in functional modules and for several subunits, it is clear in which stage they are incorporated. Also, it can be asumed that not all assembly factors for complex I have been found, because complex IV assembly requires the assistance of at least 15 assembly factors whereas the complex only contains 11 subunits. Very recently, new assembly factors of complex I have been discovered, also indicating that probably more are to be found in the near future. Even less is known about the actual function of the assembly factors, where and how do they facilitate the process of assembly. Moreover, do they actually bind several subunits and bring them to the assembly intermediate or is there another mechanism to be found?

Altogether it has to be concluded that complex I assembly is a very intricate process, as it requires the combination of 45 subunits encoded by mitochondrial and nuclear DNA. Moreover, it has been difficult to investigate the assembly since the model organism to study oxidative phosphorylation complex assembly, *Saccharomyces cerivisiae*, lacks complex I. New model organisms or other systems had to be sought and were found in *Neurospora crassa*, *Yarrowia lypolitica* and the use of human cell lines. Another difficulty can be found in the interpretation of assembly products in patient fibroblasts; it is not known if when complex I assembly is stalled, also complex I breakdown products appear. If that is the case, researchers might have looked at the breakdown products instead of assembly intermediates; therefore new systems including the tagging of subunits in non complex I affected cells might be helpful. With all the new systems introduced and the discovery of new assembly factors we are now approaching the real concept of the assembly.

References

Acin-Perez R, Bayona-Bafaluy MP, Fernandez-Silva P, Moreno-Loshuertos R, Perez-Martos A, Bruno C, Moraes CT, Enriquez JA (2004) Respiratory complex III is required to maintain complex I in mammalian mitochondria. Mol Cell 13:805–815

Albracht SP (1993) Intimate relationships of the large and the small subunits of all nickel hydrogenases with two nuclear-encoded subunits of mitochondrial NADH: ubiquinone oxidoreductase. Biochim Biophys Acta 1144:221–224

Almeida T, Duarte M, Melo AM, Videira A (1999) The 24-kDa iron-sulphur subunit of complex I is required for enzyme activity. Eur J Biochem 265:86–93

Alves PC, Videira A (1994) Disruption of the gene coding for the 21.3-kDa subunit of the peripheral arm of complex I from Neurospora crassa. J Biol Chem 269:7777–7784

Alves PC, Videira A (1998) The membrane domain of complex I is not assembled in the stopper mutant E35 of Neurospora. Biochem Cell Biol 76:139–143

Antonicka H, Ogilvie I, Taivassalo T, Anitori RP, Haller RG, Vissing J, Kennaway NG, Shoubridge EA (2003) Identification and characterization of a common set of complex I assembly intermediates in mitochondria from patients with complex I deficiency. J Biol Chem 278:43081–43088

Baba T, Ara T, Hasegawa M, Takai Y, Okumura Y, Baba M, Datsenko KA, Tomita M, Wanner BL, Mori H (2006) Construction of Escherichia coli K-12 in-frame, single-gene knockout mutants: the Keio collection. Mol Syst Biol 2:2006

Bai Y, Attardi G (1998) The mtDNA-encoded ND6 subunit of mitochondrial NADH dehydrogenase is essential for the assembly of the membrane arm and the respiratory function of the enzyme. EMBO J 17:4848–4858

Baranova EA, Holt PJ, Sazanov LA (2007) Projection structure of the membrane domain of Escherichia coli respiratory complex I at 8 A resolution. J Mol Biol 366:140–154

Berger I, Hershkovitz E, Shaag A, Edvardson S, Saada A, Elpeleg O (2008) Mitochondrial complex I deficiency caused by a deleterious NDUFA11 mutation. Ann Neurol 63:405–408

Bourges I, Ramus C, Mousson de CB, Beugnot R, Remacle C, Cardol P, Hofhaus G, Issartel JP (2004) Structural organization of mitochondrial human complex I: role of the ND4 and ND5 mitochondria-encoded subunits and interaction with prohibitin. Biochem J 383:491–499

Brandt U (2006) Energy converting NADH: quinone oxidoreductase (complex I). Annu Rev Biochem 75:69–92

Braun M, Bungert S, Friedrich T (1998) Characterization of the overproduced NADH dehydrogenase fragment of the NADH: ubiquinone oxidoreductase (complex I) from Escherichia coli. Biochemistry 37:1861–1867

Carilla-Latorre S, Gallardo ME, Annesley SJ, Calvo-Garrido J, Grana O, Accari SL, Smith PK, Valencia A, Garesse R, Fisher PR, Escalante R (2010) MidA is a putative methyltransferase that is required for mitochondrial complex I function. J Cell Sci 123:1674–1683

Carroll J, Fearnley IM, Shannon RJ, Hirst J, Walker JE (2003) Analysis of the subunit composition of complex I from bovine heart mitochondria. Mol Cell Proteomics 2:117–126

Carroll J, Fearnley IM, Skehel JM, Shannon RJ, Hirst J, Walker JE (2006) Bovine complex I is a complex of 45 different subunits. J Biol Chem 281:32724–32727

Chen R, Fearnley IM, Peak-Chew SY, Walker JE (2004) The phosphorylation of subunits of complex I from bovine heart mitochondria. J Biol Chem 279:26036–26045

Cronan JE, Fearnley IM, Walker JE (2005) Mammalian mitochondria contain a soluble acyl carrier protein. FEBS Lett 579:4892–4896

D'Autreaux B, Toledano MB (2007) ROS as signalling molecules: mechanisms that generate specificity in ROS homeostasis. Nat Rev Mol Cell Biol 8:813–824

da Silva MV, Alves PC, Duarte M, Mota N, Lobo-da-Cunha A, Harkness TA, Nargang FE, Videira A (1996) Disruption of the nuclear gene encoding the 20.8-kDa subunit of NADH: ubiquinone reductase of Neurospora mitochondria. Mol Gen Genet 252:177–183

Diaz F, Fukui H, Garcia S, Moraes CT (2006) Cytochrome c oxidase is required for the assembly/stability of respiratory complex I in mouse fibroblasts. Mol Cell Biol 26:4872–4881

Duarte M, Videira A (2000) Respiratory chain complex I is essential for sexual development in neurospora and binding of iron sulfur clusters are required for enzyme assembly. Genetics 156:607–615

Duarte M, Videira A (2009) Effects of mitochondrial complex III disruption in the respiratory chain of Neurospora crassa. Mol Microbiol 72:246–258

Duarte M, Sousa R, Videira A (1995) Inactivation of genes encoding subunits of the peripheral and membrane arms of neurospora mitochondrial complex I and effects on enzyme assembly. Genetics 139:1211–1221

Duarte M, Mota N, Pinto L, Videira A (1998) Inactivation of the gene coding for the 30.4-kDa subunit of respiratory chain NADH dehydrogenase: is the enzyme essential for Neurospora? Mol Gen Genet 257:368–375

Duarte M, Populo H, Videira A, Friedrich T, Schulte U (2002) Disruption of iron-sulphur cluster N2 from NADH: ubiquinone oxidoreductase by site-directed mutagenesis. Biochem J 364:833–839

Duarte M, Schulte U, Ushakova AV, Videira A (2005) Neurospora strains harboring mitochondrial disease-associated mutations in iron-sulfur subunits of complex I. Genetics 171:91–99

Dunning CJ, McKenzie M, Sugiana C, Lazarou M, Silke J, Connelly A, Fletcher JM, Kirby DM, Thorburn DR, Ryan MT (2007) Human CIA30 is involved in the early assembly of mitochondrial complex I and mutations in its gene cause disease. EMBO J 26:3227–3237

Efremov RG, Baradaran R, Sazanov LA (2010) The architecture of respiratory complex I. Nature 465:441–445

Fearnley IM, Carroll J, Shannon RJ, Runswick MJ, Walker JE, Hirst J (2001) GRIM-19, a cell death regulatory gene product, is a subunit of bovine mitochondrial NADH: ubiquinone oxidoreductase (complex I). J Biol Chem 276:38345–38348

Fecke W, Sled VD, Ohnishi T, Weiss H (1994) Disruption of the gene encoding the NADH-binding subunit of NADH: ubiquinone oxidoreductase in Neurospora crassa. Formation of a partially assembled enzyme without FMN and the iron-sulphur cluster N-3. Eur J Biochem 220:551–558

Fernandez-Moreira D, Ugalde C, Smeets R, Rodenburg RJ, Lopez-Laso E, Ruiz-Falco ML, Briones P, Martin MA, Smeitink JA, Arenas J (2007) X-linked NDUFA1 gene mutations associated with mitochondrial encephalomyopathy. Ann Neurol 61:73–83

Ferreirinha F, Duarte M, Melo AM, Videira A (1999) Effects of disrupting the 21 kDa subunit of complex I from Neurospora crassa. Biochem J 342:551–554

Friedrich T, Weiss H (1997) Modular evolution of the respiratory NADH: ubiquinone oxidoreductase and the origin of its modules. J Theor Biol 187:529–540

Friedrich T, Hofhaus G, Ise W, Nehls U, Schmitz B, Weiss H (1989) A small isoform of NADH: ubiquinone oxidoreductase (complex I) without mitochondrially encoded subunits is made in chloramphenicol-treated Neurospora crassa. Eur J Biochem 180:173–180

Gabaldon T, Rainey D, Huynen MA (2005) Tracing the evolution of a large protein complex in the eukaryotes, NADH: ubiquinone oxidoreductase (complex I). J Mol Biol 348:857–870

Galante YM, Hatefi Y (1978) Resolution of complex I and isolation of NADH dehydrogenase and an iron–sulfur protein. Methods Enzymol 53:15–21

Harkness TA, Rothery RA, Weiner JH, Werner S, Azevedo JE, Videira A, Nargang FE (1995) Disruption of the gene encoding the 78-kilodalton subunit of the peripheral arm of complex I in Neurospora crassa by repeat induced point mutation (RIP). Curr Genet 27:339–350

Hinchliffe P, Carroll J, Sazanov LA (2006) Identification of a novel subunit of respiratory complex I from Thermus thermophilus. Biochemistry 45:4413–4420

Hoefs SJ, Dieteren CE, Distelmaier F, Janssen RJ, Epplen A, Swarts HG, Forkink M, Rodenburg RJ, Nijtmans LG, Willems PH, Smeitink JA, van den Heuvel LP (2008) NDUFA2 complex I mutation leads to Leigh disease. Am J Hum Genet 82:1306–1315

Hoefs SJ, Skjeldal OH, Rodenburg RJ, Nedregaard B, van Kaauwen EP, Spiekerkotter U, Von Kleist-Retzow JC, Smeitink JA, Nijtmans LG, van den Heuvel LP (2010) Novel mutations in the NDUFS1 gene cause low residual activities in human complex I deficiencies. Mol Genet Metab 100:251–256

Hofhaus G, Attardi G (1993) Lack of assembly of mitochondrial DNA-encoded subunits of respiratory NADH dehydrogenase and loss of enzyme activity in a human cell mutant lacking the mitochondrial ND4 gene product. EMBO J 12:3043–3048

Hofhaus G, Attardi G (1995) Efficient selection and characterization of mutants of a human cell line which are defective in mitochondrial DNA-encoded subunits of respiratory NADH dehydrogenase. Mol Cell Biol 15:964–974

Holt PJ, Morgan DJ, Sazanov LA (2003) The location of NuoL and NuoM subunits in the membrane domain of the Escherichia coli complex I: implications for the mechanism of proton pumping. J Biol Chem 278:43114–43120

Hunte C, Zickermann V, Brandt U (2010) Functional modules and structural basis of conformational coupling in mitochondrial complex I. Science 329:448–451

Janssen R, Smeitink J, Smeets R, van den Heuvel L (2002) CIA30 complex I assembly factor: a candidate for human complex I deficiency? Hum Genet 110:264–270

Janssen RJ, Nijtmans LG, van den Heuvel LP, Smeitink JA (2006) Mitochondrial complex I: structure, function and pathology. J Inherit Metab Dis 29:499–515

Janssen RJ, Distelmaier F, Smeets R, Wijnhoven T, Ostergaard E, Jaspers NG, Raams A, Kemp S, Rodenburg RJ, Willems PH, van den Heuvel LP, Smeitink JA, Nijtmans LG (2009) Contiguous gene deletion of ELOVL7, ERCC8 and NDUFAF2 in a patient with a fatal multisystem disorder. Hum Mol Genet 18:3365–3374

Kao MC, Di BS, Nakamaru-Ogiso E, Miyoshi H, Matsuno-Yagi A, Yagi T (2005) Characterization of the membrane domain subunit NuoJ (ND6) of the NADH-quinone oxidoreductase from Escherichia coli by chromosomal DNA manipulation. Biochemistry 44:3562–3571

Kervinen M, Hinttala R, Helander HM, Kurki S, Uusimaa J, Finel M, Majamaa K, Hassinen IE (2006) The MELAS mutations 3946 and 3949 perturb the critical structure in a conserved loop of the ND1 subunit of mitochondrial complex I. Hum Mol Genet 15:2543–2552

Kirby DM, McFarland R, Ohtake A, Dunning C, Ryan MT, Wilson C, Ketteridge D, Turnbull DM, Thorburn DR, Taylor RW (2004a) Mutations of the mitochondrial ND1 gene as a cause of MELAS. J Med Genet 41:784–789

Kirby DM, Salemi R, Sugiana C, Ohtake A, Parry L, Bell KM, Kirk EP, Boneh A, Taylor RW, Dahl HH, Ryan MT, Thorburn DR (2004b) NDUFS6 mutations are a novel cause of lethal neonatal mitochondrial complex I deficiency. J Clin Invest 114:837–845

Kuffner R, Rohr A, Schmiede A, Krull C, Schulte U (1998) Involvement of two novel chaperones in the assembly of mitochondrial NADH: ubiquinone oxidoreductase (complex I). J Mol Biol 283:409–417

Kussmaul L, Hirst J (2006) The mechanism of superoxide production by NADH: ubiquinone oxidoreductase (complex I) from bovine heart mitochondria. Proc Natl Acad Sci USA 103:7607–7612

Lambert AJ, Brand MD (2009) Reactive oxygen species production by mitochondria. Methods Mol Biol 554:165–181

Lazarou M, McKenzie M, Ohtake A, Thorburn DR, Ryan MT (2007) Analysis of the assembly profiles for mitochondrial- and nuclear-DNA-encoded subunits into complex I. Mol Cell Biol 27:4228–4237

Leif H, Weidner U, Berger A, Spehr V, Braun M, van Heek P, Friedrich T, Ohnishi T, Weiss H (1993) Escherichia coli NADH dehydrogenase I, a minimal form of the mitochondrial complex I. Biochem Soc Trans 21:998–1001

Leif H, Sled VD, Ohnishi T, Weiss H, Friedrich T (1995) Isolation and characterization of the proton-translocating NADH: ubiquinone oxidoreductase from Escherichia coli. Eur J Biochem 230:538–548

Li Y, D'Aurelio M, Deng JH, Park JS, Manfredi G, Hu P, Lu J, Bai Y (2007) An assembled complex IV maintains the stability and activity of complex I in mammalian mitochondria. J BiolChem 282:17557–17562

Marques I, Duarte M, Videira A (2003) The 9.8 kDa subunit of complex I, related to bacterial Na(+)-translocating NADH dehydrogenases, is required for enzyme assembly and function in Neurospora crassa. J Mol Biol 329:283–290

Marques I, Duarte M, Assuncao J, Ushakova AV, Videira A (2005) Composition of complex I from Neurospora crassa and disruption of two accessory subunits. Biochim Biophys Acta 1707:211–220

Marques I, Ushakova AV, Duarte M, Videira A (2007) Role of the conserved cysteine residues of the 11.5 kDa subunit in complex I catalytic properties. J Biochem 141:489–493

Mathiesen C, Hagerhall C (2003) The 'antiporter module' of respiratory chain complex I includes the MrpC/NuoK subunit – a revision of the modular evolution scheme. FEBS Lett 549:7–13

McKenzie M, Ryan MT (2010) Assembly factors of human mitochondrial complex I and their defects in disease. IUBMB Life 62:497–502

Nehls U, Friedrich T, Schmiede A, Ohnishi T, Weiss H (1992) Characterization of assembly intermediates of NADH: ubiquinone oxidoreductase (complex I) accumulated in Neurospora mitochondria by gene disruption. J Mol Biol 227:1032–1042

Nouws J, Nijtmans L, Houten S, van den Brand M, Huynen M, Venselaar H, Hoefs S, Gloerich J, Kronick J, Hutchin T, Willems P, Rodenburg R, Wanders R, van den Heuvel L, Smeitink J, Vogel R (2010) Acyl-CoA dehydrogenase 9 is required for the biogenesis of oxidative phosphorylation complex I. Cell Metab 12:283–294

Ogilvie I, Kennaway NG, Shoubridge EA (2005) A molecular chaperone for mitochondrial complex I assembly is mutated in a progressive encephalopathy. J Clin Invest 115:2784–2792

Pagliarini DJ, Calvo SE, Chang B, Sheth SA, Vafai SB, Ong SE, Walford GA, Sugiana C, Boneh A, Chen WK, Hill DE, Vidal M, Evans JG, Thorburn DR, Carr SA, Mootha VK (2008) A mitochondrial protein compendium elucidates complex I disease biology. Cell 134:112–123

Palmisano G, Sardanelli AM, Signorile A, Papa S, Larsen MR (2007) The phosphorylation pattern of bovine heart complex I subunits. Proteomics 7:1575–1583

Perales-Clemente E, Fernandez-Vizarra E, Acin-Perez R, Movilla N, Bayona-Bafaluy MP, Moreno-Loshuertos R, Perez-Martos A, Fernandez-Silva P, Enriquez JA (2010) Five entry points of the mitochondrially encoded subunits in mammalian complex I assembly. Mol Cell Biol 30:3038–3047

Perier C, Tieu K, Guegan C, Caspersen C, Jackson-Lewis V, Carelli V, Martinuzzi A, Hirano M, Przedborski S, Vila M (2005) Complex I deficiency primes Bax-dependent neuronal apoptosis through mitochondrial oxidative damage. Proc Natl Acad Sci USA 102:19126–19131

Potluri P, Davila A, Ruiz-Pesini E, Mishmar D, O'Hearn S, Hancock S, Simon M, Scheffler IE, Wallace DC, Procaccio V (2009) A novel NDUFA1 mutation leads to a progressive mitochondrial complex I-specific neurodegenerative disease. Mol Genet Metab 96:189–195

Prieur I, Lunardi J, Dupuis A (2001) Evidence for a quinone binding site close to the interface between NUOD and NUOB subunits of complex I. Biochim Biophys Acta 1504:173–178

Runswick MJ, Fearnley IM, Skehel JM, Walker JE (1991) Presence of an acyl carrier protein in NADH: ubiquinone oxidoreductase from bovine heart mitochondria. FEBS Lett 286:121–124

Ryan MT, Hoogenraad NJ (2007) Mitochondrial-nuclear communications. Annu Rev Biochem 76:701–722

Saada A, Edvardson S, Rapoport M, Shaag A, Amry K, Miller C, Lorberboum-Galski H, Elpeleg O (2008) C6ORF66 is an assembly factor of mitochondrial complex I. Am J Hum Genet 82:32–38

Saada A, Vogel RO, Hoefs SJ, van den Brand MA, Wessels HJ, Willems PH, Venselaar H, Shaag A, Barghuti F, Reish O, Shohat M, Huynen MA, Smeitink JA, van den Heuvel LP, Nijtmans LG (2009) Mutations in NDUFAF3 (C3ORF60), encoding an NDUFAF4 (C6ORF66)-interacting complex I assembly protein, cause fatal neonatal mitochondrial disease, Am J Hum Genet 84:718–727

Sazanov LA, Peak-Chew SY, Fearnley IM, Walker JE (2000) Resolution of the membrane domain of bovine complex I into subcomplexes: implications for the structural organization of the enzyme. Biochemistry 39:7229–7235

Scacco S, Petruzzella V, Budde S, Vergari R, Tamborra R, Panelli D, van den Heuvel LP, Smeitink JA, Papa S (2003) Pathological mutations of the human NDUFS4 gene of the 18-kDa (AQDQ) subunit of complex I affect the expression of the protein and the assembly and function of the complex. J Biol Chem 278:44161–44167

Schafer E, Dencher NA, Vonck J, Parcej DN (2007) Three-dimensional structure of the respiratory chain supercomplex I1III2IV1 from bovine heart mitochondria. Biochemistry 46:12579–12585

Schagger H (2001) Respiratory chain supercomplexes. IUBMB Life 52:119–128

Schagger H, Pfeiffer K (2001) The ratio of oxidative phosphorylation complexes I-V in bovine heart mitochondria and the composition of respiratory chain supercomplexes. J Biol Chem 276:37861–37867

Schilling B, Aggeler R, Schulenberg B, Murray J, Row RH, Capaldi RA, Gibson BW (2005) Mass spectrometric identification of a novel phosphorylation site in subunit NDUFA10 of bovine mitochondrial complex I. FEBS Lett 579:2485–2490

Schmidt M, Friedrich T, Wallrath J, Ohnishi T, Weiss H (1992) Accumulation of the pre-assembled membrane arm of NADH: ubiquinone oxidoreductase in mitochondria of manganese-limited grown Neurospora crassa. FEBS Lett 313:8–11

Schneider D, Pohl T, Walter J, Dorner K, Kohlstadt M, Berger A, Spehr V, Friedrich T (2008) Assembly of the Escherichia coli NADH: ubiquinone oxidoreductase (complex I). Biochim Biophys Acta 1777:735–739

Schulte U, Weiss H (1995) Generation and characterization of NADH: ubiquinone oxidoreductase mutants in Neurospora crassa. Methods Enzymol 260:3–14

Schulte U, Haupt V, Abelmann A, Fecke W, Brors B, Rasmussen T, Friedrich T, Weiss H (1999) A reductase/isomerase subunit of mitochondrial NADH: ubiquinone oxidoreductase (complex I) carries an NADPH and is involved in the biogenesis of the complex. J Mol Biol 292:569–580

Sheftel AD, Stehling O, Pierik AJ, Netz DJ, Kerscher S, Elsasser HP, Wittig I, Balk J, Brandt U, Lill R (2009) Human ind1, an iron-sulfur cluster assembly factor for respiratory complex I. Mol Cell Biol 29:6059–6073

Sugiana C, Pagliarini DJ, McKenzie M, Kirby DM, Salemi R, bu-Amero KK, Dahl HH, Hutchison WM, Vascotto KA, Smith SM, Newbold RF, Christodoulou J, Calvo S, Mootha VK, Ryan MT, Thorburn DR (2008) Mutation of C20orf7 disrupts complex I assembly and causes lethal neonatal mitochondrial disease. Am J Hum Genet 83:468–478

Tuschen G, Sackmann U, Nehls U, Haiker H, Buse G, Weiss H (1990) Assembly of NADH: ubiquinone reductase (complex I) in Neurospora mitochondria. Independent pathways of nuclear-encoded and mitochondrially encoded subunits. J Mol Biol 213:845–857

Ugalde C, Janssen RJ, van den Heuvel LP, Smeitink JA, Nijtmans LG (2004a) Differences in assembly or stability of complex I and other mitochondrial OXPHOS complexes in inherited complex I deficiency. Hum Mol Genet 13:659–667

Ugalde C, Vogel R, Huijbens R, Van den Heuvel B, Smeitink J, Nijtmans L (2004b) Human mitochondrial complex I assembles through the combination of evolutionary conserved modules: a framework to interpret complex I deficiencies. Hum Mol Genet 13:2461–2472

Ugalde C, Hinttala R, Timal S, Smeets R, Rodenburg RJ, Uusimaa J, van Heuvel LP, Nijtmans LG, Majamaa K, Smeitink JA (2007) Mutated ND2 impairs mitochondrial complex I assembly and leads to Leigh syndrome. Mol Genet Metab 90:10–14

Velazquez I, Nakamaru-Ogiso E, Yano T, Ohnishi T, Yagi T (2005) Amino acid residues associated with cluster N3 in the NuoF subunit of the proton-translocating NADH-quinone oxidoreductase from Escherichia coli. FEBS Lett 579:3164–3168

Vogel RO, Janssen RJ, Ugalde C, Grovenstein M, Huijbens RJ, Visch HJ, van den Heuvel LP, Willems PH, Zeviani M, Smeitink JA, Nijtmans LG (2005) Human mitochondrial complex I assembly is mediated by NDUFAF1. FEBS J 272:5317–5326

Vogel RO, Dieteren CE, van den Heuvel LP, Willems PH, Smeitink JA, Koopman WJ, Nijtmans LG (2007a) Identification of mitochondrial complex I assembly intermediates by tracing tagged NDUFS3 demonstrates the entry point of mitochondrial subunits. J Biol Chem 282:7582–7590

Vogel RO, Janssen RJ, van den Brand MA, Dieteren CE, Verkaart S, Koopman WJ, Willems PH, Pluk W, van den Heuvel LP, Smeitink JA, Nijtmans LG (2007b) Cytosolic signaling protein Ecsit also localizes to mitochondria where it interacts with chaperone NDUFAF1 and functions in complex I assembly. Genes Dev 21:615–624

Vogel RO, van den Brand MA, Rodenburg RJ, van den Heuvel LP, Tsuneoka M, Smeitink JA, Nijtmans LG (2007c) Investigation of the complex I assembly chaperones B17.2 L and NDUFAF1 in a cohort of CI deficient patients. Mol Genet Metab 91:176–182

Weidner U, Geier S, Ptock A, Friedrich T, Leif H, Weiss H (1993) The gene locus of the proton-translocating NADH: ubiquinone oxidoreductase in Escherichia coli. Organization of the 14 genes and relationship between the derived proteins and subunits of mitochondrial complex I. J Mol Biol 233:109–122

Wessels HJ, Vogel RO, van den Heuvel L, Smeitink JA, Rodenburg RJ, Nijtmans LG, Farhoud MH (2009) LC-MS/MS as an alternative for SDS-PAGE in blue native analysis of protein complexes. Proteomics 9:4221–4228

Wittig I, Carrozzo R, Santorelli FM, Schagger H (2006) Supercomplexes and subcomplexes of mitochondrial oxidative phosphorylation. Biochim Biophys Acta 1757:1066–1072

Yamaguchi M, Belogrudov GI, Matsuno-Yagi A, Hatefi Y (2000) The multiple nicotinamide nucleotide-binding subunits of bovine heart mitochondrial NADH: ubiquinone oxidoreductase (complex I). Eur J Biochem 2672:329–336

Chapter 11
Complexes I in the Green Lineage

Claire Remacle, Patrice Hamel, Véronique Larosa, Nitya Subrahmanian, and Pierre Cardol

Abstract In land plants and green algae, mitochondria and chloroplasts were acquired sequentially through primary endosymbiotic events with a α-proteobacterium and a cyanobacterium, respectively. The inner membrane of the mitochondria harbors the enzyme complexes of the respiratory chain, the largest of them being the rotenone-sensitive NADH:ubiquinone oxidoreductase or complex I. In the thylakoid membrane of the chloroplast, besides the photosynthetic machinery, a light-independent respiratory-chain inherited from cyanobacteria drives electrons from NAD(P)H to oxygen. In most plants and algae, it comprises a homolog of bacterial complex I (NAD(P)H:plastoquinone (PQ) oxidoreductase) and a PQ oxidase (PTOX). This chapter will be thus dedicated to similarities and peculiarities of plant mitochondrial complex I compared to the well studied enzyme in mammals and fungi, as well as to the structure and role of a complex I homolog in chloroplast.

Keywords Alga • Carbonic anhydrase • Chloroplast • Higher plant • Mutant

---

C. Remacle (✉) • V. Larosa • P. Cardol
Laboratory of Genetics of Microorganisms, Institute of Botany B22, University of Liège, B-4000 Liège, Belgium
e-mail: c.remacle@ulg.ac.be

P. Hamel (✉) • N. Subrahmanian
Department of Molecular Genetics and Department of Molecular and Cellular Biochemistry, 500 Aronoff Laboratory, 318 W 12th Ave., Columbus, OH 43210, USA

Plant Cellular and Molecular Biology Graduate Program, The Ohio State University, Columbus, OH, USA
e-mail: hamel.16@osu.edu

L. Sazanov (ed.), *A Structural Perspective on Respiratory Complex I: Structure and Function of NADH:ubiquinone oxidoreductase*, DOI 10.1007/978-94-007-4138-6_11,

11.1 The Mitochondrial Complex I in Plants

11.1.1 Subunit Composition (Table 11.1)

The subunit composition of mitochondrial complex I has been intensively studied in the land plant model *Arabidopsis thaliana* and in the Chlorophycean green alga *Chlamydomonas reinhardtii*. In *Chlamydomonas*, the mitochondrial respiratory-chain complexes were separated by blue native (BN) gel electrophoresis from purified mitochondria (van Lis et al. 2003; Cardol et al. 2004). With an apparent molecular mass between 950 and 1,000 kD, complex I was then resolved into its constitutive subunits in a second dimensional SDS-gel and 30 components ranging from 7 to 77 kD could be subsequently identified by mass spectrometry analyses (Cardol et al. 2004). By searching for putative homologs of fungal (*e.g.* Videira and Duarte 2002) and mammalian (Carroll et al. 2002; Hirst et al. 2003) complex I subunits, the *Chlamydomonas* enzyme was thought to comprise at least 42 proteins. In the land plants *Vicia faba* (broad bean), *Solanum tuberosum* (potato), and *Triticum aestivum* (wheat), pioneer work in the 1990s has shown that the complex I enzyme comprised up to 30 subunits ranging from 6 to 75 kD but only few subunits were identified (Leterme and Boutry 1993; Herz et al. 1994; Combettes and Grienenberger 1999). A few years later, a similar approach to the one described for *Chlamydomonas* led to the identification of at least 39 subunits in *Arabidopsis* and in *Oryza sativa* (rice), ten of which appeared to be specific to land plants (Heazlewood et al. 2003). More recently, new attempts to characterize complex I subunit composition in *Arabidopsis* have been undertaken. A third dimensional gel electrophoresis (BN/SDS/SDS) approach developed to study complex I in the fungus *Yarrowia lipolytica* (Abdrakhmanova et al. 2004) was applied to *Arabidopsis* and enabled the subsequent identification of 42 different subunits (Meyer et al. 2008). Finally, a three step purification procedure allowed the recovery of a complex I fraction containing 49 subunits, among which one third were described as unique to plants (Klodmann et al. 2010). This procedure included (i) a mitochondrial membrane isolation step, (ii) the separation of enzymes complexes by sucrose gradient centrifugation and (iii) a cytochrome *c* affinity chromatography to remove complex III or supercomplexes. Compared to previous protocols published for the isolation of complex I from various plants that included a chromatography step (*e.g.* Leterme and Boutry 1993; Combettes and Grienenberger 1999), this procedure avoids high salt conditions and is expected, according to the authors, to preserve the integrity of complex I. Most novel subunits identified in plant complex I are small hydrophobic proteins that are probably part of the membrane domain (Klodmann et al. 2010). Some of them are highly divergent orthologs of small hydrophobic supernumerary subunits found in fungi or mammals (Gabaldon et al. 2005; Carroll et al. 2006; Morgner et al. 2008; Bridges et al. 2010; Cardol 2011). One should note that over the years the number of subunits that were identified as conserved between mammals, fungi and plants increased: 27 in 2003 (Heazlewood et al. 2003), 32–33 in 2005 (Cardol et al. 2004; Gabaldon et al. 2005), 34–37 in 2009/2010 (Huynen et al. 2009; Klodmann et al. 2010), and 41 in 2011 (Cardol 2011). This gradual discovery was fueled by the progressive

Table 11.1 Protein components of plant mitochondrial complex I

	Arabidopsis thaliana	*Chlamydomonas reinhardtii*	Mammals[e]	Fungi[f]
	Bacterial core			
1	At5g11770[a,b,c]	AAQ63698[d]	NDUFS7/PSST	NUO19.3/NUKM
2	At1g16700, At1g79010[a,b,c]	AAQ63697[d]	NDUFS8/TYKY	NUO21.3c/NUIM
3	At4g02580[a,b,c]	AAQ63695[d]	NDUFV2/24 kD	NUO24/NUHM
4	AtMg00070[a,b,c]	AAQ55457[d]	NDUFS3/30 kD	NUO30.4 (31)/NUGM
5	AtMg00510[a,b,c]	AAQ63700[d]	NDUFS2/49 kD	NUO49/NUCM
6	At5g08530[a,b,c]	AAQ63696[d]	NDUFV1/51 kD	51/NUBM
7	At5g37510[a,b,c]	AAQ73136[d]	NDUFS1/75 kD	NUO78/NUAM
8	AtMg00516[a,b,c]	AAB93446	ND1	ND1/NU1M
9	AtMg00285[b,c]	AAB93444	ND2	ND2/NU2M
10	AtMg00990[c]	AAQ55461[d]	ND3	ND3/NU3M
11	AtMg00580[b,c]	AAB93441	ND4	ND4/NU4M
12	AtMg00650	AAO61142	ND4L	ND4L/NULM
13	AtMg00513[a,b,c]	AAB93442	ND5	ND5/NU5M
14	AtMg00270[b]	AAB93445	ND6	ND6/NU6M
	Conserved supernumerary			
15	At3g08610[a,b,c]	AAS48198	NDUFA1/MWFE	NUO9.8/NIMM
16	At5g47890[b,c]	AAQ63699[d]	NDUFA2/B8	NUO10.5/NI8M
17	At2g02510[b,c]	AAS48194[d]	NDUFB3/B12	NUO10.6/NB2M
18	At5g52840[a,b,c]	AAQ73139[d]	NDUFA5/B13	NUO29.9/NUFM
19	At3g03070[c]	AAQ64639[d]	NDUFS6/13 kD A	NUO18.4/NUMM
20	At3g12260[a,b]	AAQ84469[d]	NDUFA6/B14	NUO14.8/NB4M
21	At2g42210[b]	AAS58499[d]	NDUFA11/B14.7	NUO21.3b/NUJM
22	At3g57785, At2g42310[a,b,c]	AAS48192[d]	NDUFB11/ESSS	NUO11.7/NUWM
23	At3g62790, At2g47690[a,b,c]	AAQ98888	NDUFS5/PFFD	NUO11.5/NIPM
24	At2g31490[a,b,c]	AAS48193[d]	NDUFB4/B15	NUO6.6/NUVM

(continued)

Table 11.1 (continued)

	Arabidopsis thaliana	*Chlamydomonas reinhardtii*	Mammals[e]	Fungi[f]
25	At1g04630, At2g33220[a,b,c]	AAQ64637[d]	NDUFA13/B16.6	NUO14 (13.5)/NB6M
26	At3g03100[a,b]	AAQ64638[d]	NDUFA12/B17.2	NUO13.4/N7BM
27	At2g02050[a,b,c]	AAQ73135[d]	NDUFB7/B18	NB8M
28	At5g67590[a,b]	AAQ64640[d]	NDUFS4/AQDQ	NUO21/NUYM
29	At5g18800, At3g06310[b,c]	AAQ55460	NDUFA8/PGIV	NUO20.8/NUPM
30	At4g34700	AAQ73134	NDUFB9/B22	NI2M
31	At1g49140, At3g18410[a,b,c]	AAQ55459[d]	NDUFB10/PDSW	NUO12.3/NIDM
32	At2g20360[a,b,c]	AAQ55458[d]	NDUFA9/39 kD	NUO40/NUEM
33	At5g47570[b,c]	XP_001700273	NDUFB8/ASHI	NUO20.1/NIAM
34	At1g76200[b,c]	–	NDUFB2/AGGG	NCU01436[g]
35	At4g16450[a,b,c]	AAQ64641[d]	NDUFB1/MNLL	NUO20.9/NUXM
36	At4g20150 (NDU9)[a,b,c]	AAS58501[d]	NDUFC2/B14.5B	NUO10.4
37	At4g00585[b,c]	XP_001697243[h]	NDUFC1/KFYI	NCU08300[g]/NUUM
38	At2g46540[h]	XP_001692978[h]	NDUFA3/B9	NUO9.5/NI9M
39	AAM6246[h]	AAQ73138[h]	NDUFAB1/ACPM	SDAP
40	At5g08060[h]	XP_001703194[h]	NDUFA7/B14.5A	NCU08930[g]/NUZM
41	At3g29970[h]	–	NDUFA4/MLRQ	NCU02016[h]
42	At1g67785[b,c]	–	NDUFB5/SGDH	NUO17.8
43	AAG51141[h]	–	NDUFA10/42 kD	–
44	At3g47930[a,c] L-galactono-1·4-lactone dehydrogenase	EDP08950[h]	XP_001253523[h]	NCU03188[g]
	Plant specific			
45	γ-carbonic anhydrase At5g63510, At1g19580, At3g48680· At1g47260, At5g66510[a,b,c]	AAS48196, AAS48197, AAS48195[d]	–[g]	–[g]
46	At3g07480[h]	AAS58502[d] ferredoxin-like	–	–
47	At5g14105[c]	–	–	–

48	At1g67350[b,c]	–	–	–
49	At1g68680[b]	–	–	
50	At2g27730[a,b,c]	–	–	
51	–	AAS58503[d]	–	
52	–	AAS58498[d]	–	

[a,b,c,d]Subunits identified by mass spectrometry as complex I components (Heazlewood et al. 2003; Meyer et al. 2008; Klodmann et al. 2010; Cardol et al. 2004)

[e]Accession numbers of mammalian sequences can be found in (*e.g.* Carroll et al. 2006; Hirst et al. 2003)

[f]Accession numbers for fungal sequences can be found for *Y. lipolytica* (Abdrakhmanova et al. 2004), *N. crassa* (Marques et al. 2005) and *Pichia pastoris* (Bridges et al. 2010)

[g]Although such proteins are absent from mammalian and fungal complex I, homologs are found in other eukaryotes lineages such as *Amoebozoa* (Gawryluk and Gray 2010), see text for further details

[h]Not identified by biochemical approaches as complex I component

acquisition of proteomic and genomic data in various organisms in combination to the availability of profile-to-sequence and profile-to-profile comparisons tools. At this stage, additional efforts are still needed to characterize the composition of mitochondrial complex I in distant species in order to highlight real lineage-specific or conserved subunits and decipher their role.

Nonetheless, all eukaryotic complex I enzymes investigated to date comprise approximately 25 non-core subunits, in addition to the 14/15 core subunits present in the bacterial enzyme. These additional proteins, the so-called supernumerary (or accessory) subunits, are presumed to participate in assembly, stability, or regulation, rather than in enzyme activity (*e.g.* Friedrich et al. 1998; Heazlewood et al. 2003; Marques et al. 2005; Abdrakhmanova et al. 2006; Bridges et al. 2010). Another role for non-core subunits as regulatory molecules outside the mitochondrial compartment is also emerging, as exemplified for the GRIM-19 subunit (Gene associated with Retinoic-IFN- induced mortality 19). This nucleus located protein, originally identified as a critical regulatory protein for interferon-beta and retinoic acid induced cell death (Angell et al. 2000), also localizes to mitochondrial human complex I. A GRIM19 homolog associated to complex I has been found in the green lineage (Heazlewood et al. 2003; Cardol et al. 2004) and in fungi (Abdrakhmanova et al. 2004; Bridges et al. 2010, reviewed in Remacle et al. 2008).

In *Arabidopsis*, five proteins structurally related to bacterial gamma-type carbonic anhydrases (γ-CA) have been assigned to the membrane arm of complex I (Parisi et al. 2004; Perales et al. 2004; Sunderhaus et al. 2006) while three were found in association with *Chlamydomonas* complex I (Cardol et al. 2004). Single particle electron microscopy analysis of complex I from *Polytomella* (a chloroplastless close relative of *Chlamydomonas*), *Arabidopsis*, *Zea mays* and *S. tuberosum* indicate that these γ-CA subunits could constitute a spherical domain attached to the central part of the membrane arm of complex I and exposed to the matrix (Fig. 11.1a) (Perales et al. 2005; Sunderhaus et al. 2006; Peters et al. 2008; Bultema et al. 2009). In *Arabidopsis*, γ-CA1 and γ-CA2 have also been shown to be important for complex I assembly and possibly involved in mitochondrial one-carbon metabolism (Perales et al. 2005). Overexpression of γ-CA2 in *Arabidopsis* leads to a male sterile phenotype (Villarreal et al. 2009). More recent experiments indicated that γ-CA2 trimers are capable of binding inorganic carbon (Martin et al. 2009). Since these subunits were believed to be plant-specific, it was postulated that this complex I domain might play a role in relationship to photosynthesis. However, two γ-CA were recently found in association with complex I in the amoeboid protozoon *Acanthamoeba castellanii* (Gawryluk and Gray 2010), which does not have a chloroplast and is considered to be a sister group to opisthokonts, a group of eukaryotes including metazoa and fungi (Keeling et al. 2005). γ-CA subunits are also encoded in the genomes of most eukaryotes (including the non-photosynthetic alga *Polytomella*), with the exception of opisthokonts (Gawryluk and Gray 2010). Altogether these findings suggest that this γ-CA module was lost during evolution of the opisthokont lineage and that it might play a more general role in complex I function in other eukaryotes (see Gawryluk and Gray 2010 for further discussion). In *Chlamydomonas*, a small subunit (AAS58502) was identified in complex I from

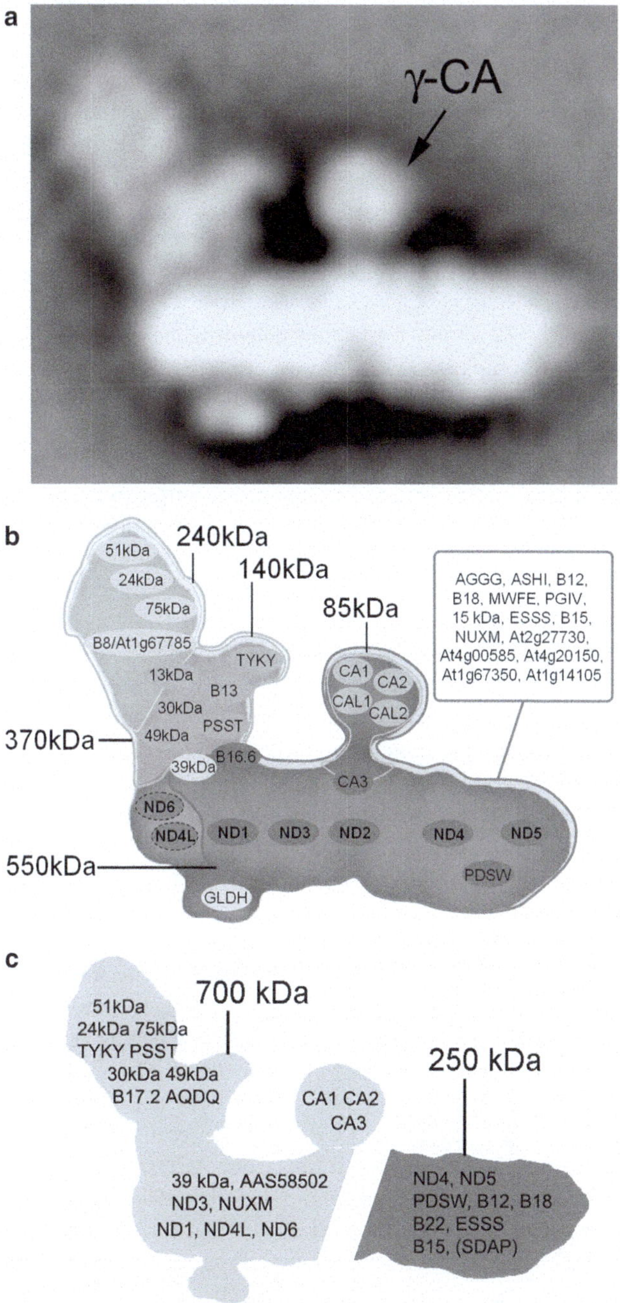

Fig. 11.1 Overall structure and subunit localization of plant mitochondrial complex I. (**a**) Projection map from *A. thaliana* complex I obtained by EM single-particle analysis (From Sunderhaus et al. 2006, Courtesy H.P. Braun and E.J. Boekema) The carbonic anhydrase domain is annotated (γ-CA). (**b**) Schematic model representation of complex I from *A. thaliana* (Modified from Klodmann et al. 2010, Courtesy H.P. Braun). (**c**) Schematic model representation of complex I from *C. reinhardtii* (Adapted from Cardol et al. 2008. With permission from Elsevier). Outline shape side-views are drawn from (Sunderhaus et al. 2006)

wild-type (Cardol et al. 2004) and in a subcomplex lacking the distal part of the membrane arm (Cardol et al. 2008). This subunit has putative homologs in land plants and shares also some similarities with ferredoxin and ferredoxin-like proteins from various sources. It was hypothesized that recruitment of a ferredoxin-like protein within plant complex I could drive electrons to the carbonic anhydrase domain (Cardol et al. 2008). But the lack of evidence of the association of such a protein with *Arabidopsis* complex I (Meyer et al. 2008; Klodmann et al. 2010) does not support this view.

In *Arabidopsis*, L -galactono-1,4-lactone dehydrogenase (GLDH) is also a structural component of complex I (Heazlewood et al. 2003; Klodmann et al. 2010). GLDH catalyses the final oxidation of galactono-$\gamma-$lactone to ascorbic acid (Ostergaard et al. 1997; Siendones et al. 1999). Its activity decreases with leaf age and is modulated by the availability of oxidized cytochrome *c* (Bartoli et al. 2000). Ascorbate, the major soluble antioxidant and redox buffer in mammals and plants, plays important roles in plant development, photoprotection, and cell expansion (*e.g.* Smirnoff 2000; Arrigoni and De Tullio 2002). Complex I from *Arabidopsis* occurs in two forms that are distinguished on the basis of their different mobility in BN-PAGE. Only the high-mobility form contains GLDH and bears an ascorbate synthesis activity that is sensitive to rotenone (Heazlewood et al. 2003). These observations pointed out towards the existence of a subpopulation of complex I whose function could regulate ascorbate synthesis by monitoring the rate of NADH-driven electron flow through complex I (Millar et al. 2003; Pineau et al. 2008). A homozygous T-DNA *Arabidopsis* mutant, deficient for GLDH, developed only when supplemented with ascorbate and was impaired in complex I content (Pineau et al. 2008). There is a GLDH protein homolog encoded in *Chlamydomonas* (Genbank accession number EDP08950), and its possible association with complex I remains to be explored.

A small fraction of the mitochondrial acyl carrier proteins (ACP) involved in the synthesis of type II fatty acids, localizing primarily in the mitochondrial matrix (Cronan et al. 2005), is associated with mammal and fungal complex I (Zensen et al. 1992; Triepels et al. 1999; Carroll et al. 2003, 2005; Schilling et al. 2005; Hinttala et al. 2005). As a consequence, depletion of mitochondrial ACP in these organisms leads to complex I impairment, in addition to a defect in fatty acid biosynthesis (Schneider et al. 1995; Feng et al. 2009; Dobrynin et al. 2010). By similarity to complex I composition in mammals and fungi, one ACP subunit has been proposed to be a part of complex I in plants and algae (Heazlewood et al. 2003; Cardol et al. 2004). In *Arabidopsis*, it has been further shown that five mitochondrial ACP isoforms exist but ACP activity occurs predominantly as two soluble isoforms (mtACP1 and mtACP2) in the mitochondrial matrix (Meyer et al. 2007). However, recent proteomic analyses of *Arabidopsis* complex I failed to detect the presence of any of the five mitochondrial ACPs (Meyer et al. 2008; Klodmann et al. 2010). Thus ACP does not seem to be a *bona fide* complex I subunit in plants and its role in complex I biogenesis awaits further experimental testing.

11.1.2 Genetic Studies

In vascular plants, the consequences of complex I defects have been mainly studied in three experimental models, *A. thaliana*, *Nicotiana sylvestris* and *Z. mays*. Mutants in nuclear genes encoding complex I subunits in *Arabidopsis* include the plant specific γ-CA2 subunit (Perales et al. 2005) discussed previously and the eukaryotic specific 18-kDa subunit (NDUFS4). Unexpectedly, this latter mutant has been isolated from a mutagenized population screened for altered responses to different stress conditions (Ishitani et al. 1997). The mutant exhibited a reduced capacity for cold acclimation and increased superoxide production, although the impact on complex I activity and assembly was not investigated (Lee et al. 2002). A T-DNA linked mutation in the same nuclear gene was then further studied (Meyer et al. 2009). This insertional mutant showed no assembly or activity of complex I and delayed germination and growth. Comparative mitochondrial proteome analysis showed that modification was restricted to the abundance of complex I subunits without significant changes in other mitochondrial proteins. Metabolite changes predominated at night and ATP level was also lower in the dark period. Overall, the decreased efficiency of ATP production by OXPHOS lead to broad rearrangements in cellular metabolism and development and to altered tolerance to abiotic stresses.

Mitochondrial mutants lacking some of the *nd* genes as a result of genome rearrangements have been described in *Z. mays* and *N. sylvestris* (tobacco). In maize, mitochondrial rearrangements yield abnormal growth called the NCS (non chromosomal stripe) phenotype, defined as striped sectors of pale-green tissues on the leaves. The NCS2 mutant, affected for both *nd4* and *nd7* genes, is male-sterile and heteroplasmic for the mutation (Marienfeld and Newton 1994). Assembly of complex I is compromised (Karpova and Newton 1999, see Sect. 11.1.3) and expression of one the three genes encoding the mitochondrial alternative oxidase (Aox2) is specifically increased (Karpova et al. 2002). In addition, chloroplast function and structure are also altered (Roussell et al. 1991). In tobacco, mitochondrial rearrangements also lead to abnormal growth and male sterility, causing the CMS (cytoplasmic male sterility) phenotype. The CMSII mutant has been extensively studied. It has been first described as having lost the *nd7* gene (Pla et al. 1995; Gutierres et al. 1997) and was next shown also to be affected in the expression of *nd1* (Gutierres et al. 1999). However, complementation with a nuclear version of *nd7* alone, which expresses a protein targeted to mitochondria restored the wild-type phenotype (Pineau et al. 2005), showing that the defect in *nd7* was the cause of the complex I defect. Similar to the maize NCS mutants, the capacity of the alternative oxidase and the amount of Aox are increased (Gutierres et al. 1997; Sabar et al. 2000), which is linked to the specific overexpression of one of the three genes encoding the mitochondrial alternative (Aox1.2) (Vidal et al. 2007; Liu et al. 2008). In parallel, the activity and the amount of alternative NAD(P)H dehydrogenases also increase contrary to the situation found in NCS maize mutants (Gutierres et al. 1997; Sabar et al. 2000). Photosynthetic efficiency is reduced and carbon assimilation under different stress conditions is modified (Sabar et al. 2000; Dutilleul et al. 2003a; Cardol et al. 2010;

Galle et al. 2010). Metabolomic studies showed an accumulation of nitrogen-rich amino acids (Dutilleul et al. 2005). Finally, the expression of many stress-related genes was changed, resulting in modified tolerance to stresses (Dutilleul et al. 2003b; Galle et al. 2010). Mitochondrial rearrangements leading to altered complex I have also been described in cucumber but the molecular nature of these rearrangements is not known (Juszczuk and Rychter 2009). Finally, several mutants have been identified in genes encoding splicing factors involved in the expression of *nd4* in *N. sylvestris* (Brangeon et al. 2000) and *A. thaliana* (Nakagawa and Sakurai 2006), and *nd1* and *nd9* in *A. thaliana* (de Longevialle et al. 2007; Jonietz et al. 2010).

In conclusion, all the plant complex I mutants present a retarded growth phenotype, sometimes male sterility, usually enhanced alternative oxidase activity and modified photosynthetic performances. Although proteomic and metabolomic studies revealed certain modifications, a general trend in the changes could not be discerned. This situation contrasts with *Chlamydomonas* complex I mutants where neither the apparent capacity of the alternative oxidase nor the photosynthetic efficiencies seem severely affected (Cardol et al. 2003).

Chlamydomonas respiratory mutants can be obtained and are viable if maintained under phototrophic conditions (where they rely on photosynthesis in the chloroplast). Mutants deficient for complex I can be easily scored on the basis of their impaired growth in the dark (Remacle et al. 2001a). Indeed, contrary to complex III or complex IV mutants that do not grow in the dark because only one active proton-pumping site is left, complex I mutants, which retain two active proton-pumping sites are still able to grow in these conditions, although at a significant lower rate (Remacle et al. 2001a). This phenotypic trait was used to screen complex I deficient mutants obtained after treatment of *Chlamydomonas* cells with acriflavine, a mutagenic agent that binds more specifically to mitochondrial DNA. The isolated mutants carry molecular lesions, usually frameshift mutations, in the mitochondrial *nd* genes (*nd1*, *nd4*, *nd5* and *nd6*). They are homoplasmic and directly amenable to biochemical studies which have led to propose a model for complex I assembly and subunit localization within the *Chlamydomonas* enzyme (see below) (Cardol et al. 2002, 2008; Remacle et al. 2001a, b). In addition, biolistic transformation of the mitochondrial genome of *Chlamydomonas* has also been employed to reconstruct mutations in the mitochondrial *nd* genes (Remacle et al. 2006).

Concerning nuclear genes, RNA interference technology allowed the isolation of mutants deficient for the ND3 and ND4L subunits (Cardol et al. 2006). These two subunits are usually encoded in the mitochondrial genome except in *Chlamydomonas* where their corresponding genes have migrated to the nucleus (Cardol et al. 2006). Recently an insertional mutagenesis was conducted to isolate loss-of-function nuclear mutations in complex I genes. Candidate complex I deficient strains were first screened on the basis of their slow growth phenotype in the dark. Then, mutants unable to assemble an active complex I were visualized by an in-gel colorimetric assay that reveals the NADH dehydrogenase activity of complex I in the presence of NADH and nitroblue tetrazolium. This simple technique is rapid and most appropriate to screen many mutants for impaired assembly of complex I (Cardol et al. 2006). To date, this screen allowed the recovery of two tagged mutants: one mutant is deficient for the

PDSW subunit (*NUOB10* gene), a non-core subunit of the distal part of the hydrophobic arm of complex I (Barbieri et al. 2011), the other is deficient for the 24 kDa subunit (*NUO5* gene), a core subunit, binding a (2Fe-2S) cluster in the soluble arm of complex I (unpublished data).

11.1.3 Internal Architecture and Biogenesis

The localization of subunits within complex I subcomplexes (or domains) has been investigated in *Chlamydomonas* and *Arabidopsis*. In *Chlamydomonas*, the analysis of mutants deficient in the synthesis of mitochondrial-encoded hydrophobic components ND1, ND4, ND5, and ND6 (Cardol et al. 2002, 2008; Remacle et al. 2001a, b, 2006), or mutants that have lost the expression of nuclear genes encoding ND3, ND4L (Cardol et al. 2006) or PDSW (Barbieri et al. 2011) led to the identification of a 200- and 700-kD subcomplexes. The 200-kD soluble subcomplex carries the rotenone insensitive NADH dehydrogenase activity and could correspond to a fragment of the matrix-exposed arm. In PDSW, ND4 and ND5-deficient cells, the 700-kD membrane-associated subcomplex also displays NADH dehydrogenase activity (Barbieri et al. 2011; Cardol et al. 2002, 2008; Remacle et al. 2006). This subcomplex is less tightly bound to the membrane than the wild-type enzyme, and is mainly composed of subunits belonging to the matrix-exposed arm (Cardol et al. 2008) (see Fig. 11.1c). In maize, the NCS2 mutant affected in ND4 also displays a partially assembled complex I loosely attached to the mitochondrial inner membrane (Karpova and Newton 1999). In *Arabidopsis*, ten subcomplexes were obtained by destabilizing the enzyme by means of various treatments (Klodmann et al. 2010). Their individual subunit composition was resolved and the major building blocks were deduced from this analysis (Fig. 11.1b). Based on the subcomplexes identified in various mutants or following destabilizing treatments, models of complex I architecture and assembly have also been proposed in *Neurospora* (Videira and Duarte 2002) and human (Hirst et al. 2003; Antonicka et al. 2003; Ugalde et al. 2004; Lazarou et al. 2009; Dieteren et al. 2008; Vogel et al. 2007a; McKenzie and Ryan 2010). Localization of conserved subunits within these subcomplexes is generally conserved among the different models for complex I assembly. Assembly models for plant complex I also include additional small modules, most remarkably a matrix-exposed domain including γ-type carbonic anhydrases (see paragraph 1.1 for further details). The current hypothesis is that complex I assembly occurs by a stepwise mechanism during which preformed modules, or assembly intermediates, are combined. The hydrogenase and the membrane modules are joined together and both are expanding through the recruitment of non-core subunits.

The biogenesis of complex I requires numerous chaperones and assembly factors, most of which are conserved in plants and algae (Table 11.2). The first two assembly factors, CIA30 and CIA84, were identified in *N. crassa*. CIA30 and CIA84 are chaperones that have been shown to be directly involved in the assembly process

Table 11.2 Assembly factors in human, higher plants and *C. reinhardtii*

Complex I assembly factor				
H. sapiens	*Higher plants*	*C. reinhardtii*	Function	Reference
NDUFAF1 (CIA30)	CIA30 (AT1G17350)	NUOFAF1	Chaperone (early assembly)	Kuffner et al. (1998)
NUBPL (Ind1)	INDL	Ind1	Assembly of Fe–S cofactors	Bych et al. (2008)
Foxred1	NP_180034 Sarcosine oxidase family protein	XP_001692123 FAD-dependent oxidoreductase	Redox reactions	Calvo et al. (2010)
C8ORF38	Os06g0104100	XP_001693265	Abundance and activity of complex I	Pagliarini et al. (2008)
C20ORF7	OsI_37783	XP_001693605	Assembly early stage of complex I	Sugiana et al. (2008)
NDUFAF2 (B17.2L)	–	–	Chaperone (late assembly)	Vogel et al. (2007b)
NDUFAF3	OsJ_32539	XP_001702394	Cooperation between NDUFAF3 and NDUFAF4 from early to late stages complex I assembly	Saada et al. (2009)
NDUFAF4 (C6ORF66)	–	XP_001701912		

of complex I, through their association with a large membrane domain in a mutant unable to assemble the holoenzyme (Kuffner et al. 1998). CIA30 is well conserved among eukaryotes and homologs were found not only in vascular plants and algae but also in humans (Table 11.2). In fact, in humans, CIA30 plays a crucial role in the early assembly of complex I and mutations in the CIA30 encoding gene are responsible for a complex I-linked mitochondrial disease (Dunning et al. 2007). To date, CIA84 seems specific to *N. crassa* and was not found in vascular plants or in algae. In 2008, Ind1 was identified as participating in the assembly of Fe–S cofactors and subunits of complex I in the yeast *Y. lipolytica* (Bych et al. 2008). As expected for such a role, Ind1 is well conserved among eukaryotes. In humans, where it is also known as NUBPL, it is critically required for the assembly of complex I with a possible role in the delivery of one or more Fe/S clusters to complex I subunits (Sheftel et al. 2009). In the past few years, the discovery of six assembly factors (C20orf7, C8ORF38, FOXRED1, NDUFAF2, NDUFAF3, and NDUFAF4) provided a significant insight into the assembly process of human complex I. Bioinformatics analysis reveals that homologs of these assembly factors are detected in algae and vascular plants, except for NDUAF2 and NDUAF4 (Table 11.2). C20orf7 is peripherally associated with the matrix side of the mitochondrial inner membrane and plays a crucial role in the early stage of complex I assembly but in a different manner than CIA30 (Sugiana et al. 2008). Knocking-down *C8orf38* in mice resulted in a reduction of both abundance and activity of complex I (Pagliarini et al. 2008). FOXRED1 is an uncharacterized protein that derives its name from a FAD-dependent oxidoreductase protein domain. A lack of FOXRED1 leads to severe complex I deficiency in humans, but its biochemical activity in the assembly process remains unclear. Four human homologs of FOXRED1 (DMGDH, SARDH, PIPOX and PDPR) perform redox reactions in amino acid catabolism, suggesting a potential link between amino acid metabolism and complex I (Calvo et al. 2010). NDUFAF2, also known as B17.2 L is a paralog of a complex I subunit (B17.2), which is not incorporated into the holoenzyme, and plays a role in a late step of the assembly/stability of complex I (Vogel et al. 2007b). Recent studies showed that NDUFAF3 and NDUFAF4 cooperate from early to late stages of complex I assembly in association with at least the highly conserved subunits NDUFS2, NDUFS3, and NDUFS8, and non-core subunit NDUFA5 (Saada et al. 2009).

Analyses of supercomplexes in plants have been investigated after solubilization with digitonin and migration in BN-PAGE (*e.g.* Eubel et al. 2004; Krause et al. 2004) and by single particle electron microscopy (*e.g.* Peters et al. 2008; Bultema et al. 2009). Different supercomplexes could be detected : $I+III_{(2)}$ (the most abundant), $III_{(2)}+IV_{(1)}$, $V_{(2)}$, $I_{(2)}+III_{(2)}$ and respirasome $I+III_{(2)}+IV_{(1)}$. In *Chlamydomonas*, solubilization of mitochondria with dodecyl-maltoside allows to detect $I+III_{(2)}$ (Cardol et al. 2008) but detailed architecture was not investigated further. As far as complex I is concerned, it is worth mentioning that complex I mutants in *Chlamydomonas* usually exhibit a higher activity of succinate:cytochrome *c* oxidoreductase (complexes II+III) (Barbieri et al. 2011; Cardol et al. 2002, 2006, 2008; Remacle et al. 2001a), which could be viewed as a compensation effect for the loss of complex I. This suggests that in some circumstances preferential association between complexes II and III exists although these two are usually not detected in supercomplexes.

11.2 Complex I in the Chloroplast

The thylakoid membranes of photosynthetic eukaryotes harbor a complex I-like enzyme, also referred to as NDH-1, that is related to cyanobacterial complex I (Battchikova and Aro 2007). Evidence for such an enzyme in the green lineage was suspected from the discovery of *ndh* genes encoding proteins with similarity to known complex I subunits in the first sequenced plastid genomes (Ohyama et al. 1986; Shinozaki et al. 1986). Based on the occurrence of the *ndh* genes in chloroplasts, NDH-1 is proposed to be present in all land plants (including ferns and mosses) (Sugiura et al. 2003; Gao et al. 2009) with the exception of some gymnosperm species (Wakasugi et al. 1994). Plastid *ndh* genes are also present in primitive green algae such as *Nephroselmis* or *Mesostigma* (Lemieux et al. 2000; Turmel et al. 1999) but absent from other microalgae such as *Chlamydomonas*, *Chlorella* or *Ostreococcus* (Wakasugi et al. 1997; Maul et al. 2002; Robbens et al. 2007). The observation that the plastid *ndh* genes are absent from the nuclear genomes of microalgae was taken as evidence that the plastid complex I was lost in such organisms (Derelle et al. 2006; Merchant et al. 2007; Palenik et al. 2007). In *Chlamydomonas* (and probably other algae or gymnosperms missing the plastid *ndh* subunits), it has been shown that type II NAD(P)H dehydrogenase enzymes operate instead of plastid complex I (Desplats et al. 2009; Jans et al. 2008; Peltier et al. 2010).

Based on electron microscopy of cyanobacterial complex I, plastid NDH-1 is proposed, similarly to bacterial and mitochondrial complex I, to display a L-shape with a hydrophobic core in the thylakoid membrane and a hydrophilic arm facing the stroma (Arteni et al. 2006). NDH-1 bearing plastids usually encode 11 NDH subunits (NdhA-NdhK), seven of which (NdhA-NdhG) are found in the membrane-embedded hydrophobic subcomplex. The counterparts of the bacterial NuoE, F and G subunits that are involved in NADH binding and oxidation (*i.e.* NADH dehydrogenase module) in bacterial/mitochondrial complex I are missing in both cyanobacterial and plastid NDH-1 (Friedrich et al. 1995). Because NADPH is the major stromal reductant, it is not clear how electrons enter plastid complex I considering that NDH-1 appears to use NADH as its preferred electron donor (Sazanov et al. 1998; Rumeau et al. 2005). However, NDH-1 might be operating as a ferredoxin-plastoquinone oxido-reductase based on the recent finding that stromal ferredoxin can act as an electron donor to NDH-1 (Yamamoto et al. 2011). The site of ferredoxin oxidation is currently unknown and it is conceivable that NDH-1 can accept electrons from several stromal donors.

It should be noted that the low abundance and instability of plastid NDH-1 has been a major challenge for the biochemical and enzymatic characterization of this protein complex (Sazanov et al. 1996, 1998). Nevertheless, a combination of genetic and biochemical approaches including the partial purification of NDH-1 led to the discovery of 18 additional nuclear-encoded structural subunits of the plastid complex I, some of which are shared with cyanobacteria while others appear

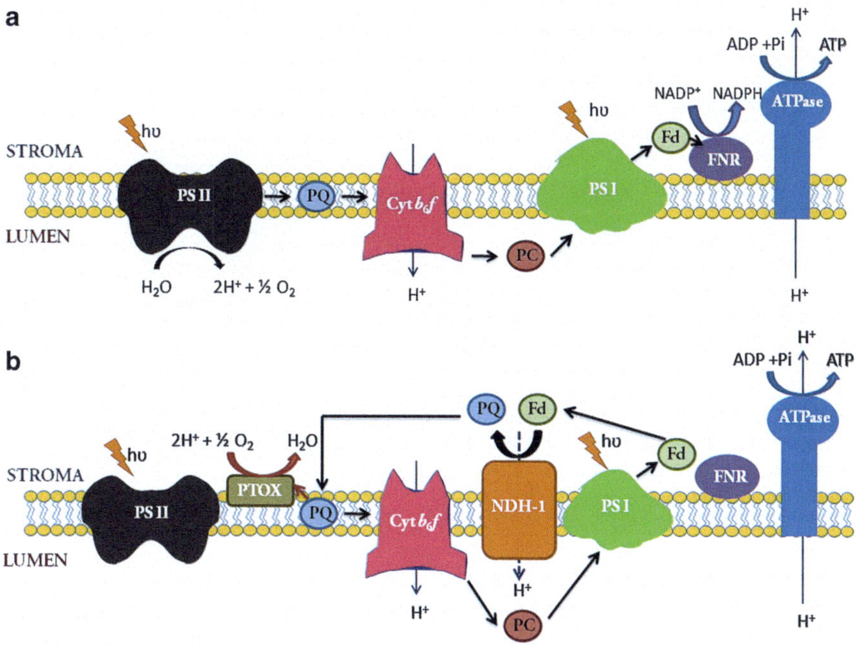

Fig. 11.2 Models for electron transfer routes in the thylakoid. (**a**) **Linear Electron Transfer**: The light dependent linear electron transfer at the thylakoid membrane starts from photo-excitation (hν) of two electrons obtained from water oxidation at Photosystem II (*PSII*). One electron is then transferred via plastoquinone (*PQ*) to cytochrome b_6f (Cyt b_6f). From Cyt b_6f, Plastocyanin(*PC*) transports the electron to Photosystem I (*PSI*) where it is again subjected to photo-excitation (hν). The electron is then received by Ferredoxin (*Fd*), which acts as the electron donor to Ferredoxin:NADP Reductase (*FNR*) for the production of NADPH. The proton gradient established by Cyt b_6f acts as the driving force for the synthesis of ATP by ATPase. (**b**) **Cyclic Electron Transfer and Chlororespiration**: In the light, cyclic electron transfer can also take place and leads to ATP synthesis with no net NADPH being produced. In this instance, the electrons flow from Fd to Plastid Complex I (NDH-1) instead of FNR. NDH-1 transfers the electrons from two Fd to one PQ. From PQ the electron directly flows into Cyt b_6f, bypassing the requirement for PSII. During chlororespiration (*arrows* represented in *brown*), the PQ pool is oxidized by a plastid terminal oxidase (*PTOX*). Chlororespiration can take place in the *dark*

to be plant specific (Rumeau et al. 2005; Munshi et al. 2006; Muraoka et al. 2006; Ishihara et al. 2007; Ishikawa et al. 2008; Majeran et al. 2008; Shimizu et al. 2008; Ishida et al. 2009; Sirpio et al. 2009a,b; Suorsa et al. 2009; Takabayashi et al. 2009; Yamamoto et al. 2011).

Genetic inactivation of several plastid-encoded subunits of NDH-1 led to the conclusion that plastid complex I is fully dispensable for plant growth under normal conditions (*e.g.* Burrows et al. 1998; Kofer et al. 1998; Shikanai et al. 1998; Horvath et al. 2000). Biochemical investigation of plastid complex I mutants showed that NDH-1 mediates chlororespiration in the dark and cyclic electron transfer around photosystem I in the light (Fig. 11.2a, b). The term “chlororespiration” was first

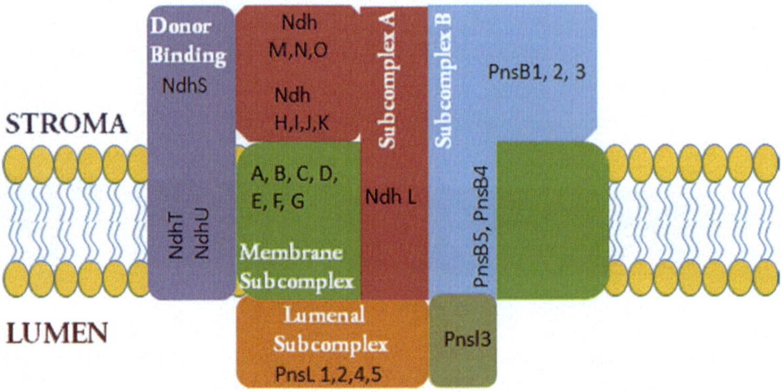

Fig. 11.3 The organization of structural subunits of Plastid Complex I. Plastid Complex I (NDH-1) is located in the thylakoid membrane of the plant chloroplast and consists of five subcomplexes: subcomplex A, subcomplex B, lumenal subcomplex, membrane subcomplex and the electron donor binding subcomplex. This diagram is based on Yamamoto et al. (2011) and Ifuku et al. (2011). We retained the terminology proposed by Ifuku et al. (2011)

coined to describe the activity of a putative electron transfer chain at the thylakoid membrane of chloroplasts to explain effects on the redox state of the plastoquinone pool (PQ) in the absence of photochemistry (Bennoun 1982). Similar to a respiratory chain, PQ was proposed to be reduced via the action of a NAD(P)H dehydrogenase and re-oxidized by oxygen via a terminal oxidase. The discovery of NDH-1 and PTOX (Plastid Terminal OXidase), a quinol oxidase related to mitochondrial alternative oxidase provided the first molecular evidence for the operation of such an electron transfer activity (Carol et al. 1999; Wu et al. 1999). It is likely that NDH-1 and PTOX constitute the components involved in chlororespiration but proof for a direct electron transfer from NDH-1 to PTOX is still missing. While loss of NDH-1 does not result in any visible phenotype, the absence of PTOX impacts carotenogenesis in leaves. Interestingly, a recent study in tomato suggests that NDH-1 controls carotenoid biosynthesis in fruit chromoplasts but not in chloroplasts (Nashilevitz et al. 2010). The NDH-1 dependent Cyclic Electron Flow around PSI (CEF) was originally defined as the transfer of electrons from stromal NADPH (that is reduced by the activity of PSI) back into the PQ pool (Rumeau et al. 2007; Peltier et al. 2010) (Fig. 11.2b). The physiological importance of NDH-1 is still unclear but several studies have highlighted the importance of the NDH-1 complex in stress conditions such as high light, drought or extreme temperatures (reviewed in Shikanai 2007; Suorsa et al. 2009; Johnson 2011). It is likely that the enzyme participates in regulating the ATP/NADPH ratio for optimal photosynthesis, a role already postulated for auxiliary electron transfer routes such as chlororespiration and CEF (Rumeau et al. 2007; Peltier et al. 2010).

The assembly of NDH-1, like its mitochondrial counterpart is proposed to proceed via modular assembly of five subcomplexes (Fig. 11.3): a membrane subcomplex, a soluble stromal subcomplex A, a membrane attached stromal facing subcomplex B, an electron donor binding subcomplex and a lumenal subcomplex

(Peng et al. 2008, 2009; Ifuku et al. 2011). The electron donor binding subcomplex contains NdhS, NdhT and NdhU. NdhS is required for the high-affinity binding of Fd to NDH-1 in an in vitro Fd-dependent PQ reduction assay and was postulated to act as the Fd-docking site in NDH-1 (Yamamoto et al. 2011). The membrane subcomplex (NdhA-NdhG, PnsB4) and the soluble subcomplex A (NdhH-NdhK, NDHM-NDHO) contain the core subunits that are also conserved in bacterial complex I. The membrane subcomplex and the soluble subcomplex A are proposed to be connected by NDHL (Shimizu et al. 2008). The subcomplex B (PnsB1, 2, 3, 4, 5) is postulated to interact with the chlorophyll binding proteins of the light-harvesting complex, LHCA5/LHCA6, which are attached to PSI (Peng et al. 2009). The identification of a novel 300 kDa complex containing NdhS, the candidate Fd-binding subunit suggests the existence of additional structural subunits, which still remain unidentified (Yamamoto et al. 2011). It is expected that yet-to-be-discovered subunit(s) containing co-factors might function in the transfer of electrons from Fd to plastoquinone. The lumenal subcomplex (PnsL1, 2, 4, 5) contains subunits that are specific to plastid complex I and is required for the stability of subcomplex A. PnsL3 is a lumenal subunit that is not part of the lumenal subcomplex but is in close interaction with subcomplex B (Ifuku et al. 2011).

In the thylakoid membranes, NDH-1 occurs as a high molecular weight complex that was shown to correspond to an NDH-1/PSI supercomplex (Lennon et al. 2003; Peng et al. 2008, 2009). In addition to the NDH-1/PSI supercomplex, three distinct subcomplexes containing NDH-1 subunits have been detected after resolution of thylakoid membranes via BN-PAGE (Ishihara et al. 2007; Peng et al. 2008; Sirpio et al. 2009a). Because the abundance of such complexes varies upon plastid differentiation, it is not clear if they correspond to assembly intermediates or subcomplexes with specialized functions. Interestingly, cyanobacteria possess at least four NDH-1 complexes that are very distinct in terms of function. Such functional versatility is achieved through a modification in subunit composition (Battchikova and Aro 2007; Ogawa and Mi 2007).

The biogenesis of NDH-1 requires several nuclear encoded proteins, some of which are involved in the splicing and editing of plastid *ndh* mRNAs while others are required at a post-translational step of the enzyme biogenesis (Suorsa et al. 2009). So far four proteins CRR1, CRR6, CRR7 and PIFI have been implicated in the assembly and/or stabilization of plastid Complex I (Munshi et al. 2005, 2006; Shimizu and Shikanai 2007; Wang and Portis 2007; Peng et al. 2010). Because there is no motif in their sequences that indicate an enzymatic activity, these components were postulated to act as NDH1-specific chaperones (CRR1, 6, 7, NDF5) or a regulator of NDH-1 activity (PIFI). One notable exception is CRR1, a stromal protein whose pyridine nucleotide binding site suggests a possible redox activity for this assembly factor (Shimizu and Shikanai 2007).

Acknowledgements Work on *Chlamydomonas* complex I was supported by a grant from the United Mitochondrial Disease Foundation (PH, CR) and FRS-FNRS 1.5.255.08 (CR), 2.4601.08 (CR), 1.5.103.10 (PC), 1.C057.09 (PC) and Action de la Recherche Concertée ARC07/12 -04 (CR). VL is supported by a FRIA fellowship. PC is a research associate from FRS-FNRS. We thank Dr. R. Lamb for critical reading of the manuscript.

References

Abdrakhmanova A, Zickermann V, Bostina M, Radermacher M, Schagger H, Kerscher S, Brandt U (2004) Subunit composition of mitochondrial complex I from the yeast *Yarrowia lipolytica*. Biochim Biophys Acta 1658:148–156

Abdrakhmanova A, Zwicker K, Kerscher S, Zickermann V, Brandt U (2006) Tight binding of NADPH to the 39-kDa subunit of complex I is not required for catalytic activity but stabilizes the multiprotein complex. Biochim Biophys Acta 1757:1676–1682

Angell JE, Lindner DJ, Shapiro PS, Hofmann ER, Kalvakolanu DV (2000) Identification of GRIM-19, a novel cell death-regulatory gene induced by the interferon-beta and retinoic acid combination, using a genetic approach. J Biol Chem 275:33416–33426

Antonicka H, Ogilvie I, Taivassalo T, Anitori RP, Haller RG, Vissing J, Kennaway NG, Shoubridge EA (2003) Identification and characterization of a common set of complex I assembly intermediates in mitochondria from patients with complex I deficiency. J Biol Chem 278:43081–43088

Arrigoni O, De Tullio MC (2002) Ascorbic acid: much more than just an antioxidant. Biochim Biophys Acta 1569:1–9

Arteni AA, Zhang P, Battchikova N, Ogawa T, Aro E-M, Boekema EJ (2006) Structural characterization of NDH-1 complexes of *Thermosynechococcus elongatus* by single particle electron microscopy. Biochim Biophys Acta 1757:1469–1475

Barbieri MR, Larosa V, Nouet C, Remacle C, Hamel PP (2011) A forward genetic screen identifies mutants deficient for mitochondrial complex I assembly in *Chlamydomonas reinhardtii*. Genetics 188:349–358

Bartoli CG, Pastori GM, Foyer CH (2000) Ascorbate biosynthesis in mitochondria is linked to the electron transport chain between complexes III and IV. Plant Physiol 123:335–344

Battchikova N, Aro EM (2007) Cyanobacterial NDH-1 complexes: multiplicity in function and subunit composition. Physiol Plant 131:22–32

Bennoun P (1982) Evidence for a respiratory chain in the chloroplast. Proc Natl Acad Sci USA 79:4352–4356

Brangeon J, Sabar M, Gutierres S, Combettes B, Bove J, Gendy C, Chetrit P, Des Francs-Small CC, Pla M, Vedel F, De Paepe R (2000) Defective splicing of the first *nad4* intron is associated with lack of several complex I subunits in the *Nicotiana sylvestris* NMS1 nuclear mutant. Plant J 21:269–280

Bridges HR, Fearnley IM, Hirst J (2010) The subunit composition of mitochondrial NADH:ubiquinone oxidoreductase (complex I) from *Pichia pastoris*. Mol Cell Proteomics 9:2318–2326

Bultema JB, Braun HP, Boekema EJ, Kouril R (2009) Megacomplex organization of the oxidative phosphorylation system by structural analysis of respiratory supercomplexes from potato. Biochim Biophys Acta 1787:60–67

Burrows PA, Sazanov LA, Svab Z, Maliga P, Nixon PJ (1998) Identification of a functional respiratory complex in chloroplasts through analysis of tobacco mutants containing disrupted plastid *ndh* genes. EMBO J 17:868–876

Bych K, Kerscher S, Netz DJ, Pierik AJ, Zwicker K, Huynen MA, Lill R, Brandt U, Balk J (2008) The iron-sulphur protein Ind1 is required for effective complex I assembly. EMBO J 27:1736–1746

Calvo SE, Tucker EJ, Compton AG, Kirby DM, Crawford G, Burtt NP, Rivas M, Guiducci C, Bruno DL, Goldberger OA, Redman MC, Wiltshire E, Wilson CJ, Altshuler D, Gabriel SB, Daly MJ, Thorburn DR, Mootha VK (2010) High-throughput, pooled sequencing identifies mutations in NUBPL and FOXRED1 in human complex I deficiency. Nat Genet. doi:10.1038/ng.659

Cardol P (2011) Mitochondrial NADH:ubiquinone oxidoreductase (complex I) in eukaryotes: a highly-conserved subunit composition highlighted by mining of protein databases. Biochim Biophys Acta. doi:10.1016/j.bbabio.2011.06.015

Cardol P, Matagne RF, Remacle C (2002) Impact of mutations affecting ND mitochondria-encoded subunits on the activity and assembly of complex I in *Chlamydomonas*. Implication for the structural organization of the enzyme. J Mol Biol 319:1211–1221

Cardol P, Gloire G, Havaux M, Remacle C, Matagne R, Franck F (2003) Photosynthesis and state transitions in mitochondrial mutants of *Chlamydomonas reinhardtii* affected in respiration. Plant Physiol 133:2010–2020

Cardol P, Vanrobaeys F, Devreese B, Van Beeumen J, Matagne RF, Remacle C (2004) Higher plant-like subunit composition of mitochondrial complex I from *Chlamydomonas reinhardtii*: 31 conserved components among eukaryotes. Biochim Biophys Acta 1658:212–224

Cardol P, Lapaille M, Minet P, Franck F, Matagne RF, Remacle C (2006) ND3 and ND4L subunits of mitochondrial complex I, both nucleus encoded in *Chlamydomonas reinhardtii*, are required for activity and assembly of the enzyme. Eukaryot Cell 5:1460–1467

Cardol P, Boutaffala L, Memmi S, Devreese B, Matagne RF, Remacle C (2008) In *Chlamydomonas*, the loss of ND5 subunit prevents the assembly of whole mitochondrial complex I and leads to the formation of a low abundant 700 kDa subcomplex. Biochim Biophys Acta 1777:388–396

Cardol P, De Paepe R, Franck F, Forti G, Finazzi G (2010) The onset of NPQ and Deltamu(H) + upon illumination of tobacco plants studied through the influence of mitochondrial electron transport. Biochim Biophys Acta 1797:177–188

Carol P, Stevenson D, Bisanz C, Breitenbach J, Sandmann G, Mache R, Coupland G, Kuntz M (1999) Mutations in the *Arabidopsis* gene IMMUTANS cause a variegated phenotype by inactivating a chloroplast terminal oxidase associated with phytoene desaturation. Plant Cell 11:57–68

Carroll J, Shannon RJ, Fearnley IM, Walker JE, Hirst J (2002) Definition of the nuclear encoded protein composition of bovine heart mitochondrial complex I. Identification of two new subunits. J Biol Chem 277:50311–50317

Carroll J, Fearnley IM, Shannon RJ, Hirst J, Walker JE (2003) Analysis of the subunit composition of complex I from bovine heart mitochondria. Mol Cell Proteomics 2:117–126

Carroll J, Fearnley IM, Skehel JM, Runswick MJ, Shannon RJ, Hirst J, Walker JE (2005) The post-translational modifications of the nuclear encoded subunits of complex I from bovine heart mitochondria. Mol Cell Proteomics 4:693–699

Carroll J, Fearnley IM, Skehel JM, Shannon RJ, Hirst J, Walker JE (2006) Bovine complex I is a complex of 45 different subunits. J Biol Chem 281:32724–32727

Combettes B, Grienenberger JM (1999) Analysis of wheat mitochondrial complex I purified by a one-step immunoaffinity chromatography. Biochimie 81:645–653

Cronan JE, Fearnley IM, Walker JE (2005) Mammalian mitochondria contain a soluble acyl carrier protein. FEBS Lett 579:4892–4896

de Longevialle AF, Meyer EH, Andres C, Taylor NL, Lurin C, Millar AH, Small ID (2007) The pentatricopeptide repeat gene OTP43 is required for trans-splicing of the mitochondrial *nad1* Intron 1 in *Arabidopsis thaliana*. Plant Cell 19:3256–3265

Derelle E, Ferraz C, Rombauts S, Rouze P, Worden AZ, Robbens S, Partensky F, Degroeve S, Echeynie S, Cooke R, Saeys Y, Wuyts J, Jabbari K, Bowler C, Panaud O, Piegu B, Ball SG, Ral JP, Bouget FY, Piganeau G, De Baets B, Picard A, Delseny M, Demaille J, Van de Peer Y, Moreau H (2006) Genome analysis of the smallest free-living eukaryote *Ostreococcus tauri* unveils many unique features. Proc Natl Acad Sci USA 103:11647–11652

Desplats C, Mus F, Cuiné S, Billon E, Cournarc L, Peltier G (2009) Characterization of Nda2, a plastoquinone-reducing type II NAD(P)H dehydrogenase in *Chlamydomonas* chloroplasts. J Biol Chem 284:4148–4157

Dieteren CE, Willems PH, Vogel RO, Swarts HG, Fransen J, Roepman R, Crienen G, Smeitink JA, Nijtmans LG, Koopman WJ (2008) Subunits of mitochondrial complex I exist as part of matrix- and membrane-associated subcomplexes in living cells. J Biol Chem 283: 34753–34761

Dobrynin K, Abdrakhmanova A, Richers S, Hunte C, Kerscher S, Brandt U (2010) Characterization of two different acyl carrier proteins in complex I from *Yarrowia lipolytica*. Biochim Biophys Acta 1797:152–159

Dunning CJ, McKenzie M, Sugiana C, Lazarou M, Silke J, Connelly A, Fletcher JM, Kirby DM, Thorburn DR, Ryan MT (2007) Human CIA30 is involved in the early assembly of mitochondrial complex I and mutations in its gene cause disease. EMBO J 26:3227–3237

Dutilleul C, Driscoll S, Cornic G, De Paepe R, Foyer CH, Noctor G (2003a) Functional mitochondrial complex I is required by tobacco leaves for optimal photosynthetic performance in photorespiratory conditions and during transients. Plant Physiol 131:264–275

Dutilleul C, Garmier M, Noctor G, Mathieu C, Chetrit P, Foyer CH, de Paepe R (2003b) Leaf mitochondria modulate whole cell redox homeostasis, set antioxidant capacity, and determine stress resistance through altered signaling and diurnal regulation. Plant Cell 15:1212–1226

Dutilleul C, Lelarge C, Prioul JL, De Paepe R, Foyer CH, Noctor G (2005) Mitochondria-driven changes in leaf NAD status exert a crucial influence on the control of nitrate assimilation and the integration of carbon and nitrogen metabolism. Plant Physiol 139:64–78

Eubel H, Heinemeyer J, Braun HP (2004) Identification and characterization of respirasomes in potato mitochondria. Plant Physiol 134:1450–1459

Feng D, Witkowski A, Smith S (2009) Down-regulation of mitochondrial acyl carrier protein in mammalian cells compromises protein lipoylation and respiratory complex I and results in cell death. J Biol Chem 284:11436–11445

Friedrich T, Steinmüller K, Weiss H (1995) The proton-pumping respiratory complex I of bacteria and mitochondria and its homologue in chloroplasts. FEBS Lett 367:107–111

Friedrich T, Abelmann A, Brors B, Guenebaut V, Kintscher L, Leonard K, Rasmussen T, Scheide D, Schlitt A, Schulte U, Weiss H (1998) Redox components and structure of the respiratory NADH:ubiquinone oxidoreductase (complex I). Biochim Biophys Acta 1365:215–219

Gabaldon T, Rainey D, Huynen MA (2005) Tracing the evolution of a large protein complex in the eukaryotes, NADH:ubiquinone oxidoreductase (complex I). J Mol Biol 348:857–870

Galle A, Florez-Sarasa I, Thameur A, de Paepe R, Flexas J, Ribas-Carbo M (2010) Effects of drought stress and subsequent rewatering on photosynthetic and respiratory pathways in *Nicotiana sylvestris* wild type and the mitochondrial complex I-deficient CMSII mutant. J Exp Bot 61:765–775

Gao L, Yi X, Yang Y-X, Su Y-J, Wang T (2009) Complete chloroplast genome sequence of a tree fern *Alsophila spinulosa*: insights into evolutionary changes in fern chloroplast genomes. BMC Evol Biol 9:130

Gawryluk RM, Gray MW (2010) Evidence for an early evolutionary emergence of gamma-type carbonic anhydrases as components of mitochondrial respiratory complex I. BMC Evol Biol 10:176

Gutierres S, Sabar M, Lelandais C, Chetrit P, Diolez P, Degand H, Boutry M, Vedel F, de Kouchkovsky Y, De Paepe R (1997) Lack of mitochondrial and nuclear-encoded subunits of complex I and alteration of the respiratory chain in *Nicotiana sylvestris* mitochondrial deletion mutants. Proc Natl Acad Sci USA 94:3436–3441

Gutierres S, Combettes B, De Paepe R, Mirande M, Lelandais C, Vedel F, Chetrit P (1999) In the *Nicotiana sylvestris* CMSII mutant, a recombination-mediated change 5′ to the first exon of the mitochondrial *nad1* gene is associated with lack of the NADH:ubiquinone oxidoreductase (complex I) NAD1 subunit. Eur J Biochem 261:361–370

Heazlewood JL, Howell KA, Millar AH (2003) Mitochondrial complex I from *Arabidopsis* and rice: orthologs of mammalian and fungal components coupled with plant-specific subunits. Biochim Biophys Acta 1604:159–169

Herz U, Schroder W, Liddell A, Leaver CJ, Brennicke A, Grohmann L (1994) Purification of the NADH:ubiquinone oxidoreductase (complex I) of the respiratory chain from the inner mitochondrial membrane of *Solanum tuberosum*. J Biol Chem 269:2263–2269

Hinttala R, Uusimaa J, Remes AM, Rantala H, Hassinen IE, Majamaa K (2005) Sequence analysis of nuclear genes encoding functionally important complex I subunits in children with encephalomyopathy. J Mol Med 83:786–794

Hirst J, Carroll J, Fearnley IM, Shannon RJ, Walker JE (2003) The nuclear encoded subunits of complex I from bovine heart mitochondria. Biochim Biophys Acta 1604:135–150

Horvath EM, Peter SO, Joet T, Rumeau D, Cournac L, Horvath GV, Kavanagh TA, Schafer C, Peltier G, Medgyesy P (2000) Targeted inactivation of the plastid *ndhB* gene in tobacco results

in an enhanced sensitivity of photosynthesis to moderate stomatal closure. Plant Physiol 123:1337–1350
Huynen MA, de Hollander M, Szklarczyk R (2009) Mitochondrial proteome evolution and genetic disease. Biochim Biophys Acta 1792:1122–1129
Ifuku K, Endo T, Shikanai T, Aro EM (2011) Structure of the chloroplast NADH dehydrogenase-like complex: nomenclature for nuclear-encoded subunits. Plant Cell Physiol 52:1560–1568
Ishida S, Takabayashi A, Ishikawa N, Hano Y, Endo T, Sato F (2009) A novel nuclear-encoded protein, NDH-dependent cyclic electron flow 5, is essential for the accumulation of chloroplast NAD(P)H dehydrogenase complexes. Plant Cell Physiol 50:383–393
Ishihara S, Takabayashi A, Ido K, Endo T, Ifuku K, Sato F (2007) Distinct functions for the two PsbP-like proteins PPL1 and PPL2 in the chloroplast thylakoid lumen of *Arabidopsis*. Plant Physiol 145:668–679
Ishikawa N, Takabayashi A, Ishida S, Hano Y, Endo T, Sato F (2008) NDF6: a thylakoid protein specific to terrestrial plants is essential for activity of chloroplastic NAD(P)H dehydrogenase in *Arabidopsis*. Plant Cell Physiol 49:1066–1073
Ishitani M, Xiong L, Stevenson B, Zhu JK (1997) Genetic analysis of osmotic and cold stress signal transduction in *Arabidopsis*: interactions and convergence of abscisic acid-dependent and abscisic acid-independent pathways. Plant Cell 9:1935–1949
Jans F, Mignolet M, Hoyoux PA, Cardol P, Ghysels B, Cuiné S, Cournac L, Peltier G, Remacle C, Franck F (2008) A type II NAD(P)H dehydrogenase mediates light-independent plastoquinone reduction in the chloroplast of *Chlamydomonas*. Proc Natl Acad Sci USA 105:20546–20551
Johnson GN (2011) Reprint of: physiology of PSI cyclic electron transport in higher plants. Biochim Biophys Acta 1807:906–911
Jonietz C, Forner J, Holzle A, Thuss S, Binder S (2010) RNA PROCESSING FACTOR2 is required for 5′ end processing of nad9 and cox3 mRNAs in mitochondria of *Arabidopsis thaliana*. Plant Cell 22:443–453
Juszczuk IM, Rychter AM (2009) BN-PAGE analysis of the respiratory chain complexes in mitochondria of cucumber MSC16 mutant. Plant Physiol Biochem 47:397–406
Karpova OV, Newton KJ (1999) A partially assembled complex I in ND4-deficient mitochondria of maize. Plant J 17:511–521
Karpova OV, Kuzmin EV, Elthon TE, Newton KJ (2002) Differential expression of alternative oxidase genes in maize mitochondrial mutants. Plant Cell 14:3271–3284
Keeling PJ, Burger G, Durnford DG, Lang BF, Lee RW, Pearlman RE, Roger AJ, Gray MW (2005) The tree of eukaryotes. Trends Ecol Evol 20:670–676
Klodmann J, Sunderhaus S, Nimtz M, Jansch L, Braun HP (2010) Internal architecture of mitochondrial complex I from *Arabidopsis thaliana*. Plant Cell 22:797–810
Kofer W, Koop HU, Wanner G, Steinmüller K (1998) Mutagenesis of the genes encoding subunits A, C, H, I, J and K of the plastid NAD(P)H-plastoquinone-oxidoreductase in tobacco by polyethylene glycol-mediated plastome transformation. Mol Gen Genet 258:166–173
Krause F, Reifschneider NH, Vocke D, Seelert H, Rexroth S, Dencher NA (2004) "Respirasome"-like supercomplexes in green leaf mitochondria of spinach. J Biol Chem 279:48369–48375
Kuffner R, Rohr A, Schmiede A, Krull C, Schulte U (1998) Involvement of two novel chaperones in the assembly of mitochondrial NADH:Ubiquinone oxidoreductase (complex I). J Mol Biol 283:409–417
Lazarou M, Thorburn DR, Ryan MT, McKenzie M (2009) Assembly of mitochondrial complex I and defects in disease. Biochim Biophys Acta 1793:78–88
Lee BH, Lee H, Xiong L, Zhu JK (2002) A mitochondrial complex I defect impairs cold-regulated nuclear gene expression. Plant Cell 14:1235–1251
Lemieux C, Otis C, Turmel M (2000) Ancestral chloroplast genome in *Mesostigma viride* reveals an early branch of green plant evolution. Nature 403:649–652
Lennon AM, Prommeenate P, Nixon PJ (2003) Location, expression and orientation of the putative chlororespiratory enzymes, Ndh and IMMUTANS, in higher-plant plastids. Planta 218:254–260
Leterme S, Boutry M (1993) Purification and preliminary characterization of mitochondrial complex I (NADH: ubiquinone reductase) from broad bean (*Vicia faba* L.). Plant Physiol 102:435–443

Liu YJ, Norberg FE, Szilagyi A, De Paepe R, Akerlund HE, Rasmusson AG (2008) The mitochondrial external NADPH dehydrogenase modulates the leaf NADPH/NADP+ratio in transgenic *Nicotiana sylvestris*. Plant Cell Physiol 49:251–263

Majeran W, Zybailov B, Ytterberg AJ, Dunsmore J, Sun Q, van Wijk KJ (2008) Consequences of C4 differentiation for chloroplast membrane proteomes in maize mesophyll and bundle sheath cells. Mol Cell Proteomics 7:1609–1638

Marienfeld JR, Newton KJ (1994) The maize NCS2 abnormal growth mutant has a chimeric nad4-nad7 mitochondrial gene and is associated with reduced complex I function. Genetics 138:855–863

Marques I, Duarte M, Assuncao J, Ushakova AV, Videira A (2005) Composition of complex I from *Neurospora crassa* and disruption of two "accessory" subunits. Biochim Biophys Acta 1707:211–220

Martin V, Villarreal F, Miras I, Navaza A, Haouz A, Gonzalez-Lebrero RM, Kaufman SB, Zabaleta E (2009) Recombinant plant gamma carbonic anhydrase homotrimers bind inorganic carbon. FEBS Lett 583:3425–3430

Maul JE, Lilly JW, Cui L, dePamphilis CW, Miller W, Harris EH, Stern DB (2002) The *Chlamydomonas reinhardtii* plastid chromosome: islands of genes in a sea of repeats. Plant Cell 14:2659–2679

McKenzie M, Ryan MT (2010) Assembly factors of human mitochondrial complex I and their defects in disease. IUBMB Life 62:497–502

Merchant SS, Prochnik SE, Vallon O, Harris EH, Karpowicz SJ, Witman GB, Terry A, Salamov A, Fritz-Laylin LK, Marechal-Drouard L, Marshall WF, Qu LH, Nelson DR, Sanderfoot AA, Spalding MH, Kapitonov VV, Ren Q, Ferris P, Lindquist E, Shapiro H, Lucas SM, Grimwood J, Schmutz J, Cardol P, Cerutti H, Chanfreau G, Chen CL, Cognat V, Croft MT, Dent R, Dutcher S, Fernandez E, Fukuzawa H, Gonzalez-Ballester D, Gonzalez-Halphen D, Hallmann A, Hanikenne M, Hippler M, Inwood W, Jabbari K, Kalanon M, Kuras R, Lefebvre PA, Lemaire SD, Lobanov AV, Lohr M, Manuell A, Meier I, Mets L, Mittag M, Mittelmeier T, Moroney JV, Moseley J, Napoli C, Nedelcu AM, Niyogi K, Novoselov SV, Paulsen IT, Pazour G, Purton S, Ral JP, Riano-Pachon DM, Riekhof W, Rymarquis L, Schroda M, Stern D, Umen J, Willows R, Wilson N, Zimmer SL, Allmer J, Balk J, Bisova K, Chen CJ, Elias M, Gendler K, Hauser C, Lamb MR, Ledford H, Long JC, Minagawa J, Page MD, Pan J, Pootakham W, Roje S, Rose A, Stahlberg E, Terauchi AM, Yang P, Ball S, Bowler C, Dieckmann CL, Gladyshev VN, Green P, Jorgensen R, Mayfield S, Mueller-Roeber B, Rajamani S, Sayre RT, Brokstein P, Dubchak I, Goodstein D, Hornick L, Huang YW, Jhaveri J, Luo Y, Martinez D, Ngau WC, Otillar B, Poliakov A, Porter A, Szajkowski L, Werner G, Zhou K, Grigoriev IV, Rokhsar DS, Grossman AR (2007) The *Chlamydomonas* genome reveals the evolution of key animal and plant functions. Science 318:245–250

Meyer EH, Heazlewood JL, Millar AH (2007) Mitochondrial acyl carrier proteins in *Arabidopsis thaliana* are predominantly soluble matrix proteins and none can be confirmed as subunits of respiratory complex I. Plant Mol Biol 64:319–327

Meyer EH, Taylor NL, Millar AH (2008) Resolving and identifying protein components of plant mitochondrial respiratory complexes using three dimensions of gel electrophoresis. J Proteome Res 7:786–794

Meyer EH, Tomaz T, Carroll AJ, Estavillo G, Delannoy E, Tanz SK, Small ID, Pogson BJ, Millar AH (2009) Remodeled respiration in ndufs4 with low phosphorylation efficiency suppresses *Arabidopsis* germination and growth and alters control of metabolism at night. Plant Physiol 151:603–619

Millar AH, Mittova V, Kiddle G, Heazlewood JL, Bartoli CG, Theodoulou FL, Foyer CH (2003) Control of ascorbate synthesis by respiration and its implications for stress responses. Plant Physiol 133:443–447

Morgner N, Zickermann V, Kerscher S, Wittig I, Abdrakhmanova A, Barth HD, Brutschy B, Brandt U (2008) Subunit mass fingerprinting of mitochondrial complex I. Biochim Biophys Acta 1777:1384–1391

Munshi MK, Kobayashi Y, Shikanai T (2005) Identification of a novel protein, CRR7, required for the stabilization of the chloroplast NAD(P)H dehydrogenase complex in *Arabidopsis*. Plant J 44:1036–1044

Munshi MK, Kobayashi Y, Shikanai T (2006) Chlororespiratory reduction 6 is a novel factor required for accumulation of the chloroplast NAD(P)H dehydrogenase complex in *Arabidopsis*. Plant Physiol 141:737–744

Muraoka R, Okuda K, Kobayashi Y, Shikanai T (2006) A eukaryotic factor required for accumulation of the chloroplast NAD(P)H dehydrogenase complex in *Arabidopsis*. Plant Physiol 142:1683–1689

Nakagawa N, Sakurai N (2006) A mutation in At-nMat1a, which encodes a nuclear gene having high similarity to group II intron maturase, causes impaired splicing of mitochondrial NAD4 transcript and altered carbon metabolism in *Arabidopsis thaliana*. Plant Cell Physiol 47:772–783

Nashilevitz S, Melamed-Bessudo C, Izkovich Y, Rogachev I, Osorio S, Itkin M, Adato A, Pankratov I, Hirschberg J, Fernie AR, Wolf S, Usadel B, Levy AA, Rumeau D, Aharoni A (2010) An orange ripening mutant links plastid NAD(P)H dehydrogenase complex activity to central and specialized metabolism during tomato fruit maturation. Plant Cell 22:1977–1997

Ogawa T, Mi H (2007) Cyanobacterial NADPH dehydrogenase complexes. Photosynth Res 93:69–77

Ohyama K, Fukuzawa H, Kohchi T, Shirai H, Sano T, Sano S, Umesono K, Shiki Y, Takeuchi M, Chang Z, Aota S-i, Inokuchi H, Ozeki H (1986) Chloroplast gene organization deduced from complete sequence of liverwort *Marchantia polymorpha* chloroplast DNA. Nature 322:572–574

Ostergaard J, Persiau G, Davey MW, Bauw G, Van Montagu M (1997) Isolation of a cDNA coding for L-galactono-gamma-lactone dehydrogenase, an enzyme involved in the biosynthesis of ascorbic acid in plants. Purification, characterization, cDNA cloning, and expression in yeast. J Biol Chem 272:30009–30016

Pagliarini DJ, Calvo SE, Chang B, Sheth SA, Vafai SB, Ong SE, Walford GA, Sugiana C, Boneh A, Chen WK, Hill DE, Vidal M, Evans JG, Thorburn DR, Carr SA, Mootha VK (2008) A mitochondrial protein compendium elucidates complex I disease biology. Cell 134:112–123

Palenik B, Grimwood J, Aerts A, Rouze P, Salamov A, Putnam N, Dupont C, Jorgensen R, Derelle E, Rombauts S, Zhou K, Otillar R, Merchant SS, Podell S, Gaasterland T, Napoli C, Gendler K, Manuell A, Tai V, Vallon O, Piganeau G, Jancek S, Heijde M, Jabbari K, Bowler C, Lohr M, Robbens S, Werner G, Dubchak I, Pazour GJ, Ren Q, Paulsen I, Delwiche C, Schmutz J, Rokhsar D, Van de Peer Y, Moreau H, Grigoriev IV (2007) The tiny eukaryote *Ostreococcus* provides genomic insights into the paradox of plankton speciation. Proc Natl Acad Sci USA 104:7705–7710

Parisi G, Perales M, Fornasari MS, Colaneri A, Gonzalez-Schain N, Gomez-Casati D, Zimmermann S, Brennicke A, Araya A, Ferry JG, Echave J, Zabaleta E (2004) Gamma carbonic anhydrases in plant mitochondria. Plant Mol Biol 55:193–207

Peltier G, Tolleter D, Billon E, Cournac L (2010) Auxiliary electron transport pathways in chloroplasts of microalgae. Photosynth Res. doi:10.1007/s11120-010-9575-3

Peng L, Shimizu H, Shikanai T (2008) The chloroplast NAD(P)H dehydrogenase complex interacts with photosystem I in *Arabidopsis*. J Biol Chem 283:34873–34879

Peng L, Fukao Y, Fujiwara M, Takami T, Shikanai T (2009) Efficient operation of NAD(P)H dehydrogenase requires supercomplex formation with photosystem I via minor LHCI in *Arabidopsis*. Plant Cell 21:3623–3640

Peng L, Cai W, Shikanai T (2010) Chloroplast stromal proteins, CRR6 and CRR7, are required for assembly of the NAD(P)H dehydrogenase subcomplex A in *Arabidopsis*. Plant J 63:203–211

Perales M, Parisi G, Fornasari MS, Colaneri A, Villarreal F, Gonzalez-Schain N, Echave J, Gomez-Casati D, Braun HP, Araya A, Zabaleta E (2004) Gamma carbonic anhydrase like complex interact with plant mitochondrial complex I. Plant Mol Biol 56:947–957

Perales M, Eubel H, Heinemeyer J, Colaneri A, Zabaleta E, Braun HP (2005) Disruption of a nuclear gene encoding a mitochondrial gamma carbonic anhydrase reduces complex I and supercomplex I+III2 levels and alters mitochondrial physiology in *Arabidopsis*. J Mol Biol 350:263–277

Peters K, Dudkina NV, Jansch L, Braun HP, Boekema EJ (2008) A structural investigation of complex I and I+III2 supercomplex from *Zea mays* at 11–13 A resolution: assignment of the carbonic anhydrase domain and evidence for structural heterogeneity within complex I. Biochim Biophys Acta 1777:84–93

Pineau B, Mathieu C, Gerard-Hirne C, De Paepe R, Chetrit P (2005) Targeting the NAD7 subunit to mitochondria restores a functional complex I and a wild type phenotype in the *Nicotiana sylvestris* CMS II mutant lacking *nad7*. J Biol Chem 280:25994–26001

Pineau B, Layoune O, Danon A, De Paepe R (2008) L-galactono-1,4-lactone dehydrogenase is required for the accumulation of plant respiratory complex I. J Biol Chem 283:32500–32505

Pla M, Mathieu C, De Paepe R, Chetrit P, Vedel F (1995) Deletion of the last two exons of the mitochondrial *nad7* gene results in lack of the NAD7 polypeptide in a *Nicotiana sylvestris* CMS mutant. Mol Gen Genet 248:79–88

Remacle C, Baurain D, Cardol P, Matagne RF (2001a) Mutants of *Chlamydomonas reinhardtii* deficient in mitochondrial complex I: characterization of two mutations affecting the *nd1* coding sequence. Genetics 158:1051–1060

Remacle C, Duby F, Cardol P, Matagne RF (2001b) Mutations inactivating mitochondrial genes in *Chlamydomonas reinhardtii*. Biochem Soc Trans 29:442–446

Remacle C, Cardol P, Coosemans N, Gaisne M, Bonnefoy N (2006) High-efficiency biolistic transformation of *Chlamydomonas* mitochondria can be used to insert mutations in complex I genes. Proc Natl Acad Sci USA 103:4771–4776

Remacle C, Barbieri MR, Cardol P, Hamel PP (2008) Eukaryotic complex I: functional diversity and experimental systems to unravel the assembly process. Mol Genet Genomics 280:93–110

Robbens S, Derelle E, Ferraz C, Wuyts J, Moreau H, Van de Peer Y (2007) The complete chloroplast and mitochondrial DNA sequence of *Ostreococcus tauri*: organelle genomes of the smallest eukaryote are examples of compaction. Mol Biol Evol 24:956–968

Roussell DL, Thompson DL, Pallardy SG, Miles D, Newton KJ (1991) Chloroplast structure and function is altered in the NCS2 maize mitochondrial mutant. Plant Physiol 96:232–238

Rumeau D, Becuwe-Linka N, Beyly A, Louwagie M, Garin J, Peltier G (2005) New subunits NDH-M, -N, and -O, encoded by nuclear genes, are essential for plastid Ndh complex functioning in higher plants. Plant Cell 17:219–232

Rumeau D, Peltier G, Cournac L (2007) Chlororespiration and cyclic electron flow around PSI during photosynthesis and plant stress response. Plant Cell Environ 30:1041–1051

Saada A, Vogel RO, Hoefs SJ, van den Brand MA, Wessels HJ, Willems PH, Venselaar H, Shaag A, Barghuti F, Reish O, Shohat M, Huynen MA, Smeitink JA, van den Heuvel LP, Nijtmans LG (2009) Mutations in NDUFAF3 (C3ORF60), encoding an NDUFAF4 (C6ORF66)-interacting complex I assembly protein, cause fatal neonatal mitochondrial disease. Am J Hum Genet 84:718–727

Sabar M, De Paepe R, de Kouchkovsky Y (2000) Complex I impairment, respiratory compensations, and photosynthetic decrease in nuclear and mitochondrial male sterile mutants of *Nicotiana sylvestris*. Plant Physiol 124:1239–1250

Sazanov LA, Burrows P, Nixon PJ (1996) Detection and characterization of a complex I-like NADH-specific dehydrogenase from pea thylakoids. Biochem Soc Trans 24:739–743

Sazanov LA, Burrows PA, Nixon PJ (1998) The chloroplast Ndh complex mediates the dark reduction of the plastoquinone pool in response to heat stress in tobacco leaves. FEBS Lett 429:115–118

Schilling B, Bharath MMS, Row RH, Murray J, Cusack MP, Capaldi RA, Freed CR, Prasad KN, Andersen JK, Gibson BW (2005) Rapid purification and mass spectrometric characterization of mitochondrial NADH dehydrogenase (complex I) from rodent brain and a dopaminergic neuronal cell line. Mol Cell Proteomics 4:84–96

Schneider R, Massow M, Lisowsky T, Weiss H (1995) Different respiratory-defective phenotypes of *Neurospora crassa* and *Saccharomyces cerevisiae* after inactivation of the gene encoding the mitochondrial acyl carrier protein. Curr Genet 29:10–17

Sheftel AD, Stehling O, Pierik AJ, Netz DJ, Kerscher S, Elsasser HP, Wittig I, Balk J, Brandt U, Lill R (2009) Human ind1, an iron-sulfur cluster assembly factor for respiratory complex I. Mol Cell Biol 29:6059–6073

Shikanai T (2007) Cyclic electron transport around photosystem I: genetic approaches. Annu Rev Plant Biol 58:199–217

Shikanai T, Endo T, Hashimoto T, Yamada Y, Asada K, Yokota A (1998) Directed disruption of the tobacco *ndhB* gene impairs cyclic electron flow around photosystem I. Proc Natl Acad Sci USA 95:9705–9709

Shimizu H, Shikanai T (2007) Dihydrodipicolinate reductase-like protein, CRR1, is essential for chloroplast NAD(P)H dehydrogenase in *Arabidopsis*. Plant J 52:539–547

Shimizu H, Peng L, Myouga F, Motohashi R, Shinozaki K, Shikanai T (2008) CRR23/NdhL is a subunit of the chloroplast NAD(P)H dehydrogenase complex in *Arabidopsis*. Plant Cell Physiol 49:835–842

Shinozaki K, Ohme M, Tanaka M, Wakasugi T, Hayashida N, Matsubayashi T, Zaita N, Chunwongse J, Obokata J, Yamaguchi-Shinozaki K, Ohto C, Torazawa K, Meng BY, Sugita M, Deno H, Kamogashira T, Yamada K, Kusuda J, Takaiwa F, Kato A, Tohdoh N, Shimada H, Sugiura M (1986) The complete nucleotide sequence of the tobacco chloroplast genome: its gene organization and expression. EMBO J 5:2043–2049

Siendones E, Gonzalez-Reyes JA, Santos-Ocana C, Navas P, Cordoba F (1999) Biosynthesis of ascorbic acid in kidney bean. L-galactono-gamma-lactone dehydrogenase is an intrinsic protein located at the mitochondrial inner membrane. Plant Physiol 120:907–912

Sirpio S, Allahverdiyeva Y, Holmstrom M, Khrouchtchova A, Haldrup A, Battchikova N, Aro EM (2009a) Novel nuclear-encoded subunits of the chloroplast NAD(P)H dehydrogenase complex. J Biol Chem 284:905–912

Sirpio S, Holmstrom M, Battchikova N, Aro EM (2009b) AtCYP20-2 is an auxiliary protein of the chloroplast NAD(P)H dehydrogenase complex. FEBS Lett 583:2355–2358

Smirnoff N (2000) Ascorbate biosynthesis and function in photoprotection. Philos Trans R Soc Lond B Biol Sci 355:1455–1464

Sugiana C, Pagliarini DJ, McKenzie M, Kirby DM, Salemi R, Abu-Amero KK, Dahl HH, Hutchison WM, Vascotto KA, Smith SM, Newbold RF, Christodoulou J, Calvo S, Mootha VK, Ryan MT, Thorburn DR (2008) Mutation of C20orf7 disrupts complex I assembly and causes lethal neonatal mitochondrial disease. Am J Hum Genet 83:468–478

Sugiura C, Kobayashi Y, Aoki S, Sugita C, Sugita M (2003) Complete chloroplast DNA sequence of the moss *Physcomitrella patens*: evidence for the loss and relocation of rpoA from the chloroplast to the nucleus. Nucleic Acids Res 31:5324–5331

Sunderhaus S, Dudkina NV, Jansch L, Klodmann J, Heinemeyer J, Perales M, Zabaleta E, Boekema EJ, Braun HP (2006) Carbonic anhydrase subunits form a matrix-exposed domain attached to the membrane arm of mitochondrial complex I in plants. J Biol Chem 281:6482–6488

Suorsa M, Sirpio S, Aro EM (2009) Towards characterization of the chloroplast NAD(P)H dehydrogenase complex. Mol Plant 2:1127–1140

Takabayashi A, Ishikawa N, Obayashi T, Ishida S, Obokata J, Endo T, Sato F (2009) Three novel subunits of *Arabidopsis* chloroplastic NAD(P)H dehydrogenase identified by bioinformatic and reverse genetic approaches. Plant J 57:207–219

Triepels R, Smeitink J, Loeffen J, Smeets R, Buskens C, Trijbels F, van den Heuvel L (1999) The human nuclear-encoded acyl carrier subunit (NDUFAB1) of the mitochondrial complex I in human pathology. J Inherit Metab Dis 22:163–173

Turmel M, Otis C, Lemieux C (1999) The complete chloroplast DNA sequence of the green alga *Nephroselmis olivacea*: insights into the architecture of ancestral chloroplast genomes. Proc Natl Acad Sci USA 96:10248–10253

Ugalde C, Vogel R, Huijbens R, Van Den Heuvel B, Smeitink J, Nijtmans L (2004) Human mitochondrial complex I assembles through the combination of evolutionary conserved modules: a framework to interpret complex I deficiencies. Hum Mol Genet 13:2461–2472

van Lis R, Atteia A, Mendoza-Hernandez G, Gonzalez-Halphen D (2003) Identification of novel mitochondrial protein components of *Chlamydomonas reinhardtii*. A proteomic approach. Plant Physiol 132:318–330

Vidal G, Ribas-Carbo M, Garmier M, Dubertret G, Rasmusson AG, Mathieu C, Foyer CH, De Paepe R (2007) Lack of respiratory chain complex I impairs alternative oxidase engagement and modulates redox signaling during elicitor-induced cell death in tobacco. Plant Cell 19:640–655

Videira A, Duarte M (2002) From NADH to ubiquinone in *Neurospora* mitochondria. Biochim Biophys Acta 1555:187–191

Villarreal F, Martin V, Colaneri A, Gonzalez-Schain N, Perales M, Martin M, Lombardo C, Braun HP, Bartoli C, Zabaleta E (2009) Ectopic expression of mitochondrial gamma carbonic anhydrase 2 causes male sterility by anther indehiscence. Plant Mol Biol 70:471–485

Vogel RO, Smeitink JA, Nijtmans LG (2007a) Human mitochondrial complex I assembly: a dynamic and versatile process. Biochim Biophys Acta 1767:1215–1227

Vogel RO, van den Brand MA, Rodenburg RJ, van den Heuvel LP, Tsuneoka M, Smeitink JA, Nijtmans LG (2007b) Investigation of the complex I assembly chaperones B17.2L and NDUFAF1 in a cohort of CI deficient patients. Mol Genet Metab 91:176–182

Wakasugi T, Tsudzuki J, Ito S, Nakashima K, Tsudzuki T, Sugiura M (1994) Loss of all ndh genes as determined by sequencing the entire chloroplast genome of the black pine *Pinus thunbergii*. Proc Natl Acad Sci USA 91:9794–9798

Wakasugi T, Nagai T, Kapoor M, Sugita M, Ito M, Ito S, Tsudzuki J, Nakashima K, Tsudzuki T, Suzuki Y, Hamada A, Ohta T, Inamura A, Yoshinaga K, Sugiura M (1997) Complete nucleotide sequence of the chloroplast genome from the green alga *Chlorella vulgaris*: the existence of genes possibly involved in chloroplast division. Proc Natl Acad Sci USA 94:5967–5972

Wang D, Portis AR Jr (2007) A novel nucleus-encoded chloroplast protein, PIFI, is involved in NAD(P)H dehydrogenase complex-mediated chlororespiratory electron transport in *Arabidopsis*. Plant Physiol 144:1742–1752

Wu D, Wright DA, Wetzel C, Voytas DF, Rodermel S (1999) The IMMUTANS variegation locus of *Arabidopsis* defines a mitochondrial alternative oxidase homolog that functions during early chloroplast biogenesis. Plant Cell 11:43–55

Yamamoto H, Peng L, Fukao Y, Shikanai T (2011) An Src homology 3 domain-like fold protein forms a ferredoxin binding site for the chloroplast NADH dehydrogenase-like complex in *Arabidopsis*. Plant Cell 23:1480–1493

Zensen R, Husmann H, Schneider R, Peine T, Weiss H (1992) De novo synthesis and desaturation of fatty acids at the mitochondrial acyl-carrier protein, a subunit of NADH:ubiquinone oxidoreductase in *Neurospora crassa*. FEBS Lett 310:179–181

Part V
Supercomplexes in Mitochondria

Chapter 12
Supramolecular Organization of the Respiratory Chain

Janet Vonck

Abstract The transmembrane protein complexes of the respiratory chain generate an electrochemical gradient over the mitochondrial inner membrane and the bacterial plasma membrane. Electrons are transferred from NADH via complex I (NADH: ubiquinone oxidoreductase) and ubiquinone to complex III (ubiquinol:cytochrome *c* oxidoreductase) and via the peripheral electron carrier cytochrome *c* and complex IV (cytochrome *c* oxidase) to the terminal acceptor, molecular oxygen. The electrochemical proton gradient is used by F_1F_o-ATP synthase to produce ATP.

The classical model of the respiratory chain envisions these complexes diffusing freely in the membrane and electron transfer occurring by random collision. This model is increasingly being challenged by studies that indicate a high level of organization of the protein complexes. After membrane solubilisation with mild detergents, supercomplexes with various combinations of complex I, III and IV and dimers of ATP synthase can be isolated. Electron microscopic studies have confirmed that these supercomplexes have a distinct architecture. A 3D reconstruction of a supercomplex from bovine heart mitochondria has shown that the ubiquinone and cytochrome *c* binding sites of the different complexes face each other, suggesting a role in substrate channeling. Formation of supercomplexes has also been shown to play a role in the assembly and stability of the complexes, especially complex I, suggesting that they constitute the functional state of the respiratory chain. Whereas the ATP synthase dimers in mitochondria were shown to form ribbons and appear to shape the cristae membranes, a higher-order organization of complexes I, III and IV has not yet been demonstrated.

J. Vonck (✉)
Department of Structural Biology, Max-Planck-Institute of Biophysics,
Max-von-Laue-Strasse 3, D-60438 Frankfurt am Main, Germany
e-mail: janet.vonck@biophys.mpg.de

L. Sazanov (ed.), *A Structural Perspective on Respiratory Complex I: Structure and Function of NADH:ubiquinone oxidoreductase*, DOI 10.1007/978-94-007-4138-6_12,

Keywords Respiratory chain • Mitochondria • Supercomplexes • Protein structure • Electron microscopy

12.1 Introduction

The respiratory chain is a major enzyme system in aerobic organisms, linking cellular metabolism to the production of ATP via the creation of a transmembrane proton gradient. The chain consists of a number of transmembrane protein complexes and small molecules which transport electrons from NADH ultimately to molecular oxygen, and in this process pump protons across the membrane. The respiratory chain is located in bacteria in the plasma membrane and in eukaryotes in the mitochondrial inner membrane, which is thought to be derived from the membrane of a bacterial endosymbiont. The complexes involved are complex I (NADH: ubiquinone oxidoreductase), complex II (succinate dehydrogenase), complex III (ubiquinol:cytochrome *c* oxidoreductase), and complex IV (cytochrome *c* oxidase) (Fig. 12.1). Electrons are transferred from complex I to complex III by ubiquinone and from III to IV by cytochrome *c*. Complex II, succinate dehydrogenase, can transfer electrons from succinate to ubiquinone, but unlike the other complexes, it does not have a proton pump function. The pumped protons reenter the cytoplasm or mitochondrial matrix via the proton channel in F_1F_o ATP synthase (complex V), which forms ATP by a rotary mechanism.

The respiratory chain complexes are large transmembrane protein complexes. Complex I is the largest complex with 14 subunits and around 550 kDa in bacteria and 45 subunits and a size of 1 MDa in mammals (Carroll et al. 2006). It oxidizes NADH and transfers two electrons to ubiquinone via flavin mononucleotide (FMN) and a series of iron-sulfur clusters, and translocates four electrons across the membrane. Complex I has long resisted structural characterization. Electron microscopy of complex I has shown that it is L-shaped in all species (Hofhaus et al. 1991; Guénebaut et al. 1997; Grigorieff 1998; Radermacher et al. 2006; Morgan and Sazanov 2008; Clason et al. 2010). The large transmembrane domain contains the proton pumping machinery, and the hydrophilic peripheral arm protruding into the matrix or cytoplasm contains the NADH binding site and all redox centers (Friedrich 1998). Electron microscopy studies have been complicated by the variability in angle between matrix and membrane arm (Radermacher et al. 2006; Clason et al. 2010), which may have functional consequences. X-ray information was first obtained for the matrix arm of complex I from *Thermus thermophilus* (Hinchliffe and Sazanov 2005; Sazanov and Hinchliffe 2006), revealing the redox centers: the primary electron acceptor FMN and nine iron-sulfur clusters, forming a pathway from the periphery to the probable quinone binding site near the membrane arm. X-ray structures of the membrane arm from *E. coli* complex I and the complete enzyme from *T. thermophilus* (Efremov et al. 2010) revealed that the membrane arm contains three homologous antiporter-like subunits with 14 transmembrane helices each, the last of which also includes a 110-Å long amphipathic helix which spans almost the whole length of the membrane arm. The structure suggests that the

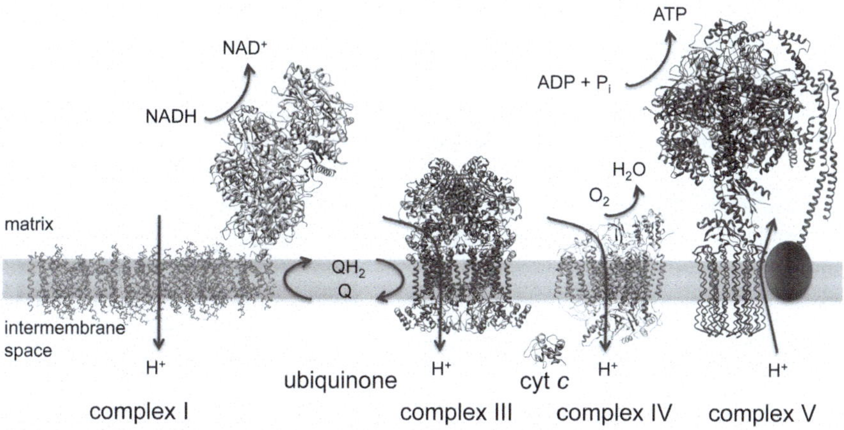

Fig. 12.1 The mitochondrial respiratory chain. The transmembrane protein complexes of the electron transport chain generate an electrochemical gradient over the mitochondrial inner membrane. NADH is oxidized to NAD^+. Electrons are transferred from NADH via complex I and ubiquinone (Q) to complex III. From there they pass through the peripheral electron carrier cytochrome *c* and complex IV to the terminal acceptor, molecular oxygen, which is reduced to water. The electrochemical proton gradient is used by complex V, F_1F_0 ATP synthase, to produce ATP. The location of the mitochondrial inner membrane is indicated. The figure was made using the UCSF Chimera package (Pettersen et al. 2004) from the Resource for Biocomputing, Visualization, and Informatics at the University of California, San Francisco (supported by NIH P41 RR-01081)

complex works like a piston, with conformational changes at the interface of the two domains inducing a motion of the long helix, tilting transmembrane helices in the three antiporter-like subunits, which results in proton transport. The membrane arm has a characteristic curved shape previously observed in EM reconstructions (Radermacher et al. 2006; Morgan and Sazanov 2008; Clason et al. 2010), although these had the wrong handedness (Efremov et al. 2010). These features are confirmed by a low-resolution structure of eukaryotic complex I from the yeast *Yarrowia lipolytica* (Hunte et al. 2010).

Complex II, succinate dehydrogenase, consists of two hydrophilic and two transmembrane subunits. The structure of the complex of *Escherichia coli* was solved by x-ray crystallography, revealing the electron transport pathway from the electron donor, succinate, to the terminal electron acceptor, ubiquinone. It was found that the redox centers are arranged in a manner that aids the prevention of reactive oxygen species (ROS) formation at the flavin adenine dinucleotide cofactor (Yankovskaya et al. 2003).

Complex III, the ubiquinol:cytochrome *c* oxidoreductase or cytochrome bc_1 complex, is a homodimeric transmembrane complex of 11 distinct subunits with large domains in the matrix, the core 1 and core 2 subunits. Crystal structures exist for complex III of several species: cow (Xia et al. 1997; Iwata et al. 1998), chicken (Zhang et al. 1998) and *Saccharomyces cerevisiae* (Hunte et al. 2000). Complex III contains three hemes per monomer, two at the cytochrome *b* subunit, heme b_L and home b_H, and one at the cytochrome c_1 subunit, and an iron-sulfur cluster at the Rieske protein. Ubiquinol is oxidized via the so-called Q cycle, which involves uptake of two protons from the matrix, release of four protons to the intermembrane space, and transfer of two electrons to two soluble cytochrome *c* molecules via heme c_1 at the intermembrane surface (Crofts 2004). A structure of the yeast enzyme with cytochrome *c* bound was also determined (Lange and Hunte 2002). Surprisingly, the substrate is only bound to one of the two cytochrome c_1 subunits.

Complex IV, cytochrome *c* oxidase, consists of four subunits in bacteria; the mammalian species has an additional nine subunits complementing the bacterial core. Crystal structures have been solved for the bacterial enzymes from *Paracoccus denitrificans* (Ostermeier et al. 1997; Iwata et al. 2002) and *Rhodobacter sphaeroides* (Svensson-Ek et al. 2002) as well as the 13-subunit complex from bovine heart mitochondria (Tsukihara et al. 1996; Muramoto et al. 2007). Complex IV contains two cytochromes (*a* and a_3) and two copper centers. It uses four reduced cytochrome *c* molecules to reduce O_2 to two water molecules. In this process, four protons are translocated across the membrane (Ferguson-Miller and Babcock 1996).

The F_1F_o ATP synthase is a two-domain structure of ~600 kDa consisting of a transmembrane, proton-translocating domain (F_o) with a minimal composition ab_2c_{10-15} in bacteria and a catalytic domain in the cytoplasm or mitochondrial matrix (F_1) with a minimal subunit composition $\alpha_3\beta_3\gamma\delta\varepsilon$. F_1F_o forms a rotary machine where proton flow through F_o causes rotation of a subcomplex $c_{10-15}\gamma\varepsilon$, inducing changes in the nucleotide binding pockets in the β subunits (Abrahams et al. 1994; Boyer 1997). The structures of different subcomplexes have been solved by x-ray crystallography, including $\alpha_3\beta_3\gamma\varepsilon c_{10}$ from the yeast mitochondrial enzyme (Stock et al. 1999), a large portion of the peripheral stalk from the mitochondrial enzyme (Kane Dickson et al. 2006) and the rotor ring from different bacterial species (Meier et al. 2006; Pogoryelov et al. 2009). The size of the ring varies from 10 to 15 *c*-subunits in different species (Stock et al. 1999; Seelert et al. 2000; Vonck et al. 2002; Pogoryelov et al. 2005; Matthies et al. 2009). The only subunit defying structural characterization is the very hydrophobic *a* subunit in the stator. Cryo-electron microscopy structures have been determined for the complete F_1F_o complex from bovine (Rubinstein et al. 2003) and yeast (Lau et al. 2008) mitochondria, which together with the fitted x-ray structures of subcomplexes give an almost complete picture of the whole complex.

This chapter reviews the organization of the complexes in the mitochondrial inner membrane and evidence for the existence of supercomplexes, with an emphasis on structural data, different models for the function of this organization and a discussion about the possibility of the presence of a higher order structure in native organelles.

12.2 The Respiratory Chain

The protein complexes of the respiratory chain were among the first membrane proteins to be isolated in the 1960s, mainly by the work of Hatefi and colleagues. It was found that enzyme activities could be separated by fractionation of membrane preparations using mild detergents, giving rise ultimately to the complexes now known as complex I–IV (Hatefi et al. 1961a). The electron transfer reactions could be reconstituted when the resulting fractions were mixed together at sufficiently high concentrations (Hatefi et al. 1961b, 1962a, b). It was assumed, however, that the complexes formed a single functional unit in the mitochondria and were present in an orderly sequence which could be disrupted by appropriate reagents (Blair 1967). In later years the random collision model of electron transport became widely accepted. According to this model, the mitochondrial inner membrane is a fluid-state rather than a solid-state membrane and all membrane proteins and redox components that catalyze electron transport and ATP synthesis are in constant and independent diffusional motion within the membrane (Hackenbrock et al. 1986). The random collision model was based on the fact that the respiratory complexes were active in isolated form and did not show association when reconstituted in liposomes (Capaldi 1982) and on the pool functions of ubiquinone (Kröger and Klingenberg 1973) and cytochrome *c* (Gupte and Hackenbrock 1988). Thus, a direct association of the respiratory chain complexes did not seem to be required for an efficient electron transfer.

In some bacteria and archaea, stable supercomplexes of the complexes III and IV were shown to exist (Berry and Trumpower 1985; Sone et al. 1987; Iwasaki et al. 1995), but these were thought to be specific for these species. Supercomplexes were not expected to be present in mitochondria and were only discovered when new preparation methods and analytical techniques were established that did not disrupt the interactions between the complexes.

12.2.1 Evidence for Supramolecular Organization: BN-PAGE

Direct indications for the existence of higher-order organization of the complexes in the mitochondrial inner membrane have come from the development of new analytic techniques like blue-native polyacrylamide gel electrophoresis (BN-PAGE) by Schägger and colleagues (Schägger and von Jagow 1991; Schägger et al. 1994; Schägger and Pfeiffer 2000; Wittig and Schägger 2007). In this method, protein complexes are solubilized by very mild detergents like digitonin that preserve the enzymatic activity as well as the protein interactions. The resulting complexes are separated by size in the presence of an anionic dye (Fig. 12.3a). Even weaker protein interactions can be preserved by electrophoresis in the absence of a dye in clear-native PAGE (CN-PAGE), although the resolution is more limited (Wittig and Schägger 2005, 2007). The complex composition is then determined by second dimension BN-PAGE with a stronger detergent, such as Triton, to dissociate the supercomplexes, and the subunit composition by second dimension SDS-PAGE.

By these methods, the existence of stoichiometric assemblies of respiratory chain complexes was shown first in the mitochondria of bovine heart and of the yeast *Saccharomyces cerevisiae* (Cruciat et al. 2000; Schägger and Pfeiffer 2000), which remain the best characterized species. Since then, supercomplexes of various compositions have been found in other species (Table 12.1).

Table 12.1 Supercomplexes in different species

		I_1III_2	$I_1III_2IV_1$	$I_1III_2IV_{2–4}$	$III_2IV_{1–2}$	I_2	I_2III_2	I_2III_4	References
Prokaryotes:	*Bacillus sp. PS3*				x				Sone et al. (1987)
	C. glutamicum				x				Niebisch and Bott (2003)
	P. denitrificans		G		x				Berry and Trumpower (1985)
									Stroh et al. (2004)
	Sulfolobus sp.				x				Iwasaki et al. (1995),
Yeast/fungi:	*S. cerevisiae*				EM[a]				Heinemeyer et al. (2007)
	P. anserina	G	G	G		G	G		Krause et al. (2004b)
	N. crassa		G	G	G[b]	G			Marques et al. (2007)
Plants:	*A. thaliana*	EM						G[c]	Eubel et al. (2003)
									Dudkina et al. (2005a)
	Asparagus	G							Dudkina et al. (2006a)
	Barley	G							Eubel et al. (2003)
	Bean	G							Eubel et al. (2003)
	Maize	G							Peters et al. (2008)
	Pea	G							Taylor et al. (2005)
	Potato	G					EM	G[c]	Eubel et al. (2003)
	Spinach	G	G		G				Krause et al. (2004a)
	Tobacco	G							Pineau et al. (2005)
Mammals:	Cow	EM	EM	G					Schäfer et al. (2006)
	Dog		G						Rosca et al. (2008)
	Human		G						Schägger et al. (2004)
	Mouse		G						Acin-Pérez et al. (2004)
	Rat		G						Dencher et al. (2007)

G found by BN-PAGE or CN-PAGE, *EM* identified by electron microscopy
[a] *S. cerevisiae* lacks complex I
[b] In mutants lacking complex I
[c] Possibly I_2III_2, see text

12.3 Supercomplexes in Different Species

12.3.1 Prokaryotes

The first supercomplexes were identified in bacteria. In the Gram-negative soil bacterium *Paracoccus denitrificans*, stable assemblies of complex III and IV were shown to exist as early as 1985 (Berry and Trumpower 1985). A supercomplex of complex III and IV was also found in the thermophilic bacterium *Bacillus* sp. PS3 (Sone et al. 1987) as well as in the soil bacterium *Corynebacterium glutamicum* (Niebisch and Bott 2003) and in the hyperthermophilic acidophilic archaeon *Sulfolobus* sp. strain 7 (Iwasaki et al. 1995), which is so far the only documented archaeal supercomplex.

The supercomplex from *P. denitrificans* contains a tightly bound cytochrome c_{552}, which has an N-terminal membrane anchor (Turba et al. 1995). Isolation in digitonin later showed that complex I was also part of a supercomplex that had the composition $I_1III_4IV_4$ (Stroh et al. 2004). Contrary to expectation, the membrane-anchored cytochrome c_{552} turned out to be not necessary for supercomplex formation, although it did stabilize the assemblies. The presence of both complex III and IV was essential for correct assembly and/or stability of complex I, and it was concluded that supercomplex formation protects the detergent-labile complex I of *P. denitrificans.* In contrast, *Escherichia coli,* the hyperthermophilic eubacterium *Aquifex aeolicus* and the yeast *Yarrowia lipolytica* all have a detergent-stable complex I, and no supercomplexes could be detected in these species (Stroh et al. 2004).

12.3.2 Yeast and Fungi

The yeast *Saccharomyces cerevisiae* does not contain complex I. Instead, several NADH dehydrogenases are present, two of which transfer electrons to ubiquinone, and with succinate as a substrate, succinate:ubiquinone oxidoreductase (complex II) is used (Boumans et al. 1998). In a study to distinguish between the solid-state and random diffusion model for the respiratory chain, inhibitor titrations by antimycin showed that neither ubiquinone nor cytochrome *c* exhibited pool behavior, showing that the respiratory chain behaves like a single functional unit (Boumans et al. 1998). This suggested the presence of supercomplexes consisting of complex III and IV plus complex II or an alternative oxidase. Subsequently, two supercomplexes from *Saccharomyces cerevisiae* were characterized biochemically, consisting of a complex III dimer and one or two copies of complex IV (Cruciat et al. 2000; Schägger and Pfeiffer 2000). Complex II was not associated with the supercomplex (Cruciat et al. 2000; Schägger and Pfeiffer 2000).

The yeast supercomplexes have been the subject of structural investigations (see below); complex IV dimers were not observed, instead one copy of complex IV was bound to each complex III monomer in the III_2IV_2 supercomplex

(Heinemeyer et al. 2007). The III_2IV_2 supercomplex also appears to interact with other proteins (reviewed in Stuart 2008), like Tim23, a protein translocase in the mitochondrial inner membrane (van der Laan et al. 2006).

In mitochondria from the filamentous fungus *Podospora anserina* supercomplexes of composition $I_1III_2IV_{0-2}$ were found in wild type strains (Krause et al. 2004b), but in mutants lacking complex IV distinct supercomplexes comprising complex I dimers (i.e. I_2 and I_2III_2) were detected. Interestingly, the mutant strains respire via an alternative pathway in which electrons are directly transferred from complex I to an alternative oxidase, and complex III is not involved in electron transfer. Its expression in high amounts and the supercomplex formation with complex I suggests a role in the stability of complex I (Krause et al. 2004b).

Various supercomplexes were also found in the filamentous fungus *Neurospora crassa* (Marques et al. 2007), including complex I dimers, I–III–IV and III–IV supercomplexes. Studies of supercomplex formation in complex I mutant strains suggested that the biogenesis of complex I is linked with its assembly into supercomplexes (Marques et al. 2007).

12.3.3 Mammals

Bovine heart mitochondria have been the model for respiratory chain studies from the start (reviewed in Hatefi 1985) and they were also the first mammalian source where supercomplexes were detected by BN-PAGE (Schägger and Pfeiffer 2000) and studied by direct structural methods (Schäfer et al. 2006, 2007). Upon solubilisation with digitonin, almost all complex I was found to be part of supercomplexes with complex III and up to four copies of complex IV (Schägger and Pfeiffer 2000). The most abundant supercomplex with 54% of all complex I is $I_1III_2IV_1$, followed by I_1III_2 with 17%, while small amounts of larger supercomplexes with 2–4 copies of complex IV can be detected (Schägger and Pfeiffer 2000) (Fig. 12.3a). Supercomplex $I_1III_2IV_1$ is also known as the respirasome. It was found to be both more active and more stable than the supercomplex lacking complex IV, suggesting a role for supercomplex formation in assembly and stabilization (Schäfer et al. 2006). Unlike complex I, complex III and especially complex IV are present in free form in significant amounts in bovine heart mitochondria (Schägger and Pfeiffer 2001).

The respirasome has also been described in other mammalian species and tissues: human skeletal muscle (Schägger et al. 2004), where supercomplex formation was found to be important for complex I assembly, human lymphoblasts (McKenzie et al. 2006), rat brain (Dencher et al. 2007), where age-related changes in abundances of supercomplexes were observed, and canine heart (Rosca et al. 2008), where a loss of supercomplexes was associated with heart failure. In mouse fibroblasts, also complex II was seen to be associated with complex I, III and IV or with III and IV (Acin-Pérez et al. 2008). The supramolecular organization of the respiratory chain thus appears to be a general and essential feature in mammalian mitochondria.

12.3.4 *Plants*

The respiratory chain in plant mitochondria is characterized by the presence of a number of alternative oxidoreductases, which catalyze electron transfer from NADH or NADPH to ubiquinone or direct electron transfer from ubiquinol to molecular oxygen, but are not involved in proton translocation (Eubel et al. 2004b). Supercomplex formation in plant mitochondria was first investigated by BN-PAGE in *Arabidopsis thaliana*, potato (*Solanum tuberosum*), bean (*Phaseolus vulgaris*), and barley (*Hordeum vulgare*) (Eubel et al. 2003). A very stable 1,500 kDa supercomplex of composition I_1III_2 was found, comprising 50–90% of complex I when solubilized by digitonin, depending on the organism, as well as smaller amounts of a 3,000 kDa complex I_2III_4. In later studies of potato tuber and stem mitochondria (Eubel et al. 2004a) and of green leaf mitochondria of spinach (Krause et al. 2004a), small amounts of supercomplexes containing complex IV, with compositions III_2IV_{1-2} and $I_1III_2IV_{1-4}$ were found, showing the presence of respirasomes also in plant mitochondria. Supercomplexes containing complex II or any of the plant-specific alternative oxidoreductases have never been observed.

The most abundant supercomplex in plants appears to be I_1III_2, which was apart from the species mentioned above also identified in tobacco (Pineau et al. 2005), pea (Taylor et al. 2005), asparagus (Dudkina et al. 2006a) and *Zea mays* (Peters et al. 2008). The very stable supercomplex I_1III_2 from *Arabidopsis thaliana* was isolated by sucrose-density-gradient ultracentrifugation and was the first supercomplex to be studied by electron microscopy and single particle analysis (Dudkina et al. 2005a).

12.4 Structural Evidence for Supercomplexes

After the first indirect indications for the existence of supercomplexes of the respiratory chain, the separation of supercomplexes on BN-PAGE gels caused a revision of the textbook view of the respiratory chain according to the random collision model. Direct structural insights in the architecture of the supercomplexes were provided more recently by the application of electron microscopy and single particle analysis (recently reviewed in Dudkina et al. 2008; Vonck and Schäfer 2009). The supercomplexes are either separated by centrifugation in sucrose density gradients or electroeluted directly from BN-gels, and then imaged by negative stain electron microscopy and subjected to single particle analysis. Several supercomplexes from mammalian, plant and yeast mitochondria have been investigated. The structural data will be discussed in this section.

12.4.1 *Supercomplex III_2IV_1 and III_2IV_2 from Yeast*

The yeast *Saccharomyces cerevisiae* lacks complex I, but BN-PAGE of its mitochondria revealed that complex III and IV form a supercomplex (Cruciat et al. 2000;

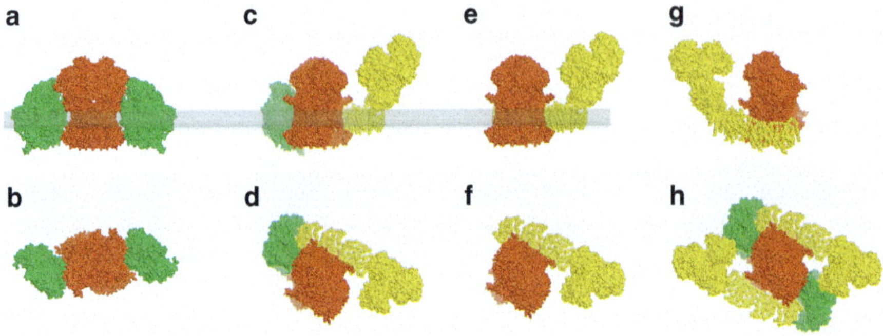

Fig. 12.2 Models for respiratory chain supercomplexes. (**a**, **b**) model for the III_2IV_2 supercomplex from yeast, after (Heinemeyer et al. 2007). Two copies of complex IV (*green*) are attached to complex III (*red*) with the convex face. (**c**, **d**) model for the bovine heart $I_1III_2IV_1$ supercomplex after (Schäfer et al. 2006; Schäfer et al. 2007). Complex IV is attached at the end of the complex I membrane arm (*yellow*) via its concave face. (**e**, **f**) Model for the bovine heart I_1III_2 supercomplex based on data in Schäfer et al. (2006) and Schäfer et al. (2007). Complex III is attached to the middle of the complex I membrane arm. (**g**) The model for the bovine heart I_1III_2 supercomplex rotated 40° along the complex I membrane arm shows complex III attached to the end of the membrane arm, as observed in plant I_1III_2 supercomplexes (Heinemeyer et al. 2007; Bultema et al. 2009). (**h**) A model for the potato $I_2III_2IV_2$ supercomplex proposed in (Bultema et al. 2009) was created by adding a second copy of complex I and complex IV to the bovine heart $I_1III_2IV_1$ supercomplex in (**c**) and (**d**). (**a**, **c**, **e**) The models in the plane of the membrane (*grey*), (**b**, **d**, **f**, **h**) the models as seen from the mitochondrial matrix. The models are built from bovine heart complex III (pdb code 1BGY) and IV (1IOC) and *Thermus thermophilus* complex I (3M9S). The figure was made using UCSF Chimera

Schägger and Pfeiffer 2000), showing that supercomplex formation is not specific for complex I. The yeast supercomplexes were subsequently studied by electron microscopy (Heinemeyer et al. 2007). The analysis revealed two complexes, III_2IV_1 and III_2IV_2, containing one or two monomers of complex IV respectively, bound to a complex III dimer (Fig. 12.2a, b). This was consistent with the sizes of the complexes of 650 and 850 kDa. Modeling of the crystal structures of complex III and IV indicated a short diffusion route for cytochrome *c*, estimated as less than 40 Å (Heinemeyer et al. 2007). The interaction of complex III and IV in the model is through the concave face of complex IV, as opposed to the proposed interaction in the bovine heart supercomplex (Schäfer et al. 2006) (Fig. 12.2c, d). Free complex IV was not detected in yeast mitochondria; all of it appears to be associated with complex III to form supercomplexes (Schägger and Pfeiffer 2000; Heinemeyer et al. 2007) and no indication of dimeric complex IV was found, either as part of a supercomplex or in isolation (Heinemeyer et al. 2007). Complex IV was shown to require the presence of complex III for assembly (Cruciat et al. 2000). The lipid cardiolipin, which is abundant in the mitochondrial inner membrane and essential for mitochondrial function (Jiang et al. 2000), appears to be essential for the formation of the supercomplexes in yeast (Schägger 2002; Pfeiffer et al. 2003; Zhang et al. 2005), and there is evidence that it acts by neutralizing the charges of lysine residues in the vicinity of the presumed interaction domain between complex III and complex IV (Wenz et al. 2009).

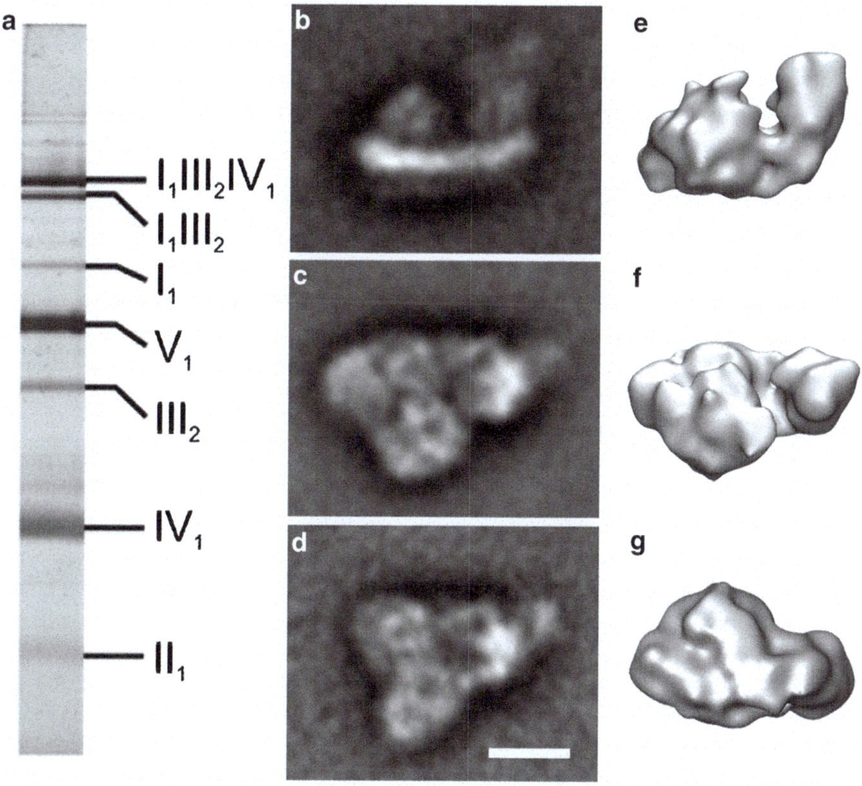

Fig. 12.3 Supercomplexes from bovine heart mitochondria. (**a**) a BN gel of solubilized bovine heart mitochondria shows apart from the individual respiratory chain complexes a strong band for a supercomplex containing complexes I, III and IV (1,700 kDa) and a weaker one for a I-III supercomplex (1,500 kDa). The highest bands represent supercomplexes with 2–4 copies of complex IV (Schäfer et al. 2006). (**b**-**d**) Electron microscopy and single particle analysis of the supercomplex bands after electroelution. (**b**) and (**c**) Show the supercomplex $I_1III_2IV_1$ in the plane of the membrane and perpendicular, respectively, and (**d**) supercomplex I_1III_2 in the same orientation as (**c**) (Schäfer et al. 2006). (**e**–**g**) the 3D reconstruction of supercomplex $I_1III_2IV_1$ (Schäfer et al. 2007), (**e**) in the plane of the membrane in an orientation corresponding to (**b**), (**f**) as seen from the matrix, corresponding to (**c**), and (**g**) as seen from the intermembrane space. The scale bar represents 10 nm

12.4.2 Supercomplex I_1III_2

In plant as well as mammalian mitochondria a supercomplex consisting of complex I and complex III (I_1III_2) has been found (Schägger and Pfeiffer 2000, 2001; Eubel et al. 2003, 2004b; Krause et al. 2004a). In plants, this supercomplex is the most abundant one (Eubel et al. 2003, 2004a). Based on BN-PAGE it has a molecular mass of ~1,500 kDa. Electron microscopy of negatively stained samples has provided structural information for the supercomplex from the plants *Arabidopsis thaliana* (Dudkina et al. 2005a), *Zea mays* (Peters et al. 2008) and potato (Bultema et al. 2009) as well as from beef heart mitochondria (Schäfer et al. 2006) (Fig. 12.3d).

In all species, a complex III dimer is associated laterally with the membrane arm of complex I, but the interaction between the two complexes differs. In the bovine supercomplex the interaction surface appears to be more extensive and complex III is attached to the middle of the complex I membrane arm, whereas in the plant structure complex III attaches to the end of the membrane arm. Lacking three-dimensional data of the supercomplexes, the interaction of complex I and III in supercomplexes appears to be species- or kingdom-specific, but the differences may be caused by different particle orientations in the negative stain images due to different shapes of the complexes (see Fig. 12.2). The x-ray structures of *E. coli*, *T. thermophilus* and *Y. lipolytica* complex I showed differences in curvature of the membrane arm (Efremov et al. 2010; Hunte et al. 2010), which may be responsible for differences in the preferred orientations of the supercomplexes on the carbon support film. Complex I from plants has several additional subunits compared to the mammalian complex (Millar et al. 2003; Parisi et al. 2004). It has a different appearance in the electron microscope than the complex from the yeast *Yarrowia lipolytica* (Radermacher et al. 2006), bovine heart or *Aquifex aeolicus* (Clason et al. 2010), showing two unique protein densities on either side of the membrane arm (Dudkina et al. 2005a; Peters et al. 2008). The spherical density on the matrix side was identified as a carbonic anhydrase complex (Sunderhaus et al. 2006), which is an enzyme unique to plant mitochondria and was previously shown to be associated with complex I. The carbonic anhydrase domain consists of three different proteins, each of which has a transmembrane domain (Sunderhaus et al. 2006) and it was seen in electron microscopy data from all plant species investigated (Dudkina et al. 2005a; Sunderhaus et al. 2006; Peters et al. 2008; Bultema et al. 2009).

Comparison of the I_1III_2 supercomplex with the more abundant, stable and active $I_1III_2IV_1$ from bovine heart mitochondria (see below) suggests that in this tissue I_1III_2 may not be physiologically relevant.

12.4.3 Supercomplex $I_1III_2IV_1$

In bovine heart mitochondria, the most abundant supercomplex identified on BN gels is $I_1III_2IV_1$, followed by I_1III_2 (Schägger and Pfeiffer 2000, 2001; Schägger et al. 2004; Krause et al. 2005; Schäfer et al. 2006) (Fig. 12.3a). Activity measurements on electroeluted samples showed that the supercomplex I_1III_2 had about twofold lower NADH dehydrogenase activity of complex I compared to the larger supercomplex, and cytochrome *c* reductase activity was hardly detectable in I_1III_2 (Schäfer et al. 2006). The larger supercomplex was also more stable than I_1III_2, which was seen on BN-PAGE and by electron microscopy to dissociate into the individual complexes (Schäfer et al. 2006). This suggests that $I_1III_2IV_1$ may be the physiologically relevant species in bovine heart mitochondria. By BN-PAGE this supercomplex, sometimes referred to as a respirasome, has also been identified in mitochondria from rat liver (Krause et al. 2005; Reifschneider et al. 2006), as well as those of plants (Eubel et al. 2004a; Krause et al. 2004a) and fungi (Krause et al. 2004b).

The complex from bovine heart mitochondria was investigated by electron microscopy and compared with the smaller supercomplex lacking complex IV (Schäfer et al. 2006) (Fig. 12.3). Thus, the location of complex IV was identified at a position distal from the complex I matrix arm. Complex IV appeared to be attached to the other complexes through the concave face, which is the dimer interface in the X-ray structure (Tsukihara et al. 1996), as opposed to the orientation in the yeast supercomplex (Heinemeyer et al. 2007).

The bovine heart supercomplex $I_1III_2IV_1$ was also the subject of the first and so far only three-dimensional reconstruction of a supercomplex (Schäfer et al. 2007). The reconstruction at a resolution of 32 Å (Fig. 12.3e, f) revealed the location of the complex III matrix domains. In the supercomplex, both complex III and IV have extensive interaction surfaces with the membrane arm of complex I and a smaller contact site with each other (Fig. 12.2c, d). The interaction between complex III and IV is much less extensive than in the yeast III_2IV_{1-2} supercomplex (Heinemeyer et al. 2007) (Fig. 12.2b) and in addition, the interaction face of not only complex IV but also complex III is different in the two species (compare Fig. 12.2b, d). The recent x-ray structure of *T. thermophilus* complex I (Efremov et al. 2010) shows a distinct curvature of the membrane arm, which fits surface features of the 3D map on the intermembrane space side (Fig. 12.3g). This shows that the handedness of the supercomplex model was determined correctly, unlike that of the complex I EM maps (Radermacher et al. 2006; Clason et al. 2010). The x-ray structures of complex III and IV and an electron microscopy map of bovine heart complex I were fitted in the map, showing that the binding sites of ubiquinone and cytochrome *c* of the different complexes were facing each other, suggesting short diffusion routes of the electron carriers (Fig. 12.4).

12.4.4 Supercomplexes $I_1III_2IV_{2-4}$

On BN gels of bovine heart mitochondria, there are several weak bands visible with a higher mass than the supercomplex $I_1III_2IV_1$ (Fig. 12.3a), which have been identified as supercomplexes of complex I and complex III with an additional 2–4 copies of complex IV (Schägger and Pfeiffer 2000). Supercomplexes of a similar composition have also been identified in plants (Eubel et al. 2004a) and in fungi (Krause et al. 2004b). Due to the low abundance of these supercomplexes it has proven difficult to determine the structure by electron microscopy, and so far it is not known if the additional copies of complex IV have a specific interaction with the other complexes, and if they occur as monomers or dimers. It has been suggested (Schäfer et al. 2006) that the interaction of complex IV with complex I in the bovine heart $I_1III_2IV_1$ supercomplex is through the interface that forms a dimer interface in the crystal structure (Tsukihara et al. 1996). If this is the case for all copies of complex IV, this would imply that complex IV occurs exclusively as monomers in the supercomplex. So far it is not clear if the presence of multiple copies of complex IV has a functional relevance.

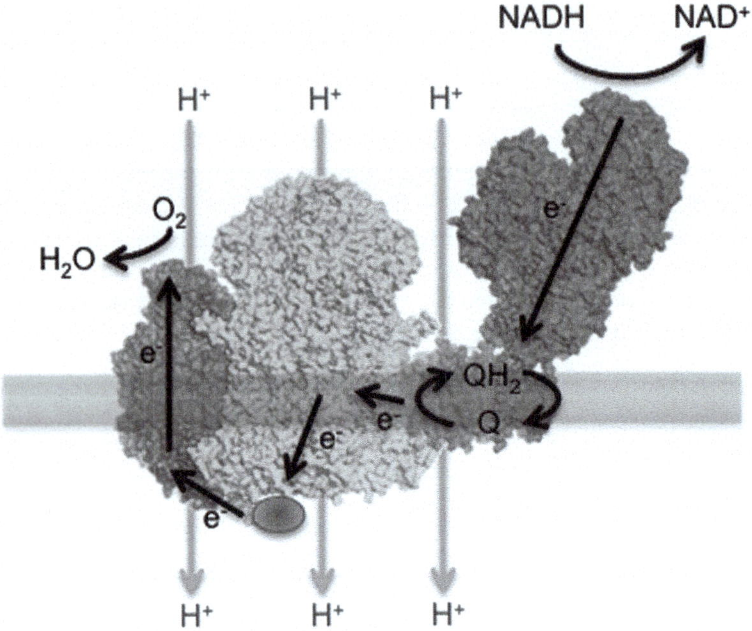

Fig. 12.4 Hypothetical electron transfer pathway in bovine heart mitochondrial supercomplex $I_1III_2IV_1$, based on the 3D structure and interpretation from (Schäfer et al. 2007). Electrons (*black arrows*) from NADH are passed through FMN and seven iron-sulfur clusters in the matrix arm of complex I to ubiquinone (Hinchliffe and Sazanov 2005; Sazanov and Hinchliffe 2006). The quinone binding site in complex I is probably located at the interface of the matrix and membrane arm (Efremov et al. 2010). Ubiquinone enters complex III (Xia et al. 1997) where some electrons enter the ubiquinone cycle and others are transferred over heme b_L and the Rieske iron-sulfur cluster to the soluble protein cytochrome *c* in the intermembrane space (see *arrow*). Cytochrome *c* transfers the electrons to subunit II of complex IV (Tsukihara et al. 1996). Complex IV transfers the electrons to the end acceptor, oxygen, to produce water. During electron transport, protons are translocated in each of the complexes from the matrix to the intermembrane space (*light arrows*), thus creating a proton gradient, which is used by ATP synthase to produce ATP

12.4.5 Supercomplexes with Two Copies of Complex I

In plants, supercomplexes with two copies of complex I were found on BN-gels, of composition I_2III_4 (Eubel et al. 2003, 2004a). An electron microscopy study of sucrose gradient fractions containing supercomplexes from potato (Bultema et al. 2009) revealed low numbers of a supercomplex with two complex I peripheral arms. This was interpreted as a $I_2III_2IV_2$ supercomplex, with a complex III sandwiched between two copies of complex I and two copies of complex IV (Fig. 12.2h). This structure is not compatible with the presence of I_2III_4, which were described in the earlier study of potato mitochondria (Eubel et al. 2004a). It is likely that these bands represented I_2III_2 instead.

In *Podospora anserina* (Krause et al. 2004b), I_2 and I_2III_2 supercomplexes were identified in the absence of complex IV, and complex I dimers were also seen in

Neurospora crassa (Marques et al. 2007). The presence of complex I dimers suggests that the structure of I_2III_2 is different in fungi than in plants, where there is no contact between the two copies of complex I (Bultema et al. 2009) (Fig. 12.2h). To date no structural information is available for the fungal supercomplexes.

12.5 Supramolecular Structure of ATP Synthase

12.5.1 Evidence for a Higher-Order Organisation of Mitochondrial ATP Synthase

The F_1F_o ATP synthase (or complex V) is responsible for the conversion of the electrochemical gradient created by the respiratory chain complexes. This complex is one of the most abundant proteins in the mitochondrial inner membrane. The F_1 particles are easily recognisable on electron micrographs of submitochondrial particles (Fernández-Morán 1962). The F_1F_o was thought to occur in monomeric form in the cristae (Capaldi 1982) based on their appearance in electron micrographs (Soper et al. 1979). A supramolecular organisation of the mitochondrial ATP synthase was first noticed in the tubular cristae of *Paramecium* by rapid-freeze deep-etch techniques (Allen et al. 1989). In these experiments, double rows of F_1 in a helical array around the cristae were found. This led to the hypothesis that the organisation of the ATP synthases was responsible for the shape of the cristae (Allen 1995). Rows of F_1's were also seen in cryo-electron tomograms of *Neurospora crassa* mitochondria (Nicastro et al. 2000) and in freeze-fracture images of yeast mitochondria (Thomas et al. 2008). The occurrence of double rather than single rows suggested that the building blocks are dimers of ATP synthase. Dimers of ATP synthases were found on BN-PA gels after solubilization of mitochondrial membranes with mild detergents, first in yeast (Arnold et al. 1998) and later in mammalian mitochondria (Schägger and Pfeiffer 2000; Krause et al. 2005; Garcia et al. 2006; Cortes-Hernandez et al. 2007; Dencher et al. 2007), higher plants (Eubel et al. 2003) and in the filamentous fungus *Podospora anserina* (Krause et al. 2004b). Higher-order oligomers which could correspond to rows of dimers were also found (Paumard et al. 2002; Krause et al. 2005; Garcia et al. 2006); a very mild separation by CN-PAGE, without the use of Coomassie stain, promotes the retention of even-numbered oligomers (4, 6 or 8) only (Wittig and Schägger 2005), again suggesting that the oligomers are formed by an association of dimers.

12.5.2 Electron Microscopy of ATP Synthase Dimers

The existence of ATP synthase dimers was proven by single-particle electron microscopy of material eluted from BN gels. Electron microscopic evidence for dimers has been obtained for mitochondria from bovine heart (Minauro-Sanmiguel et al. 2005), yeast (Dudkina et al. 2006b; Thomas et al. 2008) and the alga *Polytomella*

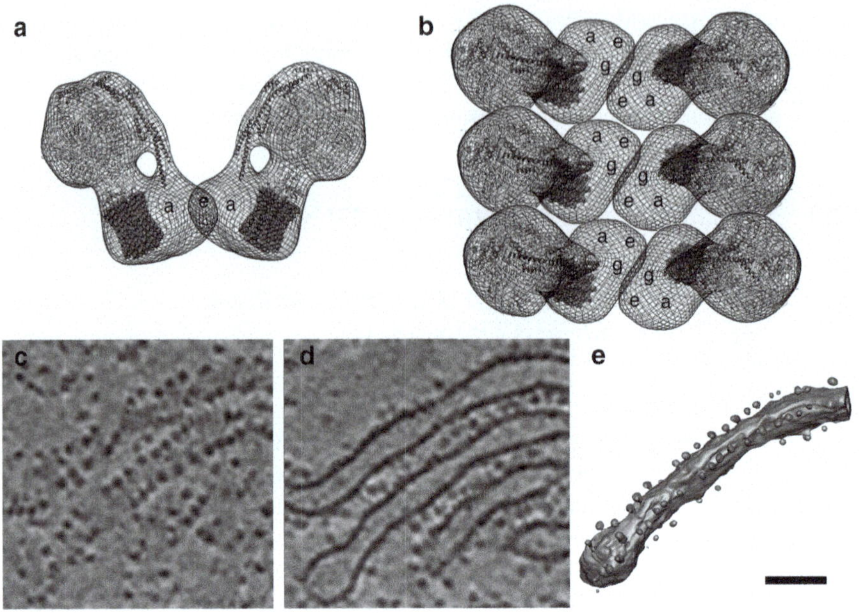

Fig. 12.5 ATP synthase dimers shape the cristae. (**a**-**b**) Hypothetical model of the supramolecular organisation of ATP synthase, based on data in Paumard et al. (2002), Arselin et al. (2003), Bustos and Velours (2005), Dudkina et al. (2005b), Minauro-Sanmiguel et al. (2005), Dudkina et al. (2006b). (**a**) Dimer seen from the membrane plane, (**b**) view of a dimer row perpendicular to the membrane. (**c**–**e**) tubular vesicles from rat liver mitochondria imaged by cryo-electron tomography. Slices of the tomogram show F_1 ATP synthase dimer rows (**c**) and a cross section of the tubes (**d**). A surface rendering of the middle tube (**e**) shows a dimer row; there is another row on the back of the tube. The scale bar is 50 nm. Tomogram courtesy of Mike Strauss and Götz Hofhaus (Figure adapted from Vonck and Schäfer 2009)

(Dudkina et al. 2005b; Cano-Estrada et al. 2010). The dimers were found to be associated through their F_o domains, which were found depending on the species at angles of 35–90° to each other, thus creating a bend in the membrane. A hypothetical arrangement of the two ATP synthases forming a dimer is shown in Fig. 12.5a, b. The angled arrangement of the F_o domains relative to each other strongly supports the notion that the association between ATP synthases is responsible for inner membrane morphology and the formation of cristae. Due to the limited resolution of the electron microscopic views, molecular details of the dimer interactions are so far unknown.

The *Polytomella* dimer is unusual in its high stability, which is probably due to the presence of a unique, 60 kDa, dimer-specific subunit in algal ATP synthase (van Lis et al. 2003, 2007). In the EM average of the *Polytomella* dimer peripheral stalks are clearly visible next to each of the F_1 units, together with an additional mass suggested to be the 60 kDa subunit (Dudkina et al. 2005b). Two 3D reconstructions of the *Polytomella* dimer were recently published, one from negatively stained images (Cano-Estrada et al. 2010) and one from subtomogram averaging of cryo-electron tomograms (Dudkina et al. 2010). The models differ in the extent of interaction between the peripheral stalks, but the main intradimer contact appears to be in the transmembrane domain.

In most of the images from yeast and bovine heart mitochondria, the peripheral stalks are not visible and may indeed be absent, as suggested by the absence of the easily recognisable OSCP subunit on top of the F_1 (Dudkina et al. 2006b). The bovine heart dimers include a smaller angle and their F_1's are close together. A bridging structure is visible connecting the bottom of the F_1's (Minauro-Sanmiguel et al. 2005). The location of the bridge coincides with the position of the inhibitor protein IF_1 in an F_1 dimer (Cabezón et al. 2003), suggesting that this protein (Minauro-Sanmiguel et al. 2005) stabilizes the dimer. Also in the bovine heart dimer, an additional structure is visible on the intermembrane space side of F_0 (Minauro-Sanmiguel et al. 2005), which is not seen in monomer preparations (Rubinstein et al. 2003). This suggests that it may represent one of the dimer-specific subunits. A likely candidate would be the hydrophilic region of subunit *e*, which is predicted to form a coiled coil on the intermembrane space side (Belogrudov et al. 1996).

The ATP synthase dimers in yeast (Dudkina et al. 2006b; Thomas et al. 2008) show relatively few structural details and a wide range of angles between monomers. However, dimers displaying 90° angles are in the majority. In one study (Dudkina et al. 2006b) it was shown that dimers that are missing one or even two F_1 parts display the same angles between the F_0 parts, which indicates that the transmembrane domains alone are responsible for the dimer interaction. The occurrence of dimers with a large and a small (35°) angle led to the suggestion that the two represent different associations within a double row: the 90° structure a "true dimer" across both rows and the 35° structure a "pseudo-dimer" made up of neighbours within a row (Dudkina et al. 2006b).

A further indication for the existence of a supramolecular arrangement of the ATP synthase is provided by an atomic force microscopy (AFM) study of inner membranes from yeast mitochondria (Buzhynskyy et al. 2007). The AFM images show staggered double rows of circular features in the membrane, in a pattern similar to that seen by freeze-etching of *Paramecium* cristae (Allen et al. 1989), which were interpreted as a view of the *c*-rings from the intermembrane space. Visible between the rings are smaller, three-domain features which the authors interpret as the F_0 subunits *a, b, e* and *g*. The *e* and *g* subunit are specific for mitochondria and are considered to be responsible for dimer formation.

12.5.3 Dimer-Specific Subunits of ATP Synthase

Separation of the subunits of ATP synthase dimers by 2D gel electrophoresis revealed the presence of three dimer-specific subunits, *e*, *g* and *k* (Arnold et al. 1998). These subunits were not essential for catalysis. Subunit *e* appears to play the central role in the dimerization process; its presence is required for the stability of subunit *g*, whereas *e* is stable in the absence of *g* (Arnold et al. 1998). Deletion mutants of *e* and *g* not only lacked ATP synthase dimers but also had abnormal mitochondria with the inner membranes forming onion-like, multilayered structures (Giraud et al. 2002; Paumard et al. 2002). The importance of *e* is also shown by its regulation by physiological status (Swartz et al. 1996; Levy and Kelly 1997). The role of the *k* subunit in dimer formation remains unclear, as deletion mutants of this subunit expressed normal levels of dimeric ATP synthase (Arnold et al. 1998).

As discussed above, the association between the two ATP synthases in a dimer occurs via the F_o part. The *e* and the *g* subunit have been shown to be F_o components each containing a single transmembrane helix (Belogrudov et al. 1996). The *e* subunit exists of an N-terminal transmembrane helix and a C-terminal hydrophilic domain with a high coiled-coil propensity located in the intermembrane space (Belogrudov et al. 1996; Arnold et al. 1997). Dimerization involves a conserved GxxxG motif in the transmembrane domains of both subunit *e* (Arselin et al. 2003) and *g* (Bustos and Velours 2005); the coiled-coil region of *e* is not essential for dimerization of the ATP synthase (Everard-Gigot et al. 2005). *e*-*g* heterodimers are responsible for the formation of ATP synthase dimers (Bustos and Velours 2005), whereas homodimers of *e* are associated with the formation of higher oligomers (Arselin et al. 2003). In a study of yeast ATP synthase, two subunits 4 (the equivalent of the bacterial peripheral stalk subunit *b*, of which there is only one copy in yeast) could be cross-linked, showing that this subunit is part of the dimer interface (Paumard et al. 2002). The peripheral stalk subunit *h* in yeast is also thought to have a role in the formation of ATP synthase dimers (Fronzes et al. 2006).

In the absence of a transmembrane proton gradient, F_1F_o would reverse its direction and act as an ATPase, pumping protons out of the mitochondrial matrix. Under these conditions the regulatory subunit IF_1 prevents ATP hydrolysis by binding to the F_1. The C-terminus of IF_1 is a dimerization domain and an IF_1 dimer binds simultaneously to two F_1's. The crystal structures of an IF_1 dimer (Cabezón et al. 2001) has been determined and also that of a bovine F_1 dimer in complex with IF_1 (Cabezón et al. 2003). Based on these structures, a regulatory role for dimer formation was proposed. However, both in yeast (Dienhart et al. 2002) and in bovine heart mitochondria (Tomasetig et al. 2002) dimerization of ATP synthase was shown to be independent of the inhibitor protein and to be mediated by the F_o domain, although the presence of IF_1 stabilizes the dimeric state of the F_1F_o in rat liver mitochondria (Garcia et al. 2006). The most likely mode of action of IF_1 is a binding to pre-existing ATP synthase dimers which are the native state of the complex.

12.5.4 ATP Synthase Dimers Shape the Cristae Membranes

A convincing three-dimensional view of the arrangement of ATP synthase on the cristae is provided by cryo-electron tomography of bovine heart and rat liver mitochondrial vesicles (Strauss et al. 2008) and more recently also of mitochondria from *Polytomella* (Dudkina et al. 2010). The F_1 heads are clearly visible in the tomograms and are arranged exclusively in double rows, forming a helical pattern around tubular vesicles (Fig. 12.5c–e) or rings around the rims of flat discs. The angles between ATP synthases in adjacent rows are approximately 90° in all species investigated, as was found for the isolated dimers of algal and yeast ATP synthase (Dudkina et al. 2005b, 2006b; Thomas et al. 2008). This shows that the dimers isolated from BN-PAGE are native and have the same conformation as in the mitochondrial inner membrane. ATP synthase dimerization appears to be specific for mitochondria; in cryo-electron tomography of chloroplasts, the ATP synthase was confined to minimally curved areas and present mainly as monomers (Daum et al. 2010). An earlier report of ATP synthase dimers detected in BN-gels from *Chlamydomonas reinhardtii* chloroplasts (Rexroth et al. 2004) could not be confirmed.

12.6 Organization of the Mitochondrial Inner Membrane

The mitochondrial inner membrane has a very high protein-to-lipid ratio of more than 3:1 by weight. The bulk of the protein in the cristae consists of the respiratory chain complexes, ATP synthase, and the ADP/ATP carrier, which exchanges the produced ATP with spent ADP in the cytoplasm (Pebay-Peyroula et al. 2003). The classical picture where these complexes are floating freely in the membrane and interacting by random collision has been revised in recent years, but the exact level of organization is still not known.

The internal membranes of a cell like the endoplasmic reticulum, Golgi apparatus, chloroplasts, and mitochondria, have complex morphologies that are maintained by highly diverse mechanisms (reviewed in Voeltz and Prinz 2007). It is clear now that the formation of cristae in the mitochondrial inner membrane is determined to a large extent, if not entirely, by the curvature in the membrane induced by dimers of ATP synthases. Two small transmembrane subunits, *e* and *g*, appear to be responsible for dimerization of the ATP synthase. The dimers form rows around the edges of the cristae, which are shaped as disks or tubes, presumably depending on the amount of lipid available per dimer. It is not known whether the dimers interact with each other actively to form rows, but it is conceivable that they come together passively to minimize the distortion of the membrane.

Much less is clear about a possible higher order organization of the respiratory chain complexes. The only indication of such a organization *in situ* appear in the rapid-freeze deep-etch study of *Paramecium* mitochondria by Allen et al., where opposite the double row of ATP synthase on tubular cristae a single row of protrusions was seen, which were interpreted as complex I (Allen et al. 1989). However, the existence of assemblies called respiratory strings has been postulated, based on the relative abundance of the different complexes and supercomplexes in different mitochondria. The model by Wittig and Schägger for bovine heart mitochondria (Wittig et al. 2006) envisions a linear aggregate of alternating complex III dimers and complex IV tetramers, with complex I bound to some of these units (Fig. 12.6a). This model takes into account the occurrence of $I_1III_2IV_4$, III_2IV_4 and IV_4 supercomplexes and the ratio of respiratory complexes I:III:IV that was determined as 1:3:6

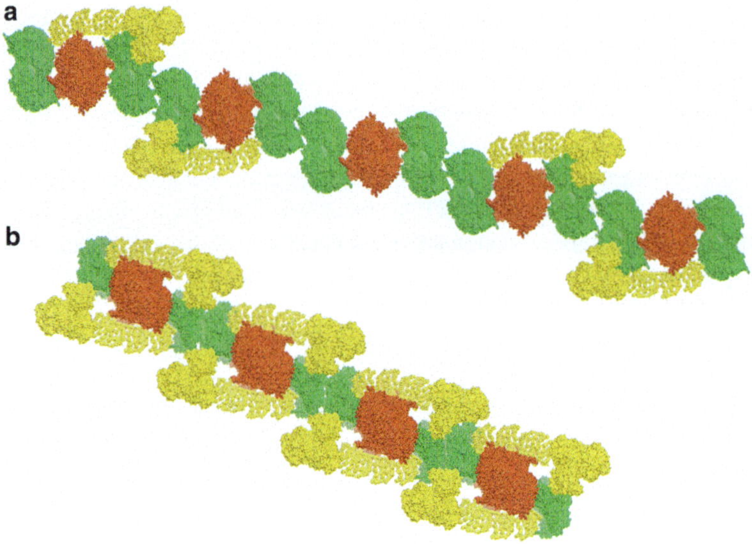

Fig. 12.6 Hypothetical models for a higher organization of respiratory chain complexes (respiratory strings). The model for bovine heart mitochondria in (**a**) is based on (Wittig et al. 2006) and is built from a linear combination of $I_1III_2IV_4$ and III_2IV_4 supercomplexes, connected via complex IV tetramers. The model for potato mitochondria in (**b**) is based on Bultema et al. (2009) and consists of $I_2III_2IV_2$ supercomplexes, connected via complex IV dimers. Complex I is shown in *yellow*, III in *red*, IV in *green*

in bovine heart mitochondria (Schägger and Pfeiffer 2001). The model by Bultema et al. for potato mitochondria (Bultema et al. 2009) is based on a single-particle electron microscopy study where among others (I_1III_2, III_2IV_1, $I_1III_2IV_1$) supercomplexes I_2III_2 consisting of two copies of complex I bound to both sides of a complex III dimer were found (Fig. 12.3h). The string has as basic unit a complex III dimer with a complex I and complex IV bound to either side, and the contacts are formed by complex IV dimers (Fig. 12.6b). Like the supercomplexes found in the two species, the respiratory string models differ considerably. Verification of the respiratory string model in general and the mammalian (Wittig et al. 2006) and plant (Bultema et al. 2009) models in particular will only be possible by the study of intact mitochondria or mitochondrial fragments. As the resolution of direct microscopy methods like cryo-electron tomography is limited to 3–8 nm, it is not to be expected that smaller proteins than the complex I matrix arm will be visualized directly. Moreover, the complex I matrix arm is similar in size to F_1, and the two are difficult to distinguish (see Fig. 12.1, which shows the bacterial complex I which is considerably smaller than the mitochondrial enzyme). It can't be ruled out that some of the features identified as ATP synthase in tomograms (Strauss et al. 2008) represented complex I.

12.7 Functions of Supercomplex Formation

12.7.1 Direct Transfer of Substrates

Several functions of the supercomplexes of the respiratory chain have been proposed. A more efficient transfer of substrates was historically the first and the most obvious role, and evidence for this function was recently reviewed by Genova et al. (2008). Before supercomplexes were isolated on BN-gels, experiments with yeast mitochondria (Boumans et al. 1998) indicated that the respiratory chain behaves as a single functional unit and suggested that complex II and alternative oxidases were part of a supercomplex with complex III and IV. Flux control experiments also indicated the presence of a I–III supercomplex in bovine heart mitochondria (Bianchi et al. 2004), although complex IV was functionally independent in this study. In these mitochondria, complex IV is in large excess and most is present in free form. Supercomplexes involving complex II were never found on BN-PAGE (Cruciat et al. 2000; Schägger and Pfeiffer 2001) and there is kinetic evidence that there is no substrate channeling from complex II to complex III and IV (Bianchi et al. 2004). As mentioned above, short diffusion distances for the electron carriers ubiquinone from complex I to complex III and cytochrome *c* from complex III to complex IV in a supercomplex were confirmed by electron microscopic analysis of isolated supercomplexes in mammalian mitochondria (Schäfer et al. 2007) (Fig. 12.4) and for cytochrome *c* also in yeast mitochondria (Heinemeyer et al. 2007).

12.7.2 Assembly and Stability of Respiratory Chain Complexes

There are many indications that in mammalian mitochondria supercomplex formation occurs concomitant with the biogenesis and assembly of the individual complexes and is essential for the assembly of complex I in particular. In one of the earliest studies of supercomplexes by BN-PAGE (Schägger and Pfeiffer 2001), it was found that in bovine heart mitochondria, almost all complex I occurs in supercomplexes, whereas complex IV is present in much larger quantities and occurs for a large part in monomeric form. The supercomplex $I_1III_2IV_1$ in bovine heart mitochondria is not only more abundant, but also more active and more stable than I_1III_2 (Schäfer et al. 2006), suggesting that complex IV not only stabilizes the other complexes but also is important for their proper function.

Several human diseases are associated with deficiencies in more than one respiratory chain complex, indicating an involvement of assembly and supercomplex formation. Dual deficiencies of complex I and IV were observed in idiopathic Parkinson patients (Benecke et al. 1993; Cardellach et al. 1993; Gu et al. 1998). Also patients with lactic acidosis syndrome displayed decreased complex I and IV activities (van Straaten et al. 2005). Patients with peripheral arteriosclerotic vascular diseases showed a combined decrease in complex I and III activity (Brass et al. 2001).

Several studies (Schägger 2002; Acin-Pérez et al. 2004; Krause et al. 2004b; Schägger et al. 2004) have shown that both complex III and IV play a role in the assembly and stability of complex I in mammalian mitochondria and that assembly errors are the cause of many mitochondrial diseases. In human skeletal muscle (Schägger et al. 2004) supercomplex formation was found to be important for complex I assembly. Genetic alterations leading to a loss of complex III prevented respirasome formation and led to the secondary loss of complex I, so that primary complex III assembly deficiencies appeared as I/III defects. Similar observations were made in mouse and human cell lines with mutations in the gene for cytochrome *b*, which is a subunit of complex III (Acin-Pérez et al. 2004). In canine heart a loss of supercomplexes was associated with heart failure, while the individual activities of the respiratory chain complexes were unchanged (Rosca et al. 2008). Toxicity studies on neuroblastoma cells showed a simultaneous inhibition of complex I and complex IV (Mazzio and Soliman 2004). In mouse fibroblasts, a requirement of complex IV for the assembly and stability of complex I was shown (Diaz et al. 2006).

Not all mutations causing supercomplex defects are localized in respiratory chain complex subunits, however. In Barth syndrome patients, respiratory chain supercomplexes were shown to be destabilized (McKenzie et al. 2006). Barth syndrome is caused by mutations in the human TAZ gene that codes for tafazzin, a putative phospholipid acyltransferase that is involved in the remodeling of cardiolipin. Cardiolipin is a unique phospholipid which is present throughout the eukaryotic kingdom and is localized in the inner membrane of mitochondria. Studies of the *S. cerevisiae crd1*Δ mutant, which lacks cardiolipin synthase, have shown that cardiolipin is essential for mitochondrial function (Jiang et al. 2000). The role of cardiolipin in supercomplex formation has been well documented in yeast; it has been shown to be essential for the stability of the III_2IV_{1-2} supercomplexes (Schägger 2002; Zhang et al. 2002; Pfeiffer et al. 2003; Zhang et al. 2005). In the crystal structure of complex III, a cardiolipin molecule is tightly bound at the periphery of the dimer (Lange et al. 2001). There is evidence that in the supercomplex it neutralizes the charges of lysine residues near the presumed interaction domain between complex III and complex IV (Wenz et al. 2009).

Although there is thus strong evidence that in mammals, complex I assembly and stability is critically dependent on the presence and proper assembly of complex III and IV, this does not appear to be the case in all species. In the fungus *Podospora anserina*, mutants have been created that lack both complex III and IV but possess active complex I (Maas et al. 2009). The authors speculate that the alternative oxidases present in these species may play a role in complex I assembly. Another possible factor is the dimerization of complex I, which so far has only been shown in filamentous fungi (Krause et al. 2004b; Marques et al. 2007). Interestingly, complex I dimers were only found in mutants lacking complex IV, and in the presence of complex IV respirasome-like supercomplexes were present (Krause et al. 2004b), suggesting that there may be alternative assembly pathways for complex I, depending on the available partners in the respiratory chain.

12.7.3 Supercomplexes and Mitochondrial Function

Mitochondria play an important role in aging and neurodegenerative diseases through the formation of reactive oxygen species (ROS) (Orrenius et al. 2007). ROS damages mitochondrial DNA, resulting in production of altered respiratory chain proteins, leading to reduced electron transfer resulting in higher ROS production in a vicious circle. It has been suggested that supercomplex formation, and especially the interaction between complex I and III, may be involved in limiting ROS production (Lenaz and Genova 2007, 2009). In rat brain, age-related changes in abundances of supercomplexes were observed, which may explain alterations in ROS production (Dencher et al. 2007). This is also consistent with the study in canine heart (Rosca et al. 2008), where an association between a loss of supercomplexes and heart failure was found. A regulatory role of supercomplexes in energy production has also been proposed (Piccoli et al. 2006): a high membrane potential may dissociate supercomplexes, resulting in a lower flux through the respiratory chain. The anionic phospholipid cardiolipin was suggested as the trigger, by acting as a sensor of pH and electrical gradient, which would alter its conformation and thus the tight binding to the respiratory chain complexes. Such a mechanism is consistent with a role of cardiolipin as a proton sink, as was proposed before (Haines and Dencher 2002).

12.8 Mechanisms of Supercomplex Formation

The specific subunits responsible for supercomplex formation have so far not been identified. In yeast, some subunits whose deletion strains contained III–IV supercomplexes could be ruled out as participants in supercomplex formation (Schägger 2002).

Complex III is only active as a dimer (Lange and Hunte 2002; Covian et al. 2004; Covian and Trumpower 2005; Gong et al. 2005), and in the EM structures of the supercomplexes, it is always present as a dimer (Dudkina et al. 2005a; Schäfer et al. 2006; Heinemeyer et al. 2007; Peters et al. 2008; Bultema et al. 2009). Complex IV is also active as a monomer, although it occurs as dimers in the crystal structure (Tsukihara et al. 1996), and a complex IV dimer has not been observed in any projection maps of supercomplexes (Schäfer et al. 2006; Heinemeyer et al. 2007). The interpretation of the orientation of complex IV in the supercomplexes from bovine heart (Schäfer et al. 2006) and yeast (Heinemeyer et al. 2007) is different: in bovine heart, complex IV is attached to the supercomplex with the concave face, which is the dimer interface in the crystal structure, whereas in yeast it is attached with the convex side. These observations are not necessarily in contradiction, because the interaction partners differ as well: in the yeast III_2IV_{1-2} supercomplex the interaction is of course IV-III, but in the mammalian $I_1III_2IV_1$ supercomplex, complex IV is attached at the end of the complex I membrane arm and has only a small interface with complex III (Schäfer et al. 2007).

Differences in crista morphology between species suggest that the details of the complex interactions in supercomplex formation may be species- or kingdom-specific. In order to better understand the functions of supercomplexes, comparisons of electron transfer rate and efficiency between the individual complexes and supercomplexes will be needed. Solving the 3D structures of more supercomplexes will be important to get information about the subunits involved in complex interactions, to understand the details of the interactions as well as the assembly pathway and the occurrence of conformational changes due to supercomplex formation. So far it is unclear, if once a supercomplex is formed it persists until degradation or if supercomplex formation is a dynamic process. In addition, mitochondria are highly dynamic structures, subject to continuous fusion and fission events, which may also be correlated with crista morphology. Advances in isolation of larger structures combined with single particle electron microscopy and studies of intact mitochondria by antibody-labeling and cryo-electron tomography will be useful to get a full picture of the structure of the electron transport chain and a better understanding of mitochondrial dynamics.

References

Abrahams JP, Leslie AG, Lutter R, Walker JE (1994) Structure at 2.8 Å resolution of F_1 ATPase from bovine heart mitochondria. Nature 370:621–628

Acin-Pérez R, Bayona-Bafaluy MP, Fernández-Silva P, Moreno-Loshuertos R, Pérez-Martos A, Bruno C, Moraes CT, Enríquez JA (2004) Respiratory complex III is required to maintain complex I in mammalian mitochondria. Mol Cell 13:805–815

Acin-Pérez R, Fernández-Silva P, Peleato ML, Pérez-Martos A, Enriquez JA (2008) Respiratory active mitochondrial supercomplexes. Mol Cell 32:529–539

Allen RD (1995) Membrane tubulation and proton pumps. Protoplasma 189:1–8

Allen RD, Schroeder CC, Fok AK (1989) An investigation of mitochondrial inner membranes by rapid-freeze deep-etch techniques. J Cell Biol 108:2233–2240

Arnold I, Bauer MF, Brunner M, Neupert W, Stuart RA (1997) Yeast mitochondrial F_1F_0-ATPase: the novel subunit e is identical to Tim11. FEBS Lett 411:195–200

Arnold I, Pfeiffer K, Neupert W, Stuart RA, Schägger H (1998) Yeast mitochondrial F_1F_0-ATP synthase exists as a dimer: identification of three dimer-specific subunits. EMBO J 17:7170–7178

Arselin G, Giraud M-F, Dautant A, Vaillier J, Brethes D, Coulary-Salin B, Schaeffer J, Velours J (2003) The GxxxG motif of the transmembrane domain of subunit e is involved in the dimerization/oligomerization of the yeast ATP synthase complex in the mitochondrial membrane. Eur J Biochem 270:1875–1884

Belogrudov GI, Tomich JM, Hatefi Y (1996) Membrane topography and near-neighbor relationships of the mitochondrial ATP synthase subunits *e, f,* and *g*. J Biol Chem 271:20340–20345

Benecke R, Strumper P, Weiss H (1993) Electron transfer complexes I and IV of platelets are abnormal in Parkinson's disease but normal in Parkinson-plus syndromes. Brain 116:1451–1463

Berry EA, Trumpower BH (1985) Isolation of ubiquinol oxidase from *Paracoccus denitrificans* and resolution into cytochrome bc1 and cytochrome c-aa3 complexes. J Biol Chem 260:2458–2467

Bianchi C, Genova ML, Parenti Castelli G, Lenaz G (2004) The mitochondrial respiratory chain is partially organized in a supercomplex assembly: kinetic evidence using flux control analysis. J Biol Chem 279:36562–36569

Blair PV (1967) Preparation and properties of repeating units of mitochondrial electron transfer. Methods Enzymol 10:208–212

Boumans H, Grivell LA, Berden JA (1998) The respiratory chain in yeast behaves as a single functional unit. J Biol Chem 273:4872–4877

Boyer PD (1997) The ATP synthase: a splendid molecular machine. Annu Rev Biochem 66:717–749

Brass EP, Hiatt WR, Gardner AW, Hoppel CL (2001) Decreased NADH dehydrogenase and ubiquinol-cytochrome c oxidoreductase in peripheral arterial disease. Am J Physiol Heart Circ Physiol 280:H603–H609

Bultema JB, Braun H-P, Boekema EJ, Kouril R (2009) Megacomplex organization of the oxidative phosphorylation system by structural analysis of respiratory supercomplexes from potato. Biochim Biophys Acta Bioener 1787:60–67

Bustos DM, Velours J (2005) The modification of the conserved GXXXG motif of the membrane-spanning segment of subunit *g* destabilizes the supramolecular species of yeast ATP synthase. J Biol Chem 280:29004–29010

Buzhynskyy N, Sens P, Prima V, Sturgis J, Scheuring S (2007) Rows of ATP synthase dimers in native mitochondrial inner membranes. Biophys J 93:2870–2876

Cabezón E, Runswick MJ, Leslie AGW, Walker JE (2001) The structure of bovine IF_1, the regulatory subunit of mitochondrial F-ATPase. EMBO J 20:6690–6696

Cabezón E, Montgomery MG, Leslie AGW, Walker JE (2003) The structure of bovine F_1-ATPase in complex with its regulatory protein IF_1. Nat Struct Biol 10:744–750

Cano-Estrada A, Vázquez-Acevedo M, Villavicencio-Queijeiro A, Figueroa-Martínez F, Miranda-Astudillo H, Cordeiro Y, Mignaco JA, Foguel D, Cardol P, Lapaille M, Remacle C, Wilkens S, González-Halphen D (2010) Subunit–subunit interactions and overall topology of the dimeric mitochondrial ATP synthase of *Polytomella* sp. Biochim Biophys Acta Bioener 1797:1439–1448

Capaldi RA (1982) Arrangement of proteins in the mitochondrial membrane. Biochim Biophys Acta 694:291–306

Cardellach F, Marti MJ, Fernandez-Sola J, Marin C, Hoek JB, Tolosa E, Urbano-Marquez A (1993) Mitochondrial respiratory chain activity in skeletal muscle from patients with Parkinson's disease. Neurology 43:2170–2172

Carroll J, Fearnley IM, Skehel JM, Shannon RJ, Hirst J, Walker JE (2006) Bovine complex I is a complex of 45 different subunits. J Biol Chem 281:32724–32727

Clason T, Ruiz T, Schägger H, Peng G, Zickermann V, Brandt U, Michel H, Radermacher M (2010) The structure of eukaryotic and prokaryotic complex I. J Struct Biol 169:81–88

Cortes-Hernandez P, Vázquez-Memije ME, Garcia JJ (2007) *ATP6* homoplasmic mutations inhibit and destabilize the human F_1F_0-ATP synthase without preventing enzyme assembly and oligomerization. J Biol Chem 282:1051–1058

Covian R, Trumpower BH (2005) Rapid electron transfer between monomers when the cytochrome bc_1 complex dimer is reduced through center N. J Biol Chem 280:22732–22740

Covian R, Gutierrez-Cirlos EB, Trumpower BH (2004) Anti-cooperative oxidation of ubiquinol by the yeast cytochrome bc_1 complex. J Biol Chem 279:15040–15049

Crofts AR (2004) The cytochrome bc_1 complex: function in the context of structure. Annu Rev Physiol 66:689–733

Cruciat C-M, Brunner S, Baumann F, Neupert W, Stuart RA (2000) The cytochrome bc_1 and cytochrome *c* oxidase complexes associate to form a single supracomplex in yeast mitochondria. J Biol Chem 275:18093–18098

Daum B, Nicastro D, Austin J, McIntosh JR, Kühlbrandt W (2010) Arrangement of photosystem II and ATP synthase in chloroplast membranes of spinach and pea. Plant Cell 22:1299–1312

Dencher NA, Frenzel M, Reifschneider NH, Sugawa M, Krause F (2007) Proteome alterations in rat mitochondria caused by aging. Annu NY Acad Sci 1100:291–298

Diaz F, Fukui H, Garcia S, Moraes CT (2006) Cytochrome *c* oxidase is required for the assembly/stability of respiratory complex I in mouse fibroblasts. Mol Cell Biol 26:4872–4881

Dienhart M, Pfeiffer K, Schaegger H, Stuart RA (2002) Formation of the yeast F_1F_0-ATP synthase dimeric complex does not require the ATPase inhibitor protein, Inh1. J Biol Chem 277:39289–39295

Dudkina NV, Eubel H, Keegstra W, Boekema EJ, Braun HP (2005a) Structure of a mitochondrial supercomplex formed by respiratory-chain complexes I and III. Proc Natl Acad Sci USA 102:3225–3229

Dudkina NV, Heinemeyer J, Keegstra W, Boekema EJ, Braun HP (2005b) Structure of dimeric ATP synthase from mitochondria: an angular association of monomers induces the strong curvature of the inner membrane. FEBS Lett 579:5769–5772

Dudkina NV, Heinemeyer J, Sunderhaus S, Boekema EJ, Braun H-P (2006a) Respiratory chain supercomplexes in the plant mitochondrial membrane. Trends Plant Sci 11:232–240

Dudkina NV, Sunderhaus S, Braun HP, Boekema EJ (2006b) Characterization of dimeric ATP synthase and cristae membrane ultrastructure from *Saccharomyces* and *Polytomella* mitochondria. FEBS Lett 580:3427–3432

Dudkina NV, Sunderhaus S, Boekema E, Braun H-P (2008) The higher level of organization of the oxidative phosphorylation system: mitochondrial supercomplexes. J Bioenerg Biomembr 40:419–424

Dudkina NV, Oostergetel GT, Lewejohann D, Braun H-P, Boekema EJ (2010) Row-like organization of ATP synthase in intact mitochondria determined by cryo-electron tomography. Biochim Biophys Acta Bioener 1797:272–277

Efremov RG, Baradaran R, Sazanov LA (2010) The architecture of respiratory complex I. Nature 465:441–447

Eubel H, Jänsch L, Braun HP (2003) New insights into the respiratory chain of plant mitochondria. Supercomplexes and a unique composition of complex II. Plant Physiol 133:274–286

Eubel H, Heinemeyer J, Braun HP (2004a) Identification and characterization of respirasomes in potato mitochondria. Plant Physiol 134:1450–1459

Eubel H, Heinemeyer J, Sunderhaus S, Braun HP (2004b) Respiratory chain supercomplexes in plant mitochondria. Plant Physiol Biochem 42:937–942

Everard-Gigot V, Dunn CD, Dolan BM, Brunner S, Jensen RE, Stuart RA (2005) Functional analysis of subunit e of the F_1F_0-ATP synthase of the yeast *Saccharomyces cerevisiae*: importance of the N-terminal membrane anchor region. Eukaryot Cell 4:346–355

Ferguson-Miller S, Babcock GT (1996) Heme/copper terminal oxidases. Chem Rev 96:2889–2907

Fernández-Morán H (1962) Low-temperature electron microscopy and X-ray diffraction studies of lipoprotein components in lamellar systems. Circulation 26:1039–1065

Friedrich T (1998) The NADH:ubiquinone oxidoreductase (complex I) from *Escherichia coli*. Biochim Biophys Acta Bioener 1364:134–146

Fronzes R, Weimann T, Vaillier J, Velours J, Brèthes D (2006) The peripheral stalk participates in the yeast ATP synthase dimerization independently of e and g subunits. Biochemistry 45:6715–6723

Garcia JJ, Morales-Rios E, Cortes-Hernandez P, Rodriquez-Zavala JS (2006) The inhibitor protein (IF1) promotes dimerization of the mitochondrial F_1F_0-ATP synthase. Biochemistry 45:12695–12703

Genova ML, Baracca A, Biondi A, Casalena G, Faccioli M, Falasca AI, Formiggini G, Sgarbi G, Solaini G, Lenaz G (2008) Is supercomplex organization of the respiratory chain required for optimal electron transfer activity? Biochim Biophys Acta Bioener 1777:740–746

Giraud M-F, Paumard P, Soubannier V, Vaillier J, Arselin G, Salin B, Schaeffer J, Brethes D, di Rago J-P, Velours J (2002) Is there a relationship between the supramolecular organization of the mitochondrial ATP synthase and the formation of cristae? Biochim Biophys Acta 1555:174–180

Gong X, Yu L, Xia D, Yu C-A (2005) Evidence for electron equilibrium between the two hemes b_L in the dimeric cytochrome bc_1 complex. J Biol Chem 280:9251–9257

Grigorieff N (1998) Three-dimensional structure of bovine NADH:ubiquinone oxidoreductase (complex I) at 22 Å in ice. J Mol Biol 277:1033–1046

Gu M, Cooper JM, Taanman JW, Schapira AHV (1998) Mitochondrial DNA transmission of the mitochondrial defect in Parkinson's disease. Ann Neurol 44:177–186

Guénebaut V, Vincentelli R, Mills D, Weiss H, Leonard KR (1997) Three-dimensional structure of NADH-dehydrogenase from *Neurospora crassa* by electron microscopy and conical tilt reconstruction. J Mol Biol 265:409–418

Gupte SS, Hackenbrock CR (1988) The role of cytochrome c diffusion in mitochondrial electron transport. J Biol Chem 263:5248–5253

Hackenbrock CR, Chazotte B, Gupte SS (1986) The random collision model and a critical assessment of diffusion and collision in mitochondrial electron transport. J Bioenerg Biomembr 18:331–368

Haines TH, Dencher NA (2002) Cardiolipin: a proton trap for oxidative phosphorylation. FEBS Lett 528:35–39

Hatefi Y (1985) The mitochondrial electron transport and oxidative phosphorylation system. Annu Rev Biochem 54:1015–1069

Hatefi Y, Haavik AG, Griffiths DE (1961a) Reconstitution of the electron transfer system I. Preparation and properties of the interacting enzyme complexes. Biochem Biophys Res Commun 4:441–446

Hatefi Y, Haavik AG, Griffiths DE (1961b) Reconstitution of the electron transfer system II. Reconstitution of DPNH – cytochrome 5 reductase, succinic-cytochrome c reductase and DPNH, succinic-cytochrome c reductase. Biochem Biophys Res Commun 4:447–454

Hatefi Y, Haavik AG, Griffiths DE (1962a) Studies on the electron transfer system XLI. Reduced coenzyme Q (QH2)-cytochrome c reductase. J Biol Chem 237:1681–1685

Hatefi Y, Haavik AG, Griffiths DE (1962b) Studies on the electron transfer system XLII. Reconstitution of the electron transfer system. J Biol Chem 237:2661–2669

Heinemeyer J, Braun HP, Boekema EJ, Kouril R (2007) A structural model of the cytochrome c reductase/oxidase supercomplex from yeast mitochondria. J Biol Chem 282:12240–12248

Hinchliffe P, Sazanov LA (2005) Organization of iron-sulfur clusters in respiratory complex I. Science 309:771–774

Hofhaus G, Weiss H, Leonard K (1991) Electron microscopic analysis of the peripheral and membrane parts of mitochondrial NADH dehydrogenase (complex I). J Mol Biol 221:1027–1043

Hunte C, Koepke J, Lange C, Rossmanith T, Michel H (2000) Structure at 2.3 Å resolution of the cytochrome bc_1 complex from the yeast *Saccharomyces cerevisiae* co-crystallized with an antibody Fv fragment. Structure 8:669–684

Hunte C, Zickermann V, Brandt U (2010) Functional modules and structural basis of conformational coupling in mitochondrial complex I. Science 329:448–451

Iwasaki T, Matsuura K, Oshima T (1995) Resolution of the aerobic respiratory system of the thermoacidophilic archaeon, *Sulfolobus* sp. strain 7. I. The archaeal terminal oxidase supercomplex is a functional fusion of respiratory complexes III and IV with no *c*-type cytochromes. J Biol Chem 270:30881–30892

Iwata S, Lee JW, Okada K, Lee JK, Iwata M, Rasmussen B, Link TA, Ramaswamy S, Jap BK (1998) Complete structure of the 11-subunit mitochondrial cytochrome bc_1 complex. Science 281:64–71

Iwata S, Ostermeier C, Ludwig B, Michel H (2002) Structure at 2.8 Å resolution of cytochrome *c* oxidase from *Paracoccus denitrificans*. Nature 376:660–669

Jiang F, Ryan MT, Schlame M, Zhao M, Gu Z, Klingenberg M, Pfanner N, Greenberg ML (2000) Absence of cardiolipin in the crd1null mutant results in decreased mitochondrial membrane potential and reduced mitochondrial function. J Biol Chem 275:22387–22394

Kane Dickson V, Silvester JA, Fearnley IM, Leslie AGW, Walker JE (2006) On the structure of the stator of the mitochondrial ATP synthase. EMBO J 25:2911–2918

Krause F, Reifschneider NH, Vocke D, Seelert H, Rexroth S, Dencher NA (2004a) "Respirasome"-like supercomplexes in green leaf mitochondria of spinach. J Biol Chem 279:48369–48375

Krause F, Scheckhuber CQ, Werner A, Rexroth S, Reifschneider NH, Dencher NA, Osiewacz HD (2004b) Supramolecular organization of cytochrome *c* oxidase- and alternative oxidase-dependent respiratory chains in the filamentous fungus *Podospora anserina*. J Biol Chem 279:26453–26461

Krause F, Reifschneider NH, Goto S, Dencher NA (2005) Active oligomeric ATP synthases in mammalian mitochondria. Biochem Biophys Res Commun 329:583–590

Kröger A, Klingenberg M (1973) Further evidence for the pool function of ubiquinone as derived from the inhibition of the electron transport by antimycin. Eur J Biochem 39:313–323

Lange C, Hunte C (2002) Crystal structure of the yeast cytochrome *bc*$_1$ complex with its bound substrate cytochrome *c*. Proc Natl Acad Sci USA 99:2800–2805

Lange C, Nett JH, Trumpower BH, Hunte C (2001) Specific roles of protein-phospholipid interactions in the yeast cytochrome *bc*$_1$ complex structure. EMBO J 20:6591–6600

Lau WCY, Baker LA, Rubinstein JL (2008) Cryo-EM structure of the yeast ATP synthase. J Mol Biol 382:1256–1264

Lenaz G, Genova ML (2007) Kinetics of integrated electron transfer in the mitochondrial respiratory chain: random collisions vs. solid state electron channeling. Am J Physiol Cell Physiol 292:1221–1239

Lenaz G, Genova ML (2009) Structural and functional organization of the mitochondrial respiratory chain: a dynamic super-assembly. Int J Biochem Cell Biol 41:1750–1772

Levy FH, Kelly DP (1997) Regulation of ATP synthase subunit e gene expression by hypoxia: cell differentiation stage-specific control. Am J Physiol Cell Physiol 272:C457–C465

Maas MFPM, Krause F, Dencher NA, Sainsard-Chanet A (2009) Respiratory complexes III and IV are not essential for the assembly/stability of complex I in fungi. J Mol Biol 387:259–269

Marques I, Dencher NA, Videira A, Krause F (2007) Supramolecular organization of the respiratory chain in *Neurospora crassa* mitochondria. Eukaryot Cell 6:2391–2405

Matthies D, Preiß L, Klyszejko AL, Muller DJ, Cook GM, Vonck J, Meier T (2009) The c_{13} ring from a thermoalkaliphilic ATP synthase reveals an extended diameter due to a special structural region. J Mol Biol 388:611–618

Mazzio EA, Soliman KF (2004) Effects of enhancing mitochondrial oxidative phosphorylation with reducing equivalents and ubiquinone on 1-methyl-4-phenylpyridinium toxicity and complex I-IV damage in neuroblastoma cells. Biochem Pharmacol 67:1167–1184

McKenzie M, Lazarou M, Thorburn DR, Ryan MT (2006) Mitochondrial respiratory chain supercomplexes are destabilized in Barth Syndrome patients. J Mol Biol 361:462–469

Meier T, Ferguson SA, Cook GM, Dimroth P, Vonck J (2006) Structural investigations of the membrane-embedded rotor ring of the F-ATPase from *Clostridium paradoxum*. J Bacteriol 188:7759–7764

Millar AH, Mittova V, Kiddle G, Heazlewood JL, Bartoli CG, Theodoulou FL, Foyer CH (2003) Control of ascorbate synthesis by respiration and its implications for stress responses. Plant Physiol 133:443–447

Minauro-Sanmiguel F, Wilkens S, Garcia JJ (2005) Structure of dimeric mitochondrial ATP synthase: novel F_0 bridging features and the structural basis of mitochondrial cristae biogenesis. Proc Natl Acad Sci USA 102:12356–12358

Morgan DJ, Sazanov LA (2008) Three-dimensional structure of respiratory complex I from *Escherichia coli* in ice in the presence of nucleotides. Biochim Biophys Acta Bioener 1777:711–718

Muramoto K, Hirata K, Shinzawa-Itoh K, Yoko-O S, Yamashita E, Aoyama H, Tsukihara T, Yoshikawa S (2007) A histidine residue acting as a controlling site for dioxygen reduction and proton pumping by cytochrome *c* oxidase. Proc Natl Acad Sci USA 104:7881–7886

Nicastro D, Frangakis AS, Typke D, Baumeister W (2000) Cryo-electron tomography of *Neurospora* mitochondria. J Struct Biol 129:48–56

Niebisch A, Bott M (2003) Purification of a cytochrome *bc*$_1$-*aa*$_3$ supercomplex with quinol oxidase activity from *Corynebacterium glutamicum*. Identification of a fourth subunit of cytochrome *aa*$_3$ oxidase and mutational analysis of diheme cytochrome *c*$_1$. J Biol Chem 278:4339–4346

Orrenius S, Gogvadze V, Zhivotovsky B (2007) Mitochondrial oxidative stress: implications for cell death. Annu Rev Pharmacol Toxicol 47:143–183

Ostermeier C, Harrenga A, Ermler U, Michel H (1997) Structure at 2.7 Å resolution of the *Paracoccus denitrificans* two-subunit cytochrome *c* oxidase complexed with an antibody F_V fragment. Proc Natl Acad Sci USA 94:10547–10553

Parisi G, Perales M, Fornasari M, Colaneri A, Schain N, Casati D, Zimmermann S, Brennicke A, Araya A, Ferry J, Echave J, Zabaleta E (2004) Gamma carbonic anhydrases in plant mitochondria. Plant Mol Biol 55:193–207

Paumard P, Vaillier J, Coulary B, Schaeffer J, Soubannier V, Mueller DM, Brethes D, di Rago J-P, Velours J (2002) The ATP synthase is involved in generating mitochondrial cristae morphology. EMBO J 21:221–230

Pebay-Peyroula E, Dahout-Gonzalez C, Kahn R, Trézéguet V, Lauquin GJ-M, Brandolin G (2003) Structure of mitochondrial ADP/ATP carrier in complex with carboxyatractyloside. Nature 426:39–44

Peters K, Dudkina NV, Jänsch L, Braun H-P, Boekema EJ (2008) A structural investigation of complex I and I + III_2 supercomplex from *Zea mays* at 11–13 Å resolution: assignment of the carbonic anhydrase domain and evidence for structural heterogeneity within complex I. Biochim Biophys Acta Bioener 1777:84–93

Pettersen EF, Goddard TD, Huang CC, Couch GS, Greenblatt DM, Meng EC, Ferrin TE (2004) UCSF Chimera: a visualisation system for exploratory research and analysis. J Comput Chem 25:1605–1612

Pfeiffer K, Gohil V, Stuart RA, Hunte C, Brandt U, Greenberg ML, Schägger H (2003) Cardiolipin stabilizes respiratory chain supercomplexes. J Biol Chem 278:52873–52880

Piccoli C, Scrima R, Boffoli D, Capitanio N (2006) Control by cytochrome *c* oxidase of the cellular oxidative phosphorylation system depends on the mitochondrial energy state. Biochem J 396:573–583

Pineau B, Mathieu C, Gérard-Hirne C, Paepe RD, Chétrit P (2005) Targeting the NAD7 subunit to mitochondria restores a functional complex I and a wild type phenotype in the *Nicotiana sylvestris* CMS II mutant lacking nad7. J Biol Chem 280:25994–26001

Pogoryelov D, Yu J, Meier T, Vonck J, Dimroth P, Müller DJ (2005) The c_{15} ring of the *Spirulina platensis* F-ATP synthase: F_1/F_0 symmetry mismatch is not obligatory. EMBO Rep 6:1045–1052

Pogoryelov D, Yildiz Ö, Faraldo-Gómez JD, Meier T (2009) High-resolution structure of the rotor ring of a proton-dependent ATP synthase. Nat Struct Mol Biol 16:1068–1073

Radermacher M, Ruiz T, Clason T, Benjamin S, Brandt U, Zickermann V (2006) The three-dimensional structure of complex I from *Yarrowia lipolytica*: a highly dynamic protein. J Struct Biol 154:269–279

Reifschneider NH, Goto S, Nakamoto H, Takahashi R, Sugawa M, Dencher NA (2006) Defining the mitochondrial proteomes from five rat organs in a physiologically significant context using 2D blue-native/SDS-PAGE. J Proteome Res 5:1117–1132

Rexroth S, Meyer zu Tittingdorf JMW, Schaßmann HJ, Krause F, Seelert H, Dencher NA (2004) Dimeric H^+-ATP synthase in the chloroplast of *Chlamydomonas reinhardtii*. Biochim Biophys Acta 1658:202–211

Rosca MG, Vazquez EJ, Kerner J, Parland W, Chandler MP, Stanley W, Sabbah HN, Hoppel CL (2008) Cardiac mitochondria in heart failure: decrease in respirasomes and oxidative phosphorylation. Cardiovasc Res 80:30–39

Rubinstein JL, Walker JE, Henderson R (2003) Structure of the mitochondrial ATP synthase by electron cryomicroscopy. EMBO J 22:6182–6192

Sazanov LA, Hinchliffe P (2006) Structure of the hydrophilic domain of respiratory complex I from *Thermus thermophilus*. Science 311:1430–1436

Schäfer E, Seelert H, Reifschneider NH, Krause F, Dencher NA, Vonck J (2006) Architecture of active mammalian respiratory chain supercomplexes. J Biol Chem 281:15370–15375

Schäfer E, Dencher NA, Vonck J, Parcej DN (2007) Three-dimensional structure of the respiratory chain supercomplex $I_1III_2IV_1$ from bovine heart mitochondria. Biochemistry 44:12579–12585

Schägger H (2002) Respiratory chain supercomplexes of mitochondria and bacteria. Biochim Biophys Acta 1555:154–159

Schägger H, Pfeiffer K (2000) Supercomplexes in the respiratory chain of yeast and mammalian mitochondria. EMBO J 19:1777–1783

Schägger H, Pfeiffer K (2001) The ratio of oxidative phosphorylation complexes I-V in bovine heart mitochondria and the composition of respiratory chain supercomplexes. J Biol Chem 276:37861–37867

Schägger H, von Jagow G (1991) Blue native electrophoresis for isolation of membrane protein complexes in enzymatically active form. Anal Biochem 199:223–231

Schägger H, Cramer WA, von Jagow G (1994) Analysis of molecular masses and oligomeric states of protein complexes by blue native electrophoresis and isolation of membrane protein complexes by two-dimensional native electrophoresis. Anal Biochem 217:220–230

Schägger H, de Coo R, Bauer MF, Hofmann S, Godinot C, Brandt U (2004) Significance of respirasomes for the assembly/stability of human respiratory chain complex I. J Biol Chem 279:36349–36353

Seelert H, Poetsch A, Dencher NA, Engel A, Stahlberg H, Müller DJ (2000) Proton-powered turbine of a plant motor. Nature 405:418–419

Sone N, Sekimachi M, Kutoh E (1987) Identification and properties of a quinol oxidase super-complex composed of a bc1 complex and cytochrome oxidase in the thermophilic bacterium PS3. J Biol Chem 262:15386–15391

Soper JW, Decker GL, Pedersen PL (1979) Mitochondrial ATPase complex. A dispersed, cytochrome-deficient, oligomycin-sensitive preparation from rat liver containing molecules with a tripartite structural arrangement. J Biol Chem 254:11170–11176

Stock D, Leslie AG, Walker JE (1999) Molecular architecture of the rotary motor in ATP synthase. Science 286:1700–1705

Strauss M, Hofhaus G, Schröder RR, Kühlbrandt W (2008) Dimer ribbons of ATP synthase shape the inner mitochondrial membrane. EMBO J 27:1154–1160

Stroh A, Anderka O, Pfeiffer K, Yagi T, Finel M, Ludwig B, Schägger H (2004) Assembly of respiratory complexes I, III and IV into NADH oxidase supercomplex stabilizes complex I in *Paracoccus denitrificans*. J Biol Chem 279:5000–5007

Stuart R (2008) Supercomplex organization of the oxidative phosphorylation enzymes in yeast mitochondria. J Bioenerg Biomembr 40:411–417

Sunderhaus S, Dudkina NV, Jänsch L, Klodmann J, Heinemeyer J, Perales M, Zabaleta E, Boekema EJ, Braun H-P (2006) Carbonic anhydrase subunits form a matrix-exposed domain attached to the membrane arm of mitochondrial complex I in plants. J Biol Chem 281:6482–6488

Svensson-Ek M, Abramson J, Larsson G, Tornroth S, Brzezinski P, Iwata S (2002) The X-ray crystal structures of wild-type and EQ(I-286) mutant cytochrome *c* oxidases from *Rhodobacter sphaeroides*. J Mol Biol 321:329–339

Swartz DA, Park EI, Visek WJ, Kaput J (1996) The e subunit gene of murine F_1F_0-ATP synthase – Genomic sequence, chromosomal mapping and diet regulation. J Biol Chem 271:20942–20948

Taylor NL, Heazlewood JL, Day DA, Millar AH (2005) Differential impact of environmental stresses on the pea mitochondrial proteome. Mol Cell Proteomics 4:1122–1133

Thomas D, Bron P, Weimann T, Dautant A, Giraud M-F, Paumard P, Salin B, Cavalier A, Velours J, Brèthes D (2008) Supramolecular organization of the yeast F_1F_0-ATP synthase. Biol Cell 100:591–601

Tomasetig L, Di Pancrazio F, Harris DA, Mavelli I, Lippe G (2002) Dimerization of F_0F_1 ATP synthase from bovine heart is independent from the binding of the inhibitor protein IF_1. Biochim Biophys Acta 1556:133–141

Tsukihara T, Aoyama H, Yamashita E, Tomizaki T, Yamaguchi H, Shinzawa-Itoh K, Nakashima R, Yaono R, Yoshikawa S (1996) The whole structure of the 13-subunit oxidized cytochrome c oxidase at 2.8 Å. Science 272:1136–1144

Turba A, Jetzek M, Ludwig B (1995) Purification of *Paracoccus denitrificans* cytochrome c_{552} and sequence analysis of the gene. Eur J Biochem 231:259–265

van der Laan M, Wiedeman N, Mick DU, Guiard B, Rehling P, Pfanner N (2006) A role for Tim21 in membrane-potential-dependent preprotein sorting in mitochondria. Curr Biol 16:2271–2276

van Lis R, Atteia A, Mendoza-Hernandez G, Gonzalez-Halphen D (2003) Identification of novel mitochondrial protein components of *Chlamydomonas reinhardtii.* A proteomic approach. Plant Physiol 132:318–330

van Lis R, Mendoza-Hernández G, Groth G, Atteia A (2007) New insights into the unique structure of the F_0F_1-ATP synthase from the chlamydomonad algae *Polytomella* sp. and *Chlamydomonas reinhardtii*. Plant Physiol 144:1190–1199 van Straaten HLM, van Tintelen JP, Trijbels JMF, van den Heuvel LP, Troost D, Rozemuller-Kwakkel JM, Duran M, de Vries LS, Schuelke M, Barth PG (2005) Neonatal lactic acidosis, complex I/IV deficiency, and fetal cerebral disruption. Neuropediatrics 36:193–199

Voeltz GK, Prinz WA (2007) Sheets, ribbons and tubules – how organelles get their shape. Nat Rev Mol Cell Biol 8:258–264

Vonck J, Schäfer E (2009) Supramolecular organization of protein complexes in the mitochondrial inner membrane. Biochim Biophys Acta Mol Cell Res 1793:117–124

Vonck J, Krug von Nidda T, Meier T, Matthey U, Mills DJ, Kühlbrandt W, Dimroth P (2002) Molecular architecture of the undecameric rotor of a bacterial Na^+-ATP synthase. J Mol Biol 321:307–316

Wenz T, Hielscher R, Hellwig P, Schägger H, Richers S, Hunte C (2009) Role of phospholipids in respiratory cytochrome bc_1 complex catalysis and supercomplex formation. Biochim Biophys Acta Bioener 1787:609–616

Wittig I, Schägger H (2005) Advantages and limitations of clear-native PAGE. Proteomics 5:4338–4346

Wittig I, Schägger H (2007) Electrophoretic methods to isolate protein complexes from mitochondria. Methods Cell Biol 80:723–741

Wittig I, Carrozzo R, Santorelli FM, Schägger H (2006) Supercomplexes and subcomplexes of mitochondrial oxidative phosphorylation. Biochim Biophys Acta Bioener 1575:1066–1072

Xia D, Yu CA, Kim H, Xia J, Kachurin AM, Zhang L, Yu L, Deisenhofer J (1997) Crystal structure of the cytochrome bc1 complex from bovine heart mitochondria. Science 281:64–71

Yankovskaya V, Horsefield R, Törnroth S, Luna-Chavez C, Miyoshi H, Léger C, Byrne B, Cecchini G, Iwata S (2003) Architecture of succinate dehydrogenase and reactive oxygen species generation. Science 299:700–704

Zhang ZL, Huang LS, Shulmeister VM, Chi YI, Kim KK, Hung LW, Crofts AR, Berry EA, Kim SH (1998) Electron transfer by domain movement in cytochrome bc_1. Nature 392:677–684

Zhang M, Mileykovskaya E, Dowhan W (2002) Gluing the respiratory chain together. Cardiolipin is required for supercomplex formation in the inner mitochondrial membrane. J Biol Chem 277:43553–43556

Zhang M, Mileykovskaya E, Dowhan W (2005) Cardiolipin is essential for organization of complexes III and IV into a supercomplex in intact yeast mitochondria. J Biol Chem 280:29403–29408

A Structural Perspective on Complex I

Current knowledge of the structure and function of respiratory complex I (NADH:ubiquinone oxidoreductase).

Edited by Leonid Sazanov

Complex I (NADH:ubiquinone oxidoreductase) is the first and largest enzyme of the respiratory chain in mitochondria and plays a central role in cellular energy production. Complex I dysfunction has been implicated in many human neurodegenerative diseases, including Parkinson's. Mutations in complex I subunits are among the most common human genetic disorders known. Until recently, in the absence of structural information, complex I was the least understood enzyme of the respiratory chain. How it couples electron transfer between NADH and quinone to the translocation of protons across the membrane was a complete enigma.

Recent years have been marked by spectacular progress in the elucidation of complex I's structure, finally allowing the interpretation of a wealth of data accumulated in about 50 years since the discovery of the enzyme. Understanding of the mechanism of this large molecular machine, involving long-range conformational changes, is beginning to emerge. This book is the first devoted entirely to complex I. It contains chapters written by leaders in the field, covering a wide range of topics from the structure of the complex, properties of its many redox centers, subunit composition, mutagenesis studies, evolution of the enzyme and on to the current understanding of the coupling mechanism and the molecular basis for human pathologies.

Features:

- concise and authoritative review of current state-of-the-art in research on respiratory complex I
- the first book entirely devoted to complex I, a central enzyme in bioenergetics
- recent breakthroughs in structure elucidation finally allow understanding of the mechanism
- mutations leading to common and severe human disorders are discussed

L. Sazanov (ed.), *A Structural Perspective on Respiratory Complex I: Structure and Function of NADH:ubiquinone oxidoreductase*, DOI 10.1007/978-94-007-4138-6,

Index

L. Sazanov (ed.), *A Structural Perspective on Respiratory Complex I: Structure and Function of NADH:ubiquinone oxidoreductase*, DOI 10.1007/978-94-007-4138-6,

The manufacturer's authorised representative in the EU is Springer Nature Customer Service Centre GmbH, Europaplatz 3, 69115 Heidelberg, Germany. If you have any concerns regarding our products, please contact ProductSafety@springernature.com

Printed and bound by CPI Group (UK) Ltd, Croydon, CR0 4YY
25/09/2026
02222899-0002